CITIES OF THE GLOBAL SOUTH READER

The *Cities of the Global South Reader* adopts a fresh and critical approach to the field of urbanization in the developing world. The *Reader* incorporates both early and emerging debates about the diverse trajectories of urbanization processes in the context of the restructured global alignments in the last three decades. Emphasizing the historical legacies of colonialism, the *Reader* recognizes the entanglement of conditions and concepts often understood in binary relations: first/third worlds, wealth/poverty, development/underdevelopment, and inclusion/exclusion. By asking: "Whose city? Whose development?" the *Reader* rigorously highlights the fractures along lines of class, race, gender, and other socially and spatially constructed hierarchies in global South cities. The *Reader*'s thematic structure, where editorial introductions accompany selected texts, examines the issues and concerns that urban dwellers, planners, and policy makers face in the contemporary world. These include the urban economy, housing, basic services, infrastructure, the role of non-state civil society-based actors, planned interventions and contestations, the role of diaspora capital, the looming problem of adapting to climate change, and the increasing spectre of violence in a post 9/11 transnational world.

The *Cities of the Global South Reader* pulls together a diverse set of readings from scholars across the world, some of which have been written specially for the volume, to provide an essential resource for a broad interdisciplinary readership at undergraduate and postgraduate levels in urban geography, urban sociology, and urban planning as well as disciplines related to international and development studies. Editorial commentaries that introduce the central issues for each theme summarize the state of the field and outline an associated bibliography. They will be of particular value for lecturers, students, and researchers, making the *Cities of the Global South Reader* a key text for those interested in understanding contemporary urbanization processes.

Faranak Miraftab is Professor in the Department of Urban and Regional Planning at the University of Illinois, Urbana-Champaign, where she teaches on globalization and transnational planning and coordinates the department's international programs and activities for its undergraduate and graduate degree programs. Miraftab's research concerns the global and local contingencies involved in the formation of the city and citizens' struggle to access urban space and socio-economic resources.

Neema Kudva is Associate Professor in the Department of City and Regional Planning at Cornell University. She directs the International Studies in Planning Program (ISP) and is faculty lead of the Nilgiris Field Learning Center, a collaborative interdisciplinary project of Cornell University and the Keystone Foundation, India. Kudva's research is in two areas: the institutional structures that undergird planning and development at the local level, and contemporary urbanization, particularly issues related to small cities and their regions.

THE ROUTLEDGE URBAN READER SERIES

Series editors

Richard T. LeGates
Professor Emeritus of Urban Studies and Planning, San Francisco State University

Frederic Stout
Lecturer in Urban Studies, Stanford University

The Routledge Urban Reader Series responds to the need for comprehensive coverage of the classic and essential texts that form the basis of intellectual work in the various academic disciplines and professional fields concerned with cities and city planning.

The readers focus on the key topics encountered by undergraduates, graduate students, and scholars in urban studies, geography, sociology, political science, anthropology, economics, culture studies, and professional fields such a city and regional planning, urban design, architecture, environmental studies, international relations, and landscape architecture. They discuss the contributions of major theoreticians and practitioners and other individuals, groups, and organizations that study the city or practice in a field that directly affects the city.

As well as drawing together the best of classic and contemporary writings on the city, each reader features extensive introductions to the book, sections, and individual selections prepared by the volume editors to place the selections in context, illustrate relations among topics, provide information on the author, and point readers towards additional related bibliographic material.

Each reader contains:

- Between thirty and sixty *selections* divided into five to eight sections. Almost all of the selections are previously published works that have appeared as journal articles or portions of books.
- A *general introduction* describing the nature and purpose of the reader.
- *Section introductions* for each section of the reader to place the readings in context.
- *Selection introductions* for each selection describing the author, the intellectual background and context of the selection, competing views of the subject matter of the selection, and bibliographic references to other readings by the same author and other readings related to the topic.
- One or more plate sections and illustrations at the beginning of each section.
- An index.

The series consists of the following titles:

THE CITY READER

The City Reader: Fifth edition – an interdisciplinary urban reader aimed at urban studies, urban planning, urban geography, and urban sociology courses – is the *anchor urban reader*. Routledge published a first edition of *The City Reader* in 1996, a second edition in 2000, a third edition in 2003, and a fourth edition in 2007. *The City Reader* has become one of the most widely used anthologies in urban studies, urban geography, urban sociology, and urban planning courses in the world.

URBAN DISCIPLINARY READERS

The series contains *urban disciplinary readers* organized around social science disciplines and professorial fields: urban sociology, urban geography, urban politics, urban and regional planning, and urban design. The urban disciplinary readers include both classic writings and recent, cutting-edge contributions to the respective disciplines. They are lively, high-quality, competitively priced readers which faculties can adopt as course texts and which also appeal to a wider audience.

TOPICAL URBAN ANTHOLOGIES

The urban series includes *topical urban readers* intended both as primary and supplemental course texts and for the trade and professional market. The topical titles include readers related to sustainable urban development, global cities, cybercities, and city cultures.

INTERDISCIPLINARY ANCHOR TITLE

The City Reader: Fifth edition
Richard T. LeGates and Frederic Stout (eds)

URBAN DISCIPLINARY READERS

The Urban Geography Reader
Nick Fyfe and Judith Kenny (eds)

The Urban Politics Reader
Elizabeth Strom and John Mollenkopf (eds)

The Urban and Regional Planning Reader
Eugénie Birch (ed.)

The Urban Sociology Reader: Second edition
Jan Lin and Christopher Mele (eds)

The Urban Design Reader: Second edition
Michael Larice and Elizabeth Macdonald (eds)

TOPICAL URBAN READERS

The City Cultures Reader: Second edition
Malcolm Miles, Tim Hall with Iain Borden (eds)

The Cybercities Reader
Stephen Graham (ed.)

The Global Cities Reader
Neil Brenner and Roger Keil (eds)

The Sustainable Urban Development Reader: Third edition
Stephen M. Wheeler and Timothy Beatley (eds)

Cities of the Global South Reader
Faranak Miraftab and Neema Kudva (eds)

Forthcoming:
The City Reader: Sixth edition
Richard T. LeGates and Frederic Stout (eds)

For further Information on The Routledge Urban Reader Series please visit our website:
http://www.routledge.com/articles/featured_series_routledge_urban_reader_series/
or contact

Andrew Mould
Routledge
2 Park Square
Milton Park
Abingdon
Oxon, OX14 4RN
England
andrew.mould@routledge.co.uk

Richard T. LeGates
Department of Urban Studies and Planning
San Francisco State University
1600 Holloway Avenue
San Francisco, CA 94132
(510) 642-3256
dlegates@sfsu.edu

Frederic Stout
Urban Studies Program
Stanford University
Stanford, CA 94305-2048
fstout@stanford.edu

Cities of the Global South Reader

Edited by

Faranak Miraftab

and

Neema Kudva

LONDON AND NEW YORK

First published 2015
by Routledge
2 Park Square, Milton Park, Abingdon, Oxon OX14 4RN

and by Routledge
711 Third Avenue, New York, NY 10017

Routledge is an imprint of the Taylor & Francis Group, an informa business

British Library Cataloguing in Publication Data
A catalogue record for this book is available from the British Library

Library of Congress Cataloging in Publication Data
Cities of the global South reader / edited by Faranak Miraftab,
Neema Kudva. — 1 Edition.
pages cm. — (Routledge urban reader series)
Includes bibliographical references and index.
1. Urbanization—Developing countries. 2. Cities and towns—Growth.
I. Miraftab, Faranak, editor. II. Kudva, Neema, editor.
HT384.D44C584 2014
307.7609172'4—dc23 2014022360

ISBN: 978-0-415-68226-8 (hbk)
ISBN: 978-0-415-68227-5 (pbk)
ISBN: 978-1-315-75864-0 (ebk)

Typeset in Amasis and Berthold Akzidenz Grotesk
by Keystroke, Station Road, Codsall, Wolverhampton

Printed and bound by CPI Group (UK) Ltd, Croydon, CR0 4YY

Contents

Figures

Tables

Acknowledgements

Our debts on this project are many. Dick LeGates has been a wonderful series editor who answered dozens of questions promptly and provided us with detailed instructions on how best to manage a project of this ambition and scope. Our editor at Routledge, Andrew Mould, shepherded the project through an extensive review process and toward publication over an extended period. We are grateful to the anonymous reviewers of the volume's proposed content and structure for their helpful comments and suggestions. We also owe a very special thanks to editorial assistants Faye Leerink and Sarah Gilkes, who managed to find hard-to-reach people to get copyright permissions and handled a host of problems with good humor and patience. Several student assistants have helped us with a range of tasks over the years from preparing the proposal, gathering articles, preparing bibliographies, scanning articles, and chasing copyright permissions. Many thanks at UIUC to Azad Amir-Ghassemi, Sofia Sianis, and Anjali Krishnan, and at Cornell to Fan Fan Zhao, Ke Tong, and Shinae Park. We are also indebted to our artist colleagues and friends, Iftikhar Dadi at Cornell and Pradeep Dalal, who introduced us to the artists whose work brings yet another perspective to the cities included in the *Reader.* They helped us curate the images that we have included. Many thanks also to the artists who worked with us to deliver images for the *Reader.* A special thanks to artist and architect Nasrin Nawab for producing the base map of Tehran that Nicolas Grefenstette at Cornell worked on further, and to Nasroulah Kasraian and Ali Madanipour for making their photos of Tehran available to us.

This book would have never seen publication if it hadn't been for the editing and wordsmithing skills of James Kilgore, who carefully read through everything we wrote, emphasized the need for conceptual clarification, and flagged points of confusion and potential contradictory interpretations. Jim is an educator in the field, a skilled editor, and prolific author; we could not have been in better hands. We also thank Terri Gitler, who copy-edited the manuscript and caught everything that the rest of us missed.

We drew on many experiences to pull this *Reader* together, some more directly than others. Our experience teaching two large undergraduate classes at our home campuses shaped the book's structure: UP185, Cities in a Global Perspective, at UIUC and CRP 101 (later CRP 1101), Global Cities, at Cornell. Our lectures made their way into the thematic introductions even as our classes improved immeasurably because of our collaboration. The *Reader* also draws on two projects that were aimed at improving our courses by bringing in a variety of voices–Virtual Cities, a collaboration between Neema Kudva and Bill Goldsmith at Cornell (2003–2005) and multipliCities in collaboration with Ken Salo (UIUC), Sophie Oldfield (University of Cape Town), and Keith Pezzoli (University of California, San Diego) (2008–2011).

We relied heavily on each other through the process, playing equal roles in editing this volume and writing the introductions collaboratively–and note this to dispel speculations on who between us put in more work or less effort. Our work was made easier by the fact that there was always another person with whom to argue and think through an issue. For that and for companionship that made the process of this publication so much more pleasant, we thank each other. Last but not least we thank our spouses and children, who not only patiently put up with us while we worked on this project but also supported us in many different ways–intellectually and practically. Faranak has a bundle of thanks for her muses and inspiration Ken, Rahi, and Omeed, and is grateful for everlasting encouragement by her parents, siblings, and close friends. Neema thanks her many friends who fed and listened to her in Ithaca, India, and Boulder, particularly Jim Bull and Erin Mulligan for publishing and editorial advice; her parents and sister for making space for her work; and David, Kieran, and Mira for good humoredly putting up with her and her constantly shifting schedule. We dedicate this book to our children in reverse birth order: Mira Diana Kudva Driskell, Omeed Miraftab Salo, Rahi Miraftab Salo, and Kieran Kudva Driskell, with the hope that they will better understand the cities where they live.

Introduction

Figure 1 Detail from "Master Plan" at the multimedia installation *TRASH* done collaboratively with Chintan: Environmental Research and Action Group, an NGO that works with trash pickers in Delhi, India.
Source: © Vivian Sundaram, 2009.

Editors' Introduction to the Volume

Many significant shifts in thematic and geographic focus have taken place since the advent of the field of urbanization in the developing world. The "Third World cities" of the first two development decades were megacities and primate cities produced by rural–urban migration and synonymous with large swathes of squatter and slum housing. Urban scholarship of this period drew heavily on the Latin American experience. Asia's resurgence in urban and development scholarship started in the 1970s with the rise of the city-states Singapore and Hong Kong, and the newly industrializing countries of East Asia. Today the Asian giants, China and India, along with a range of other significant economies such as South Africa, Brazil, Russia and Indonesia, dominate a conversation that includes a wide variety of issues, from governance arrangements, gendered perspectives, and infrastructure development to the role of non-state civil society-based actors, diaspora capital, and the looming problem of adapting to climate change. Equally important, the emergence of the EU and the disintegration of the Soviet Union restructured global alignments and brought understudied global regions such as the non-EU bloc of Eastern European countries and the Central Asian republics into view. Along with these events came the increasing specter of violence in a post-9/11 transnational world and the little-analyzed fallout from the near collapse of financial markets in 2008.

Our reality is a planet where more than half the population lives in cities and towns of various sizes, and the bulk of future rapid urban growth is expected to occur in Africa and Asia. It is also an urban reality where the "third world" is a space of finely grained differentiation, not coterminous with national or urban boundaries, that is as present in the ghettos and enclaves of North America and Western Europe as in the slums of Latin America, Africa, or Asia. The spirit of change grips some places, which hurtle in directions that provide exemplars for the rest of the world, while other places seem to stand still or even decline.

This complex context is one starting point for this *Reader*. The other starting point is the pedagogic challenges we face as educators based at US institutions who focus on understanding global inequalities through the study of cities and urbanization. We struggle with the problem of engaging students in communities that lie at a distance, metaphorically or literally, from the places we inhabit on a daily basis. The majority of our students have little experience of urban communities outside their safe middle-class experience, regardless of whether they have lived in the US, Western Europe, or various cities of the global South. For many of them, the unfamiliar cities and communities in this *Reader* are either places of exotica and islands of hypermodernity or dismal homogenous spaces of poverty and backwardness where terrorists, danger, and strange others lurk. How do we help explain the complexity of life in these places? How do we bring alive the heterogeneity of cities embedded in different historical trajectories despite the media construction of the "third world" and poverty as a single block of backwardness and despair? And how do we convey these cities as places of challenge and great promise, where people are active agents shaping their future as opposed to being passive victims of difficult circumstances waiting to be rescued by the development machine?

These questions become yet more challenging when joined to the conceptual and political commitments we make in our work, and in this *Reader*. We seek to challenge widely prevalent categories of cities and models of development, dominant paradigms of planning, and taken-for-granted notions of political democracy, inclusion, and citizenship. We seek to give voice to scholars and activists from, or with roots in, the global South.

We recognize that gross global inequalities and the persistent production and reproduction of those inequalities lie at the root of many thematic issues discussed in this *Reader*; the questions "for whom?" and "by whom?" structure the *Reader's* thematic introductions.

For us, cities are not just nodes on transaction networks, or physical collections of built form specific to a context and global movement, or a mix of cultures over time. We are in the tradition of a long line of scholars and researchers who view cities as social processes, as political assemblages in which formal and informal institutions of governance are forged and continue to be shaped as polities change over time. Various processes impacting societies shape cities, ranging from changing migration patterns and large-scale population movements to changes in geopolitical power and the technologies of infrastructure, communication, and manufacturing.

What has emerged from contemporary urban reality and our pedagogical and scholarly commitments is a unique interdisciplinary anthology that is wide-ranging yet focused in scope. The *Cities of the Global South Reader* introduces students to a burgeoning body of work on the cities of the global South, and to the ways in which scholars and researchers have sought to conceptualize and understand processes of city building and maintenance in the global South. Students who work their way through this volume will emerge with a clear sense of the leading voices and key debates in the world of urban development in the global South.

Many anthologies and volumes have been published on urbanization and development. Some focus solely on categories like "Global Cities" or "Colonial Port Cities." Others cover a region or country: "Chinese Cities," or "Contestations in Cities of X region," for example. Still others are theme-based, addressing topics like "Innovation Hubs" or "Healthy Cities" that cut across regions and categories. These books sometimes include detailed case studies from different cities and are enormously useful in building knowledge on cities of the global South; however, our *Reader* adds another dimension. It is structured across categories, themes, and regions to offer cross-cultural and multiregional perspectives on key challenges of urbanization, development and planning in the global South. As a reader and not a textbook, it draws on the classic literature in the field as well as contemporary scholarship, including a number of articles from leading scholars especially commissioned for this book. We combine this literature with thematic introductions written by the editors to outline the intellectual history and lineage of ideas that dominate the study of cities in the global South. Thus, the *Reader* combines the scope and coherence of a textbook with the multiple voices and diversity that is the strength of an edited anthology.

The thematic introductions at the opening of each of the twelve sections of this *Reader* are key to its intellectual and pedagogic mission. By offering a comprehensive overview of key debates on each theme, they help the reader to place the debates (and article selections) in perspective, understand regional specificities and commonalities, and they introduce the broader political economic and institutional context that shapes practices and policies. These section introductions also lay out the intellectual history of each theme, from the relationship between development and urbanization to studies of housing, infrastructure, and the transfer of knowledge between places. They foreground critical debates that have shaped forms of action from above through formal institutions and policies, from below through collective grassroots movements, and from several directions simultaneously through the interventions of private actors in the institutional realms of the market and civil society.

The primary audiences of this interdisciplinary *Reader* are course instructors and students enrolled in a range of fields: global studies, development studies, urban studies, international studies, geography, urban planning and other disciplines and programs that afford students the opportunity to adopt a global focus. Moreover, we are teaching not only in programs and institutions across the US, Canada, the UK, and Australia, but also in other international locations where the medium of instruction is English. A secondary audience includes English-speaking scholars and others interested in understanding the cities they live in, work in or visit. But even as we designed this volume keeping our intended audiences in mind and in the face of the challenges and realities described earlier, we made additional choices in selections and structuring themes. These choices are critical to further understanding the *Reader* as an intellectual project. We group and describe these in three areas below.

OUR CHOICE OF TERMINOLOGY: THE GLOBAL SOUTH

In titling this anthology *Cities of the Global South Reader* we recognize that we are inevitably tying ourselves and this book to a particular category, "the global South," and thus abetting the creation of hierarchies, as Jenny Robinson's critique has made clear. (See the article selection in the *Reader* for more details.) For us, however, the term "global South" does not signify singular and self-contained territories and categories constructed oppositionally in terms of first *or* third world, global South *or* North cities. As one of our authors, David Satterthwaite pointed out to us, his and Jorge Hardoy's choice of the term "third world" in 1991 was made to invoke the French use of the term "third estate" as synonymous with people and not the "first estate" which signified nobility, and in conscious opposition to other terms like "less-developed" or "developing" countries. Our use of "global South" also does not signify discrete territories hierarchically positioned among a range of categories of cities that have dominated urban studies: global, metropolitan, peripheral, large, small, Third World, First World, global South, or global North. We recognize the entanglement of the Third and First Worlds, global South and North, the metropole and the colonies even as we seek to understand, in the words of Edward Said, their overlapping territories and intertwined histories (1994). We recognize the Third World within the First World; the global South in North America and the European Union; what we call multipliCities the multiple worlds within each city.

In choosing the term global South, however, we do wish to emphasize a shared heritage of recent colonial histories in the global peripheries. These have constituted hybrid societies distinct from those that colonized others and also different from majority white-settler societies such as the US, Canada, and Australia, where indigenous populations were decimated. This shared heritage, combined with the post-World War II experience of development to "alleviate" poverty, has resulted in unique trajectories of sociopolitical and economic development across the global South. As a conceptual construct, therefore, we contend that "global South" offers a useful frame of reference by acknowledging the colonial past and a more recent shared development history. For this reason, we also include China as part of the global South in this *Reader*. This was a complicated choice, as China defies categories. Largely driven by a massive urban transformation along its eastern coast, the Chinese have moved more people out of poverty faster than any nation in history. Yet China continues to struggle with sharp regional inequities, rural–urban differentials on a range of development indicators, and millions of "floating population" migrants who lack basic rights in the cities to which they have illegally moved. In addition, China's rise and transformation constitute an important chapter in the urbanization story of our globalized, transnational world.

THEMATIC STRUCTURE OF THE READER

This *Reader* has twelve sections organized in five parts with each section containing two to four article selections prefaced by a thematic introduction. At its core, this *Reader* focuses on surfacing injustices manifest in urban poverty, inequality, and exclusions based on gender, race, class and ethnic identities, among other markers of social hierarchies. In structuring the *Cities of the Global South Reader* we refuse to treat these core issues in self-contained sections, choosing instead to highlight these issues across all thematic sections. In doing so, we recognize the entangled co-dependence of wealth/poverty, development/underdevelopment, first/third worlds, privilege/powerlessness, and various forms of inclusion/exclusion. The sections themselves echo topics that have long been central to the scholarship on cities (examples include urban history, housing, infrastructure, and citizenship) even as they pick up on other topics that have emerged more recently, such as risk, basic services, governance, and participation.

ARTICLE SELECTIONS WITHIN THEMATIC SECTIONS

No anthology, however ambitious, can cover every region, every theme, or every position in every pertinent debate. Pragmatic reasons, such as a large and varied scholarship in English on urban experiences in African, Asian, and Latin American cities have led us to focus on these parts of the world even as we try to maintain a balance between the regions and remain inclusive of Southern voices. Our choice is not meant to imply that these three regions are more important or that the scholarship carries more weight than work in other languages or work that is published on other regions, including post-socialist Eastern Europe and Central Asia or the Caribbean.

Nor do our selections suggest exclusive authority for individual authors. Our primary aim has been to offer the platform of this volume to academic and grassroots voices who speak to the core issues we wish to highlight. In this vein, we have included excerpts from a range of texts that have made a mark in shaping larger debates over time. Some did so by being provocative and fomenting critique, others by introducing new ways of looking at and understanding processes in the cities of the global South. We have, however, not included excerpts from policy papers produced by government agencies or international development institutions, regardless of how influential these papers were in shaping debates or silencing individual voices around urban development, although they are mentioned in the articles and in the Editors' introductions. In our selections we have also faced pragmatic limitations. Some superb and influential texts were not included because creating short, clear excerpts was difficult. In other cases the writing style was not easily accessible to our intended audiences.

Our analytic and pragmatic concerns have led us, in some cases, to address key issues by commissioning leading scholars and emerging voices to contribute original pieces to our work. These articles add yet another important dimension to this *Reader* and to the field more broadly. For example, Ali Madanipour's article on Tehran uses first-person narratives to highlight multiple experiences of city dwellers through the lens of class, gender, ethnic, and spatial positions. The synoptic articles by Anthony King on colonialism, Michael Goldman on development, and Richard Harris on housing capture the broader intellectual and institutional landscape of shifting debates–a task no single existing article could achieve. Commissioned articles by Sudeshna Mitra on transnational capital and by Sahar Khamis and Katherine Vaughn on cyberactivism, as well as the introduction to cities and risk by Andrew Rumbach, are charged with discussing recent phenomena and currently unfolding processes. Together these commissioned pieces and excerpts of classic and contemporary published material, supplemented by the thematic introductions we have drafted, offer readers a perspective that focuses on the production of cities as seen through a historical, interdisciplinary lens that brings in regional variation and uncovers processes that generate the energy and inequalities that mark contemporary city-worlds.

HOW TO USE THIS *READER*

Each section is designed to stand alone even as the authors draw connections and note relations between and across themes. Each thematic section, however, is best understood when the introduction is read along with the article selections. To optimize the instructional use of this *Reader*, we urge instructors to assign the article excerpts along with the companion section introduction. Adopters can use this *Reader* in various ways: by assigning all twelve section introductions, for example, to offer their students an extensive overview of debates that shaped the field or by focusing on selected themes. Although each section does stand alone, we stress that it makes the most sense when the introduction and article excerpts are read together.

We now turn to the organization of the *Reader* and the specifics of its use. As noted earlier, the *Reader* is organized in five parts and twelve sections. Each section opens with the editors' introduction to frame the issues and guide the readers through the selected or commissioned texts. **Part I, *The City Experienced***, makes use of narratives and representations. Here we see that there are multiple cities and city experiences within the same city depending on the position of the citizens. We refer to this as the multipliCity. **Part II, *Making the "Third World" City***, historicizes the city-making process, detailing the role of colonial experience and post-World War II development in shaping the city. Understanding the "historical underpinnings" of every city (as Lefebvre would have it) is important as we seek to comprehend how they work today. **Part III, *The City Lived***, captures the experience of cities by demonstrating the shifts in migratory flows, and the ways in which work and housing articulate issues that were central to early "third world" urbanization literature and continue to be central to contemporary city life. **Part IV, *The City Environment***, shows how infrastructure; the violence that emerges from poverty, privatization, and war; and resilience in the face of both everyday disasters and threats associated with climate change define the city environment. **Part V, *Planned Interventions and Contestations***, brings attention to themes of governance, participation, and citizenship. The selected readings clustered thematically around these issues focus on the complexities and the shifting constellation of relationships among key actors that shape cities and development processes: the state (at different levels), the private companies, and the civil society organizations. This part takes a critical look at the contestations among key actors involved in shaping cities everywhere. The final and concluding section of the *Reader* is dedicated to the ongoing debate around the ways in which expert knowledge, traveling transnationally and often hierarchically, shapes the city and seeks to mediate the relationships among these potentially contesting actors.

The article text selections are reproduced essentially as they appeared in the original. We have not changed original wording of older selections that might strike some modern readers as antiquated, tried to improve on writing in the original that we found awkward, or provided summaries of deleted sections. Where we have edited selections, material omitted is indicated by square brackets [. . .]. Our own explanations of terms are also enclosed by square brackets [ ]. We include bibliographic references and additional notes only where the selections refer to them. In the interests of accessibility, we have not included explanatory endnotes unless they are vital to the meaning of the text. At the author's request, the selection written by Manuel Castells included in Part V has not been excerpted. Madhav Badami and Ahmed Kanna assisted us in preparing excerpts of their work. Manuel, Madhav and Arjun allowed us free use of their work, for which we are thankful.

And so we come to the end of this introduction and to three years of preparation and writing. We would like to leave you, our readers, with the thought that the third world and the global South are both "here" where we live, and "there," in the cities that we study from afar. We conclude by pointing to the ways in which binary concepts such as developed/developing countries, First/Third World and global North/South (as used here) remain problematic and, ironically, useful.

REFERENCE

Said, E. (1994). *Culture and Imperialism*, New York: Vintage.

Figure 2 First World, Third World?
Source: © Chapatte, 2013.

PART I

The city experienced

Figure 3 Iranian cinema mirrors Tehrani lives.
Source: © Ali Madanipour.

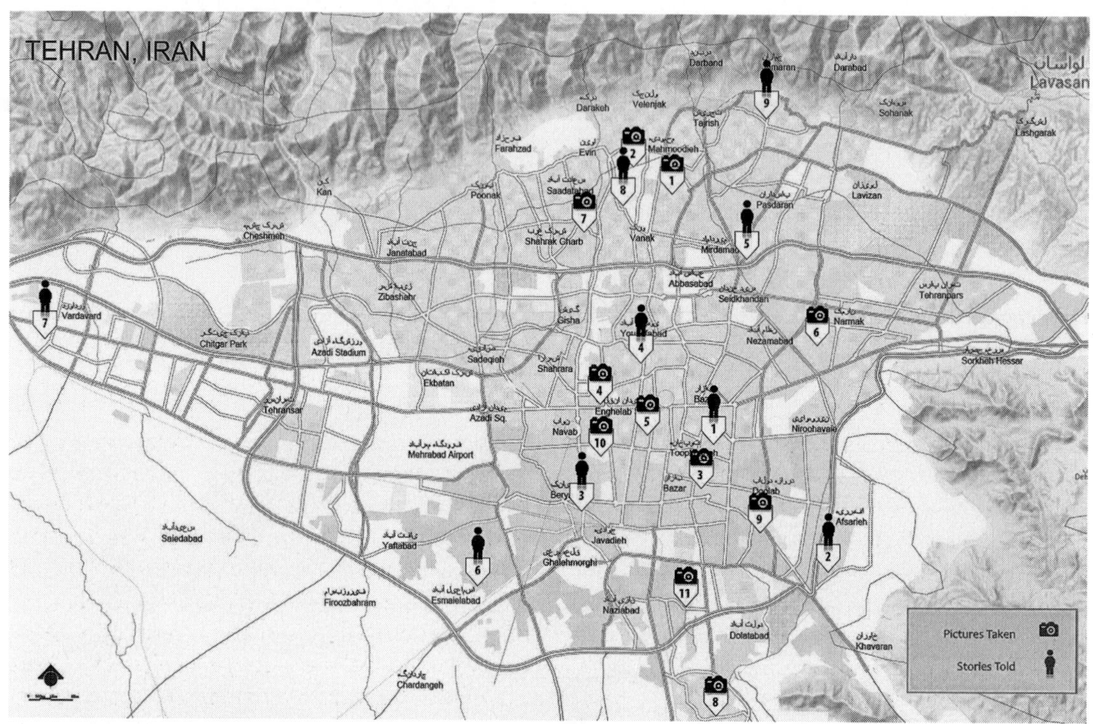

Figure 4 Map of Tehran showing locations of pictures and stories.
Source: © Faranak Miraftab and Neema Kudva.

"Urban Lives: Stories from Tehran"[1]

Essay written for *Cities of the Global South Reader* (2015)

Ali Madanipour

How can we acquire a complete profile of the city? Is it at all possible to arrive at such a picture? Whatever approach we take, we confront a different scene. Whoever we talk to will tell us a different story. Do we dismiss these stories as "subjective" and try to build up a completely "objective" model of reality? But we know that even this objective story is a narrative told from a perspective, embedded in a set of contexts and experiences. We can look at the patterns of the city's economy, its social life, the way it is managed, the way its space is organized, and some of the problems it faces. We can draw a sketch of the city along these lines and add more details to render a more accurate picture of the city. Whatever we do, however, will not capture a complete portrait of the city in its rich diversity and complexity of meaning. If we took any of these approaches, something very essential would be missing: the stories and the voice of the people who live in the city.

Below we hear the stories of a variety of people from one particular city: Tehran, the capital of Iran and one of the major cities of the global South. Like any other city, Tehran is a dense tapestry of relationships and meanings, developed and constantly changed by its inhabitants in their daily lives. Compiled below are nine stories of men and women, of young and old, of rich and poor—a cross-section of the population in Tehran. By hearing these stories we hope you begin to get a sense of the city. Obviously these nine people don't tell the whole story of Tehran, as there is no end to the diversity of these stories and each can fill many books. But they do offer us some glimpses of what lies beneath the skin of the city from different perspectives.

However, before you read these personal accounts, some historical content might help to make the narratives more understandable. For this purpose, three events are important to keep in mind: the 1979 revolution which overthrew the monarchy (the Shah or the King) and brought into place an Islamic republic where religious codes of conduct were reinforced in the public arena; the war with Iraq which started shortly after that and lasted through the 1980s; and the political instability of Afghanistan to this day beginning with the Russian occupation in 1979 (hence the presence of Afghani refugees in the city). These events intertwine with the daily experiences of pleasure and pain in the city and how residents have attributed meaning to their urban lives. Here are some scenes and a handful of possible monologues from among millions of Tehrani voices.

[Editors' Note: Along with these stories we also include a series of pictures taken in Tehran and a map of the city that shows the approximate locations of the pictures and the stories. The pictures do not belong to the stories. But combined we hope the keyed-in locations will help you navigate the diverse experiences of the city of Tehran.]

PORTRAITS FROM A LABYRINTH

1

I am a young man. I used to be a teacher in Kabul, Afghanistan, but had to run away because of the civil war in my country. Here I have had to work as a construction worker. It is very hard and nothing in my previous life had prepared me for this job. I wish I could go back to Kabul and live an ordinary, more dignified life. But things are worse back home, with all the continuous fighting. Here some people don't like Afghanis. They think we take work opportunities away from the locals and get paid less for the same work. None of them appreciate that we get paid less for often harder jobs, which even Iranian workers are not prepared to do, like digging and cleaning the sewage wells. Some times when I go to the street corner, where we are picked up by the clients, we get jeered by the Iranian workers who were more expensive and could not find a job. Others think we are too violent, that on the evidence of only a few incidents, often among ourselves.

I live with six other Afghani workers in a small section of an old, crumbling house in the middle of Tehran. It is awful to live like that, but I have no other alternative. These houses look a bit like some places in Afghanistan. Some parts of Tehran have huge buildings with fancy shapes I had only seen in the movies. One good thing about being in the middle of the city is that a lot of things happen here. Our rent is low and we are together. There are so many people from everywhere that nobody bothers us about our documents. I don't wear my traditional costume so that I don't stand out in the street. If I don't get a job for a day, I just wander around in the city center, looking at thousands of passers-by and at the street vendors, shops, and places. I can get some rest in a mosque or go to one of many cinemas nearby. I like these cinemas, although in this country they don't show many Indian films, which are very popular back home.

Despite all this, for me Tehran is an alien place. I am Shiite and my mother tongue is Persian, the same as most people here. Even my Dari accent is not a problem because many people have different accents here. There are some Iranian people who even don't speak Persian. But these don't make me feel at home. Things are very different here and I am not comfortable with most things. My jobs don't last long and I have to look for new ones. I know some friends who work in a neighborhood on the outskirts of the city, where many new buildings are being built. They live on the site and if I am desperate I can go and stay with them. But this line of work cannot go on forever. I wish to get married and start a family, but there are no prospects. Most Afghanis who have come to Iran are men and I don't know any local people. I have not heard from my parents for a while, that is ever since one of my clan came over and said they were rather well but suffering from the civil war.

2

I am a middle-aged woman, living in Gharchak, in the southern outskirts of Tehran. I go out to the city every day, working as a domestic helper, cleaning, washing, cooking, etc. I am originally from a village in Khurasan. My husband did not own agricultural land and there was nothing to do in the village apart from working on other people's land. He thought it'd be better to do the same in Tehran and earn much more. Therefore, we moved to Tehran, where we had some relatives who had come before. My husband learned to work as a plasterer. He was often violent and was addicted to drugs. He was killed in an accident on the construction site. After him, I have had to bring up our eight children on my own. I had rented a room in Afsariyeh, southwest city. Life in that house was really difficult. Several families lived there each in one room round a courtyard. In our village, houses are much more spacious and you don't have strangers as house mates. But here, with so many people in a house, things were too bad. One of my daughters who was often ill had to fast during the month of Ramadan because of pressure from neighbors. They were very nosy. Thank God we are out of that house now. My elder son managed to buy a piece of land in Gharchak and with the help of his cousins built a home for us.

I take a bus to the city every day. All my time in this city I have had to travel to various places where my clients live. I have had a few good clients for many years. They appreciate that a trustworthy domestic helper is not easy to find. Before the revolution, it was easier to find work. People even imported domestic help from the Philippines. I rarely go back to my village. There is nothing there for me. I left when I was very young and my children are now my most important relatives. I have always had to work to earn my living and support my children. I have had some help from various people, like some of my clients, for my children. But that was never enough. God knows how I got used to this huge city, which has no beginning and no end, after coming from our little village.

3

I am a middle-aged housewife, living with my husband in Amiriyeh in the south of Tehran. We are both originally from Shiraz and have come to Tehran many years ago. When our children were still young, we used to go back to Shiraz every summer. We go back rarely now, as the children are no longer interested, fewer of our close relatives are alive or live there, and traveling is not easy. We have lived in this house for nearly thirty years. My two youngest children were born here. I go out of the house every day, for shopping. I know all the shopkeepers in our street and they know me. In the past, they were friendlier and looked after us. Nowadays they seem to have so many customers that they don't care much about any of them. At the time of the war rations, the butcher and the grocer and many others all felt like the God on earth. I'm glad those war days are gone. The city would go completely dark and you could hear these missiles landing in around the city. One fell not very far from here. You keep your nerves for the first few, but afterwards you get seriously worried, although I believe in fate and used to think that if I was destined to be hit by one of Saddam's missiles, I would be hit no matter what I do to prevent it. Many families I know have suffered somehow, or have lost one or more members in the years of revolution and war. The war days are gone now but most subsidies are also gone. Everything is now far more expensive and it is very difficult for us to afford a decent standard of living.

The city has become much noisier and dirtier. We used to sleep on our roof in the summer, enjoying the cool breeze from the mountains. Now if we sleep outside, in the morning our sheets will have become dirty with chunks of soot. I constantly wash and clean everything, but can't do much about our carpets and walls, which get dirty very quickly after each cleaning and redecorating. Our little pool in the courtyard is now almost always dirty, even the day after we have changed the water. Only a few years ago, the water would stay clean much longer.

My children have all grown up and live and work in different parts of Tehran. They come and visit most weekends, on Fridays. The only time I go out of this neighborhood is to visit them or our relatives. Once I knew almost everyone who lived in this street. Now very few of them have remained. Most people we used to know have moved out to other parts of the city, many to the north. Some of their children have gone overseas. We used to have a colonel and a judge as neighbors. Their houses had large gardens and old trees. Now those large houses are mostly gone. One household has sold its courtyard, to be turned into a new house. Another has been completely re-developed. All the houses used to be one or two stories high. From our rooftop, we could see a number of beautiful domes and minarets and many trees. Now our views are blocked by taller and taller buildings, all apartments. That is why we have now more neighbors and more cars parked on the street. Now our neighbors are lower down the social scale, small shopkeepers and minor government employees. We don't know many of them and have no relationships with them. My children insist that we should sell and go to some better place in the north of the city. I have refused to go. I still feel more comfortable here. I can walk to all the shops and every kind of shop is nearby. Elsewhere, shops may be far. Basically, I don't like to leave this neighborhood. I like my home. It is not the best in the world but I like it. I don't want to have to learn to live in a new place.

4

I am a young single man. I work in an office and live on my own in my apartment in the upper floor of my parents' house, although I rely heavily on my mother's cooking. I drive to work every day, which takes me half an hour. I try to avoid being an example of the Donkey's Theorem. A donkey walks the shortest possible distance between two points: from where it is to where the food lies. I try to use a different route every day, just to make it a bit more interesting. Some routes are more congested than the others. In some winter days the travel time can double easily.

I have heard some of my friends complaining that Tehran has no identity, that it is a gray and uninteresting place. I don't really think so. I think Tehran, like every other city, has its interesting and uninteresting places. Some people talk about how neighbors don't know each other and how the city alienates people. I have to say, I don't want to know my neighbors and don't want them to know me. I want to come and go freely as I wish. It is worse after the revolution, as there are more restrictions on personal freedom and it is as if you are always being watched. You always have to have a good excuse for what you do.

If you ask me, people in Tehran have no real attachment to their neighborhoods. They are constantly moving and changing their places. I live now in Yousefabad and may go to Shemiran if I find a good apartment, although I will have a problem persuading my parents. Admittedly, some areas have specific characters. Especially some Armenian areas have a very nice, almost European feeling. There is always a lovely smell of coffee in the air. I go to Cafe Naderi for a cup of coffee any time I pass by, especially in the summer evenings when you can sit outside in its cool garden. I've heard that in the past, that is, before the revolution, they used to have live bands in summer evenings. I love summer evenings in the city, in the wide pavements of Naderi Street, or Vali Asr Street north of Vali Asr Square, or the Tajrish area. The sun is setting so the weather is not as hot, but you still have the daylight of the long summer days. People have come out of their homes for a walk, or for shopping, or are going back home from work. All the shops are open and everything seems to be very relaxed. Summer evenings are nice even in the old neighborhoods, although not much of them are left now. The voice of *muezzin* is echoed in the roofed streets of the bazaar, or the narrow streets of residential areas, where you find people have sprayed water in front of their doors, to make them cool and inviting. People also water their plants and trees inside their courtyards, which helps cooling the air, although every year at the beginning of the summer there is constant warning about the dangers of water shortage. There is another warning at the beginning of each summer: about women's dress. Many women find it really difficult to cover themselves fully in the heat of the summer and this as you know is against the Islamic codes.

5

I am a young woman. I was born in Tehran and have lived here all my life. I have just started university, which is in the middle of the city, which means I have to take a bus every day from my home in Pasdaran, in the north of the city. This is a big change for me. Before this, I just walked every day to my high school. I had to study very hard to pass the university's entrance test, in which hundreds of thousands across the country participate. It is very competitive and you have to work really hard to get in, unless you are in some sort of privileged position of having fought or lost a close relative in the war. That is why I have not been doing much else or going anywhere. My only entertainment has been going to the movies with friends from school or going out with my parents visiting our relatives or occasionally going to a restaurant. The highlight of the year is usually the Persian New Year at the beginning of spring, when we visit friends and relatives and they visit us. Oh yes, and we sometimes go to the Caspian Sea in the summers, which I like very much, although I hate swimming with my clothes on.

At high school my movements were very limited and I didn't know much of the city. Now I have to move around on my own and it is both exciting and a bit scary. Buses are convenient, as men and women are separated on them. Women's section is always quieter than men's, I guess because women don't go out as much. At school we were all girls but here at the university there are boys too, although our contacts with them are rather limited and somehow under scrutiny. There are also strict rules about our dress. My mother and I wear scarves only when we go outside, not in front of friends and family.

I intend to finish university and then find work, because I want to be independent. After that I will think about marriage. My parents are very good and don't try to persuade me to marry as soon as possible, as some parents do.

The university is in a large site in the middle of the city. There are many bookshops in front of the main entrance and I take a walk there any time that I can, both for trying to find the books I need and for fun. Just next to the university is the Enghelab Square, which is always very busy. It is always full of buses, taxis, cars, and crowds of people moving in all directions. There are all sorts of people there. That is why I try to avoid it if I can. My brother says Tehran has become like a big village, because there are so many rural people in the city. He says they come to the city and ruin the character of the place, since they do not know how to behave and live in a city. He says all these peasants should be sent back to their villages. I find my brother's views on this too harsh.

6

I am a middle-aged man. I have a small shop in the south of the city, working on my own as a cobbler.

I have a large family, living in a small house nearby. We have been in this house for many years. I walk to my shop every day. I come home for a lunch break and some rest. Apart from my shop and my home, I rarely go elsewhere. My earnings are rather limited, but I have no complaints. What I earn is destined for me, no matter what I do. That is why I am so content. If I find myself wanting something, I say to myself, this is the material self and has to be controlled. Somebody asked a Sufi, "Where do you stand in this world?" He answered, "Nothing happens in this world without my consent, not even a leaf falling off a tree." He was asked, "How can you make such a claim?" He answered, "Because I am completely happy with all the events in the world, as I know that God is in charge. I am not concerned with the material well-being." Some devout people are very interested in appearances and procedures of religion. I am more interested in what lies behind words and appearances, and try to see through them.

My eldest son wants to get married. This is why he has been pushing to redevelop our house. Previously, it was a one-story house with a small courtyard. Our kitchen was in the basement, but washing and a lot of other housework were done in the courtyard. Now, he has pulled down the old house and has built a three-story building, with one separate apartment on each floor. He and his bride will occupy the top floor. I don't mind these changes. It would be very difficult for him to afford buying a place elsewhere and it is good for us in that the family stays together, although we can't provide enough space for all our children to stay with us after getting married.

Some days, especially Mondays and Thursdays, I have some visitors here at home. They are from all walks of life and often come for recharging their spiritual batteries. We drink tea, smoke cigarettes, talk, and read classical Persian poetry, whose major works are often inspired by mysticism.

7

I am a public sector employee, living with my wife and three children in our home in a township near Karaj, in the western suburbs of Tehran. I have to commute to Tehran every day, which means spending a long time on the road waiting for and on the bus. My income is so low that I am constantly under stress with desperation. I feel very ashamed in front of my family, as I can't provide a decent standard of living for them. My children are still young and have many needs that I am unable to meet. This is why I am very unhappy, but I don't know what to do. This is a problem of people on fixed incomes who get squeezed when prices go up like this. Traders and businessmen just adjust their charges. We are stuck. Some of my friends moonlight, mainly running a taxi service with their car. Others spend all their time buying subsidized goods and then selling them in the black market. Some even resort to corruption. None of these options is open to me.

We managed to buy this house with a lot of problems, saving for a long time, selling my wife's jewelry, and borrowing from our relatives. We ran out of money in the middle of construction, so we had to postpone some of the work, including the facade, which is now left in rough brick. I hope to finish the work in the future whenever I can afford it. A lot of other buildings in this township are in this state of half completion. We used to rent a small flat in the south of the city but moved out here to have a place of our own. We could not afford to buy a place inside Tehran. There are many others in this township who are like us. Others are newcomers from towns and villages. I would say people here are friendlier with each other than where we were in the city. Many are from the same town or village and stick to each other for support. But there are many problems here in this township: the school is not good, our streets are not paved, our electricity connections are erratic, and we don't have a park. Luxuries such as telephone lines are out of the question. There are only a few shops catering for a large number of people. We have to queue for everything. My wife spends most of her time in the queues to buy our basic needs.

I have heard that people kill themselves and their children out of desperation. Sometimes I feel that I am on the verge of doing just that, although my religious beliefs prevent me from committing a sin like that. But what can I do? I feel very old, tired, and hopeless.

8

I am a retired man, living in Elahiyeh, north city. I used to be a senior civil servant before the revolution. After all those years of working, my pension is ridiculously small, only enough to cover my telephone bill. I wonder how some people can live on their pensions. Luckily I have some real estate in Tehran and on the Caspian coast, which helps me earn a living. My house is large, to be honest, too large for me and my wife. Our two children live in California. Last time we were out there to visit them, my wife was keen to see if we could stay there, to be near them. We reckoned if we sold this house, we could buy a good house over there and put some money in the bank to keep us going. I can easily sell this house for one million US dollars. But we both thought it would be hard for us to live in another country at this age. Our children go to work and we are not sure how much time they could spend with us. Moreover, our standard of life here is definitely better, despite all the problems of the past twenty years. We have a gardener and a cleaner, which I am not sure if I can afford over there. Some of my contemporaries who fled the country have lived in the West comfortably. Others have had quite a miserable time. I think I am better off staying here in these last days of my life and be properly buried.

Most days I go out for a walk in the park or visit my friends. I enjoy reading history, especially of the nineteenth century, which is a fascinating period of our history. Some of my friends think we are still under the spell of the conspiracies which were shaped against us in the nineteenth century. Before it was banned, we had a satellite dish installed, which was very entertaining with many channels showing good movies and news. Some people have now hidden their satellite dish inside their houses. I decided it wasn't worth the consequences. I rarely go to the city as the air is so polluted. Some days children and the elderly are specifically asked to stay at home. Our part of the city has been a very good neighborhood for many years. Houses are all large, with good sized courtyards and gardens. But these days things are changing with high rise buildings cropping up everywhere. Just imagine next door a large garden disappears and you are confronted with all these new flats looking into your house. Suddenly your street becomes crowded with noisy cars and people. I don't know what I would do if this happened to me. In one or two cases people have been able to effectively protest against the building project. But mostly the municipality wins, because they and the developers make a lot of money out of it. If it happened to me, maybe I would sell at a very good price and buy a flat. It is not easy after all to keep a large house, cleaning is one problem, security is another. Last year a friend of mine was burgled in the middle of the night while all the family was asleep. The burglars had cut the iron bars of the living room window and had taken away his very precious carpet. Some of these high rise buildings have proper security guards and check all the visitors. Maybe I'd buy one of those flats anyway.

9

I am a businessman, living with my wife and two young children in Darabad, right next to the northern mountains. In the winter, it is difficult to move about, for the snow here is much heavier than the rest of the city. But the compensation is cleaner air and cooler summers, although these days air pollution is everywhere in the city, and the winters are warmer. We are a traditional family with strong religious beliefs. I work with my father, in our factory that makes parts for the car industry. It is located in the far west of the city, towards Karaj, so I travel long distances every day. Many of my friends work in manufacturing, but the industry has suffered a lot from economic sanctions, as foreign credit has been difficult to get and sometimes we have been forced to buy what we need with cash. I work long hours, but my income is good and I own a part of the company. Soon I will take over from my father, who is getting rather old.

It is very common to receive jokes, in particular political jokes, as text messages. But I also use my mobile phone for a better purpose: constantly checking the price of gold and the rate of currency exchange. Some years ago, I spent a large part of my savings on buying gold coins, which I have not regretted, as gold has been going up all the time. My conversations with friends and family revolve around the prices of things: gold, foreign currency, property, cars, as well as everyday necessities like meat and fruit. These days things are very expensive, but there is also a lot of money around. There are many stories about people who are fabulously wealthy, but I have not been close enough to see how they live and what they do. Apparently, a lot of them keep their fortunes outside the country, but Porsche sells more cars in Tehran than in any other city in the Middle East, and I've heard of ice creams covered with edible golden leaves sold for US$300.

My father and I spend a good deal on charities, especially during religious ceremonies. In recent years, my wife and I have also been to many foreign holidays in Thailand, Russia, and Europe. But we also spend some weekends in our holiday villa on the Caspian Sea coast, on the northern side of the Alburz mountains. I also bought an apartment in Dubai, which has lost a lot of its value since their property crash, but my wife likes to go shopping in Dubai, so we keep it for now.

My father-in-law is a well-off merchant, importing medical devices for hospitals. He has good connections with the bazaar in the city center, where all the merchants used to concentrate and may still be the heart of the city's economy. He used to have an office there himself, but he now works from an office in the north-central part of the city. He tells us about merchants who have a tiny office of 2 by 3 meters in the bazaar, furnished only with a desk and a telephone, but making millions each day by just buying and selling imported goods, without even seeing the goods. Although a new stock exchange has opened, the bazaar still works in the more traditional ways. With a new metro line and the pedestrianization of its front street, the bazaar has revived as a retail center, being much more widely accessible than before and full of shoppers.

Some of my friends have left Iran, but I have never thought about leaving the country and living abroad. I love it here. Despite all the political troubles, I am still optimistic about the future. Some people with radical ideas are just naïve or idealistic, and there is also a lot of mismanagement and vested interests. But I don't think any ordinary person here has the appetite for another revolution or war. People just want peaceful improvement, rather than violent change.

Figure 5 Luxury apartments in the northern parts of Tehran.
Source: © Ali Madanipour.

Figure 6 Tehran's traditional bazaar has kept some of its appeal as a retail center.
Source: © Ali Madanipour.

Figure 7 Political news avidly followed at a street kiosk.
Source: © Ali Madanipour.

Figure 8 Tehran's central areas boosted with pedestrianization and the expansion of the metro network.
Source: © Ali Madanipour.

Figure 9 Street vendor serving warm beetroot.
Source: © Ali Madanipour.

Figure 10 The city of 12 million people is growing upwards.
Source: © Ali Madanipour.

Figure 11 A street in south Tehran.
Source: © Nasroulah Kasraian and Hamideh Zolfaghari.

Figure 12 A tea house in south Tehran.
Source: © Nasroulah Kasraian and Hamideh Zolfaghari.

Figure 13 A cobbler's shop in central Tehran.
Source: © Nasroulah Kasraian and Hamideh Zolfaghari.

Figure 14 Apartments in south Tehran.
Source: © Nasroulah Kasraian and Hamideh Zolfaghari.

You have now "heard" the stories of these nine residents of Tehran. We hope these stories give you an idea of the complexity and diversity of the city—of how people live, what kind of work they do, what sorts of relationships and values they cherish. These stories can form a starting point, a launching pad into a much more detailed and comprehensive view of the cities of the global South.

NOTE

1 This article draws subitatially from A. Madanipour (1998) *Tehran: The Making of a Metropolis*. West Sussex: Wiley.

PART II

Making the "Third World" city

Figure 15 *Proposal for a Vietnamese Landscape.*
Source: © Tuan Andrew Nguyen, 2006.

SECTION 1
Historical Underpinnings

Editors' Introduction

We start this introduction by briefly looking at the field of urban history and the conceptualizations of the city used by historians. Following this, we describe "historical underpinnings," a term that we borrow from Henri Lefebvre (1974[1991]) to emphasize the spatial and material basis of social relations that shape the urban experience through time. The article selections focus on a particular period in history when European colonial empires dominated, an underpinning that is shared by most of the global South and shapes its cities in unique ways.

The city has long been the subject of historical study as people sought to understand how cities emerged and thrived. However, as a field of study, urban history came into its own in the early to mid-twentieth century as the United States and Europe became predominantly urban societies. Modes of analysis often used to study cities across the global South developed in the North in the context of then dominant social science concepts and theories. This began to change as urban history drew on other academic fields and subfields on the study of empire, colonization, development, industrialization, immigration, housing, rights-based struggles, and social movements (Maylam, 1994). Today urban history has become an extensive and diverse field of study.

In this introduction we will focus on three aspects that are important to cities of the global South: (i) whose history is being uncovered and recorded, and by whom; (ii) how do historians theorize the city when they write the "history of the city"; and (iii) how is history used to understand and theorize the Southern city–an issue of importance to those of us interested in understanding the cities this *Reader* covers.

KEY ISSUES

Whose history?

Understanding whose history is being recorded and by whom is important to evaluating any historical account. This is as true of historical and archaeological records of ancient urban societies in Mesopotamia, China, Egypt, or the Indian subcontinent as it is of accounts produced more recently across the global South. Here we are interested in the period that starts with early European colonialism. Historical accounts typically have one of three distinct conceptual foci. First, there are the celebratory Eurocentric accounts that privilege European modernity and emerge from the idea of European superiority in the age of colonial expansion. In this perspective, colonial expansion brings enlightenment, progress, and change to the passive, immobile worlds of tribal and caste solidarities. Histories written from this perspective, whether by travelers, colonial administrators and scholars, or by the colonized elite remain important sources for understanding cities. Second, there are the counter-narratives of nationalist historiographers. These histories aim to create a culturally coherent account that places the nation at the center of the narrative, whether it is India, South Africa, Mexico, or Indonesia. Finally, there are the critical radical histories written in response to both colonialist and nationalist perspectives. These subaltern histories or histories from below focus on presenting the perspectives of those who do not hold privilege or power within the system. Regardless of which schools of historiography they engage, these radical investigations demonstrate a commitment to uncovering how power works to make the city in different moments across time.

Theorizing the city

A theoretical concept of the city and how it works is central to writing any "history of the city", even if historians do not always express such concepts overtly.

Early urban historians frequently conceptualized the city as a *single entity, a geographically discrete place, and a physical artifact*, often within a larger system of cities (the metropolis, the region) and clearly distinguished from the rural. These historical accounts defined settlements along a trajectory that moved from rural to small town, the peri-urban fringe and exurbs to the suburbs and then into the city itself. But even in thinking about cities as discrete entities, there are two explanations for their origin and growth (or death). The first tends to focus on the city as *an economic entity, a central locus in an economic system* of trade, manufacturing, and commerce. This is connected to or separate from the idea of the city as *an expression of political relations and power*, an important node in systems of governance. Understanding the nature of urban economic viability, for example, highlights how the fortunes of each city are linked closely to their regions and how they change over time. Just as importantly, it allows historians to focus on the importance of location, on trade and shipping routes, and on networks of production, circulation, and consumption.

By contrast, understanding cities as spatial articulations of political relations and power highlights the evolution of contemporary democracy, dissent, and rule. Examples of cities that project political power include metropoles at the center of empire such as Rome, Constantinople, Beijing, Delhi, London, or Washington, DC. The political relations perspective also allows historians to analyze cities as places that offer an opportunity for the public to come together to deliberate and voice dissent. Thus, in these histories we find people organizing, speaking up, and fighting for their rights in the streets, squares, parks, and neighborhoods. We may find workers, women, immigrants, and minorities in action, whether in the organized movements of the nineteenth century or the more recent Arab Spring and Occupy movements.

Using these three conceptualizations of cities allows historians to construct the story of a single city or use the stories of single cities as exemplars of an age which serve to trace trajectories of urbanization processes. There are several studies that help students of the city understand the relationship of the city to region, to broad urbanization processes, the successes and failures of policies, and the spatial dimensions of social and political control over time. These include Lewis Mumford's widely read *The City in History: Its Origins, Its Transformations and Its Prospects* (1961), Spiro Kostof's two volume opus *The City Shaped* (1991) and *The City Assembled* (1992), and Arnold Toynbee's heavily critiqued *Cities on the Move* (1970). There are also works that opened up questions on the historical roots of colonialism and globalization, such as Janet Abu-Lughod's *Before European Hegemony* (1989). Other writers have used a more personal form of thinking about urban futures while drawing idiosyncratically on urban historiography. These include influential writings on the city by architects and planners who mine the past to imagine the visionary city of the future, from Ebenezer Howard's (1902) "garden cities" and Corbusier's "radiant city" (1935) both of which continue to assert significant influence, to Constantine Doxiadis's "ecumenopolis" (1968, 1974).

But cities are much more than discrete, imageable entities of clear shape and form, just as they are much more than containers that allow life to flourish. Lewis Mumford captured the shift that was taking place in approaches to understanding cities in a seminal article *What is a City?* (1937) when he wrote: "The city in its complete sense, then, is a geographic plexus, an economic organization, *an institutional process*, a theater of social action, and an aesthetic symbol of collective unity" (1937: 94, italics ours).

Mumford's view (shaped by Patrick Geddes, the Scottish biologist turned regionalist after whom Mumford later named his son) was radical in its presentation of the idea of the region within which the city is embedded, and in its early articulation of the idea of constitutive social and cultural processes. Since then, in the last seventy years or so, sociocultural and political interpretations of the city have become richer and more nuanced. Two important factors have been central to this process. First, the ascendancy of Marxist and, later, postmodern and poststructuralist urban theory in the second half of the twentieth century brought new conceptual frameworks and understandings. Second, the many changes in the technologies of movement and communication as well as shifting geopolitical alignments, have led to deeper analyses of the economic, political, and social nature of cities. A new urgency to theorize contemporary cities emerged when the world's urban population crossed the 50 percent mark around 2010, with most future growth predicted to occur in cities of the global South.

Using history to understand the city

In an essay focused on the United States, *What Good is Urban History?*, the sociologist Charles Tilly described urban history as the "quintessential social history" with cities as the "privileged sites for study of the interaction between large social processes and routines of local life" (1996: 703, 704). He also pointed out that historians as careful scrutinizers of archives, documents, narratives, and artifacts tend to be cautious theorizers, leaving theorizing the city through history to sociologists like Tilly, geographers, anthropologists, planners, economists, and other commentators. The broadest and most compelling theorizations of the city, however, draw on a historical understanding to allow us to read the urban structures and processes we inherit and to make sense of the social and political relations that produce and are produced by life in the city.

The influential Chicago School of Sociology played a huge role in theorizing the North American city of the early twentieth century. They created a model that influenced urban studies more broadly for several decades. Marxist critiques were crucial in the reconceptualization of the contemporary city and two theorists were essential to this effort: the geographer David Harvey and the sociologist Manuel Castells. Harvey's prolific writing, including *Social Justice and the City* (1973) and *The Urbanization of Capital* (1985), shaped not just Marxist geography as a field, but the entire field of urban studies. Harvey emphasized a dialectical relationship between the city as process and as object or thing when he wrote, "We should focus on processes rather than things and we should think of things as products of processes" (1997:21). Manuel Castells's writings, building on the work of his teacher, Alain Touraine, were key in explaining the transformation of the city. His classic *The Urban Question* (1977) framed the political economy of contemporary cities, while his subsequent volume, *The City and the Grassroots* (1983), stressed the role of social movements and resistance. Later Castells turned his attention to the technologies of the Internet and the digital age in *The Information Age Trilogy* (1996, 1997, 2000). A rich scholarship has emerged employing these and other related theoretical foundations to explain the contemporary urban world in relation to broader processes of political-economic change and globalization. Understanding these urban constitutive processes helped identify sites for potential action for those who chose to be political activists, community organizers, or policy and design interventionists.

History thus both provides lessons in the possible paths that urban development can take, and is the underpinning to the city we experience today. As we noted earlier, we borrow the term "underpinning" from Lefebvre who emphasized that every social relation needed a spatial material underpinning and noted (1974[1991]: 403) that "'something' always survives or endures." His work, like Harvey's and Castells's that followed him, was an early exploration of the production of that "something"–space. Yet Lefebvre did not fetishize space, arguing instead for a critical historical analysis of the dialectical relation between space and social relations, which could explain the changing urban experience of contemporary cities. While we cannot discuss Lefebvre, Harvey, or Castells in more detail in this section on historical underpinnings, we highlight two articles that focus on the colonial experience that provide a central underpinning for theorizing cities across the global South. The colonial period is also the starting point for this *Reader* and in some sense for the category of the global South, an idea we discussed in more detail in the introduction to this *Reader.*

ARTICLE SELECTIONS

There are two selections for this section. The first is by the sociologist Anthony King, renowned for his work on the production of colonial and postcolonial built environments (1976, 1984, 2004). The second selection is from the work of geographer, Doreen Massey, well known for her writing on reconceptualizing space and articulating its relationship to politics (see *Space, Place and Gender* (1994) and *For Space* (2005)). King drew on his large oeuvre to write this article for the *Reader*, while Massey's article comes from a three-volume textbook that she and her collaborators prepared for a course, *Understanding Cities*, at The Open University in the UK. The ideas they explore here are examples of how history is used to understand the city. Both selections focus on the colonial experience, which is central to understanding most contemporary cities of the global South.

King's path-breaking book on colonial Delhi, published in 1976 and researched while he was teaching at the Indian Institute of Technology at New Delhi, changed the landscape of urban studies in the global South in important ways. It marked a break from both the imperial colonial and the nationalist traditions, while opening up questions of cultural and social processes in the production of built form. Several other publications followed including *The Bungalow* (1984) and *Spaces of Global Cultures* (2004). In the piece included here, King addresses the ways in which historical forms of colonialism–specifically in the two long waves of expansion from 1500–1800 CE, and from 1800–1925 CE–have had a powerful influence and shaped the nature of urbanization and the development of what were long called "third world cities." It is a synoptic piece, drawing on a large literature to describe the ways in which colonialism is critical to understanding contemporary cities across the postcolonial global South. King introduces the reader to the patterns and forms of colonial urban development that persist and shape contemporary places. He indicates the importance of understanding how our context shapes us and produces particular forms of scholarship–making us blind to some ideas while remaining open to others. Most importantly King emphasizes the centrality of understanding local conditions noting that even as colonial settlements across Asia and Africa may share similar features, they differ according to the "geography and history of the settlement" itself (page 36)–a point that we emphasize as well.

Massey explores the "internal intensities of individual cities," their specific histories but places the city in the world, in a wider set of interconnections demonstrating how this can potentially reorder our understanding of the city and its relationships. More importantly, given Harvey's distinction between the city as thing and as process, she speaks to the notion of the city as "open intensity." For Massey, cities are first and foremost places of meeting, of encounters, and they are open, continually changing. It is from this idea that everything else flows. Thus, when Massey looks at a city in history, she sees multiple histories, a mixture that evolves over time in the geographic sense of expressing both the specificity of a place and its changing links with the world beyond. To quote Massey, in each moment of its history the city is thus a "focus of a wider geography" (page 43). The city is thus a focus for interconnections–within itself and to the world around. She goes on to explain that mixing demands the maintenance of difference through contestations, where the spaces and enclaves created by city dwellers and by the rulers are fundamental to the expression and organization of this mixing. Crucially for Massey cities are essentially open places of wider interconnections. Through writing the history of Mexico City, the focus of the selection we present here, Massey forefronts the continual process of negotiation and renegotiation that construct the city as thing.

Figure 16 Empires through the ages.
Source: © Zapiro.

REFERENCES

Abu-Lughod, J. (1989) *Before European Hegemony: The World System AD 1250–1350.* New York: Oxford University Press.

Castells, M. (1977) *The Urban Question: A Marxist Approach.* Cambridge, MA: MIT Press.

Castells, M. (1983) *The City and the Grassroots: A Cross-Cultural Theory of Urban Social Movements.* Berkeley: University of California Press.

Castells, M. (1996) *The Rise of the Network Society*. Malden, MA: Blackwell.

Castells, M. (1997) *The Power of Identity*. Malden, MA: Blackwell.

Castells, M. (2000) *End of Millennium.* Oxford: Blackwell.

Le Corbusier. (1935) *La Ville Radieuse* (The Radiant City). Boulogne: Editions de L'Architecture d'Aujour'hui.

Doxiadis, C. A. (1968) *Ekistics: An Introduction to the Science of Human Settlements.* New York: Oxford University Press.

Doxiadis, C. A. (with Papaionnou, J. C.) (1974) *Ecumenopolis: The Inevitable City of the Future.* New York: W. W. Norton.

Harvey, D. (1973) *Social Justice and the City*. Baltimore, MD: Johns Hopkins University Press.

Harvey, D. (1985) *The Urbanization of Capital: Studies in the History and Theory of Capitalist Urbanization.* Baltimore, MD: Johns Hopkins University Press.

Harvey, D. (1997) "Contested cities: Social process and spatial form." In N. Jewson and S. McGregor (Eds.), *Transforming Cities: Contested Governance and New Spatial Divisions.* London, New York: Routledge, pp. 19–27.

Howard, E. (1902) *Garden Cities of To-morrow* (3rd edn). London: S. Sonnenschein and Co. (originally published in 1898 as *To-morrow: A Peaceful Path to Real Reform*).
King, A. (1976) *Colonial Urban Development: Culture, Social Power, and Environment.* London: Routledge & Kegan Paul.
King, A. (1984) *The Bungalow: The Production of a Global Culture.* London: Routledge & Kegan Paul.
King, A. (2004) *Spaces of Global Cultures: Architecture, Urbanism, Identity.* London: Routledge.
Kostof, S. (1991) *The City Shaped: Urban Patterns and Meanings Through History.* Boston: Little, Brown.
Kostof, S. (1992) *The City Assembled: The Elements of Urban Form Through History.* Boston: Little, Brown.
Lefebvre, H. (1974[1991]) *The Production of Space.* Cambridge, MA: Blackwell.
Massey, D. (1994) *Space, Place, and Gender.* Minneapolis: University of Minnesota Press.
Massey, D. (1998) "Cities interlinked." In D. Massey, J. Allen, and S. Pile (Eds.), *City Worlds.* London: Routledge.
Massey, D. (2005) *For Space.* London: Sage.
Maylam, P. (1994) "Explaining the apartheid city: 20 years of South African urban historiography." *Journal of Southern African Studies*, 21(1): 19–38.
Mumford, L. (1937) "What is a city?" *Architectural Record*, LXXXII: 58–62. (Reprinted in R. LeGates, and F. Stout (Eds.) *The City Reader* (1991). London and New York: Routledge.)
Mumford, L. (1961) *The City in History: Its Origins, Its Transformations, and Its Prospects.* New York: Harcourt, Brace & World.
Tilly, C. (1996) "What good is urban history?" *Journal of Urban History*, 22: 702–719.
Toynbee, A. (1970) *Cities on the Move.* New York: Oxford University Press.

"Colonialism and Urban Development"

Essay written for *Cities of the Global South Reader* (2015)

Anthony D. King

INTRODUCTION

My aim in this essay is to address the ways in which historical forms of colonialism have influenced the nature of urbanization and, especially, the development of cities in the "Third World." I use imperialism to refer to the imposition of the power of one state over the people and territories of another, frequently by military force. Where imperialism originates in the metropole, what happens in the colonies resulting from economic, political, and cultural control and domination is colonialism, even though this can take different forms (Loomba 1998). As far as planning and urbanism are concerned, colonialism is also about the forms of dominance that particular colonialisms take (AlSayyad 1992). Despite its power, colonialism is rarely totally hegemonic. Colonial influences and practices have invariably been resisted as well as accommodated; vernacular cultures have been reinvigorated as well as transformed (King 1995).

Colonialism has existed throughout history but the main focus of this essay is on the overseas expansion of Europe from the fifteenth century onwards, into the Americas, Asia, the Middle East, and Africa. This took various forms from significant numbers of people permanently occupying territory such as the settler colonies of southern Africa, Australasia or the Americas, or fewer colonists exploiting the resources and indigenous labor of other lands. At the height of European and Japanese imperialism before the First World War, imperial powers held some 85 percent of the earth's territory as colonies, protectorates, dependencies, dominions, or commonwealths.

There have been two long waves of colonial expansion and contraction, the first, from 1500 to 1800, the second, from 1800 to 1925. Each related to long wave cycles in the world economy. The major colonizers were Spain, Portugal, and the Netherlands, principally between 1500 and 1750, and France and Great Britain, from 1600 to 1925. In this second period, from 1870, Belgium, Germany, Italy, Japan, and the USA entered as "late-comers." In what we may call the "first competitive era" the spheres encompassed by colonization included Iberian America, the greater Caribbean, Northern America, African ports, Indian ports, the East Indies, and, in the second era, the Indian Ocean islands, Australasia, interior India, Indo-China, interior Africa, the Mediterranean, Pacific Ocean islands, Chinese treaty ports and Arabia (Taylor 1985).

Regardless of its form, colonialism has been a major historical force shaping settlements and cities, patterns of urbanization more generally, and the development of built forms and architecture.

COLONIAL PORT CITIES

Unlike the earlier empires of China or the Moghuls, which mainly covered territorially contiguous lands (Darwin 2007), European empires of the so-called "Age of Discovery" were the outcome of the overseas trading and military intentions of maritime powers. The colonial port cities of these empires were built on the profits (and often labor) derived from slavery, tobacco, sugar, precious metals, tea, coffee, and other commodities. They were eventually to grow into enormous urban conglomerations and cities which today dominate the coastal regions of the world's continents: New York, New Orleans, Rio de

Janeiro, Buenos Aires, Lagos, Cape Town, Colombo, Mumbai, Kolkata, Chennai, Singapore, Hong Kong, and Shanghai. While in many cases built up from hubs of an earlier ocean trade, all developed as prominent colonial port cities (Basu 1985). The rapid growth of these coastal cities in many cases left the inland regions largely undeveloped with poor infrastructure and gross gender imbalances. Later, Western technologies, particularly the railway, reinforced this process of urbanization, linking the interior of continents to the ports, draining out the economic surplus to the metropole and, as in India, providing colonial armies with rapid transport facilities to sites of potential political unrest. Yet in both European and Ottoman empires, the railway also created economic and political opportunities and networks (Celik 2008). Burgeoning port cities became the sites of social, political and cultural exchange as well as exploitation.

Many of these colonial ports were to develop into major cities and became the first colonial capitals. The powerful legacy of colonialism in Africa, for example, is shown by the fact that in 2010, in 46 of the continent's 53 independent states, the primary city in the urban hierarchy is the colonial capital, major port, or port capital. Over half of the 50 largest cities on the continent were cities of those historical types (Myers 2011). They were the sites of government, administration, commerce, and the proselytizing activities of both Western and local missionaries. In many cases, these cultural and social melting pots are the origin of the multiethnic, culturally hybridized cities of today.

Decades, or sometimes centuries, later, when what were once colonies became independent states, in some instances the national capital was transferred from these early port cities to an inland site. Here, such changes provided a new space for the symbolic construction, through planning, architecture, and urban design, of a national cultural identity and the development of a national polity; in Brazil, from Rio de Janeiro to Brasilia; in Nigeria, from Lagos to Abuja; in Tanzania, from Dar-es-Salaam to Dodoma. In India, such a move, from coastal Calcutta to geographically central Delhi, had—ironically—already been taken in 1912 by the imperial government. For newly independent nation-states, shifting capitals has often entailed massive population upheavals, enormous expenditure, and the diversion of resources from the pressing needs of development.

COLONIAL URBAN DEVELOPMENT

> The colonial world is a world divided into compartments of native quarters and European quarters, of schools for natives and schools for Europeans; in the same way we need not recall apartheid in South Africa . . . The colonial world is a world cut in two. The dividing line, the frontiers are shown by barracks and police stations. In the colonies it is the policeman and the soldier who are the official, instituted go-betweens, the spokesmen of the settler and his rule of oppression . . . The settler's town is a strongly built town, all made of stone and steel . . . The town belonging to the colonized people, or at least the native town, the Negro village, the medina, the reservation, is a place of ill fame . . . It is a world without spaciousness; men live on top of each other . . . The native town is a crouching village, a town on its knees, a town wallowing in the mire. (Fanon 1968: 37–39)

> The major metropolis in almost every newly-industrial country is not a single unified city, but, in fact, two quite different cities, physically juxtaposed but architecturally and socially distinct. These dual cities have usually been a legacy from the colonial past. It is remarkable that such a phenomenon has remained almost unstudied. (Abu-Lughod 1965: 429)

These quotations, the first by Martiniquan nationalist writer, Frantz Fanon, conveying the sense of oppression and injustice experienced by the colonized in North Africa, the second, by American sociologist Janet Abu-Lughod, describing Cairo and other cities in developing countries, each underscore the idea of segregation and duality as the central features of what has come to be known as "the colonial city."

As Osterhammel (1997) and others have pointed out, the oft-repeated emphasis on duality has tended to oversimplify the socio-spatial realities as well as the representation of the colonial city itself. If we focus only on the geographical and spatial features, the form may differ depending on whether the colonial city is a port, inland town, hill station, major capital, military base, or transport hub. These geographical questions of location, site, and space are only some of many characteristics which have been suggested as features and/or determinants of colonial urban form (Basu 1979, 1985; King 1985; Pacione 2001: 437).

Other supposed features of the colonial city include "external origins," "pluralistic structure," "racial segregation," and "hybridized culture." For Osterhammel, the key feature is not separation but rather the opposite: mixing, pluralism, and hybridization.

Some typology may help us to understand the physical and spatial characteristics of different European colonial cities around the world. Such a typology would distinguish between societies and territories according to population size, nature of economy and culture, and extent of urbanization. Other criteria would include motives for colonization, the number and degree of permanence of the colonizing population, the degree of coercion exercised over the indigenous inhabitants, including complete enslavement or even annihilation.

Linked to the question of motivation is the resource being sought and how this is determined, over time, by changes in the technological and cultural appraisals of the metropolitan power. At least ten possibilities affected the physical, spatial, and social form which the city developed. Thus, where an indigenous settlement already existed, the colonists have the following choices, depending on their intentions, number, requirements, and the stage of colonization:

1 the site and accommodation are occupied with little or no modifications (Zanzibar);
2 the site and accommodation are occupied but modified and enlarged (as with many small inland colonial administrative centers);
3 the existing settlement is razed and built over (Mexico City);
4 the site and accommodation are incorporated into a new planned settlement (Batavia);
5 the new settlement is built separate from but close to the existing one (New Delhi);
6 the existing settlement is ignored and a new one built at a distance from it (Rabat).

Where there is no previous indigenous settlement, the new foundation can be built as follows:

1 for the colonists only—non-colonists (i.e. indigenous inhabitants or "intervening groups") remain outside, providing their own settlement and accommodation (New York);
2 for the colonists only; no other permanent settlement by non-colonial groups is permitted (Sydney);
3 for the colonists but with a separate location and accommodation for indigenous and intervening groups (Nairobi);
4 for the colonists and all (or some) of the intervening and indigenous groups in the same area (Kingston, Jamaica).

Given the spatial limitations, this essay is particularly focused on colonial developments in South Asia and, to a lesser extent, Africa.

THE COLONIAL CITY: A COMPLEX CONCEPT

While an early reference to the colonial city was made by American anthropologists Robert Redfield and Milton Singer in their classic article on "The Cultural Role of Cities" (Redfield and Singer 1954), the first serious attempt to theorize the notion was by American geographer Ronald Horvarth (1969). The emergence of this task was prompted by a growing dissatisfaction with existing comparative urban theory and its excessive focus on Western industrial experience and the dominance of the Chicago School.

Another impetus, however, came from predominantly Western-based urban professionals, planners and architects concerned with urban development planning following the end of colonial rule (United Nations 1971). Between these often inter-related sources, a more theoretically informed understanding of the idea of both colonial and postcolonial cities emerged. This understanding recognized that, despite the political independence of some seventy nations in the decade or two after 1945, the persistence of colonial urban structures—represented by the grossly unequal social and spatial divisions in major cities and the unequal distribution of resources, including access to housing, schools and health facilities—was an affront to the democratic aspirations of newly independent nations. (See quotation by Abu-Lughod 1965 above.) This "colonial space" of the European part of the city (including its distance from the indigenous settlement) was based not just on the standards of the "modern city" in the West but also on the demands of a colonial elite, who held an inflated representation of those standards. In many postcolonial cities, decades after independence, while the urban areas allocated to indigenous people under colonialism crumbled under the pressure from rapid migration, indigenous elites had moved into, and were still living in, the spaces occupied by their colonial predecessors.

Recent studies of the colonial city phenomenon, however, by a new generation of scholars, suggest that colonialism is too often seen as a hegemonic, unifying category, which reduces differences and ambiguities. While generalizations can be valuable, they also obscure important differences between both times and places.

For instance, Indonesian architect and scholar, Abidin Kusno, writing in 2000, asked

> to what extent have studies centered on European imperialism themselves "colonized" ways of thinking about colonial and postcolonial space . . . the standpoint or focus from which these works are written still tends to be that of Europe. (Kusno 2000: 6)

Kusno's position here is characteristic of what, from the late 1980s, has become known as a "post-colonial critique," an analysis of colonial discourse that questions "Western knowledge's categories and assumptions" (Young 1990: 11). The intention, in Chakrabarty's (1992) incisive phrase, is to "provincialize Europe" by placing critical emphasis on indigenous rather than colonial perspectives and accounts. Studies from the 1990s have shifted the focus of research on the colonial city by making more extensive use of vernacular language sources and shifting their intellectual position.

For example, as far as the understanding of colonial Delhi is concerned, recent studies of "old" Delhi have used a range of vernacular sources to criticize the rigid binary structures ("colonial/colonized," "traditional/modern," etc.), within which the development of New Delhi has been portrayed. They have shown how space between the two cities was less a strict divide and more an opportunity, which the indigenous population utilized to their own advantage (Hosagrahar 2005). Similarly, Yeoh's (1996) study of colonial Singapore provides evidence to show how far the Chinese community, rather than being confined in the colonial city, maintained control over their own spaces. Writing about nineteenth-century Calcutta, often represented in European texts and maps as being rigidly divided between a "white town" and a "black town," Chattopadhyay (2005) has argued that the supposed boundaries between these two places were essentially "blurred." In some cases, the very existence of two such places is unrecognized in local accounts and perceptions. Rather, the terms, "black town" and "white town" were used by the British to "sustain an imperial narrative of difference and European superiority" (ibid: 77); boundaries were more fluid than the imperial narrative would suggest. These and other studies (Chopra 2011; Glover 2008) show that generalizing about "the colonial city" (King 1985; Pacione 2001) needs to be accompanied by knowledge of specific local cases.

COLONIALISM AND URBAN PLANNING

Town and city planning largely developed in Europe and North America in the early years of the twentieth century. They were a reaction to the rapid and chaotic growth of towns and cities caused by capitalist industrialization. Even before the formal acknowledgement of "town and city planning," many states had already passed legislation and taken measures to "discipline" urban development. Public health issues, particularly the spread of diseases, often precipitated measures to regulate the layout of streets, control noxious industries, ensure a minimum supply of open space, and establish "Improvement Trusts" to oversee building practices. Such steps were a major shift from earlier "grand schemes" of urban design associated with aristocratic, monarchical, or religious power.

In the colonies, the urban assumptions of industrial capitalism combined with cultural practice to produce particular spatial forms well before a specifically named "professional" expertise of "town planning" arose. In understanding this development, particularly in the British colonies, three phases are crucial:

1. A phase up to the early twentieth century when settlements, camps, towns and cities were laid out according to codes and principles, designed to ensure military and political dominance.
2. A second phase which began in the early twentieth century. This coincided with the development of a formally established practice of "town planning" theory, legislation, ideology, and professional knowledge. In this phase, the Master Plan was the key concept.
3. A third, postcolonial, and neo-colonial phase following independence, which tends to continue colonial practices.

Town planning was also exported to the colonies. However, this exportation took place in a context of economic relations that were shaped primarily by the needs of the colonizing power. In town planning terms, this led to what sociologist Manuel Castells (1977) termed "dependent urbanization." This meant that in metropolitan society urbanization went hand in hand with industrialization, whereas in the colonized territories urbanization took place in the absence of industrialization. This is a powerful illustration of the systemic nature of the imperial–colonial relationship. Urban planning in the colonies was part of this process of "dependent urbanization." The evolution of urban systems (and regional planning) and the organization of urban space in colonial society can thus be accounted for by "the internal (and especially) the external distribution of power" (Friedmann and Wolff 1982: 13–14).

In this context, the Master Plan became the instrument of power for controlling and disciplining the occupation of land. In the metropole, the Master Plan aimed to bring order to the disorder and chaos of the market system, simultaneously marginalizing the poor. In the colonies, the Master Plan, as a map, similarly excluded disorder, the poor, and especially, squatter settlements which, officially, "did not exist." At the micro level, planning exercised control over street traders and informal housing, its ultimate object, through "clearance policies," being one of "spatial purification" (Dupont 2011).

Nonetheless, there were local populations who met urban planning initiatives with both resistance and accommodation. As various authors have shown (Nasr and Volait 2003; Yeoh 1996; Hosagrahar 2005) when planning is "exported" from the metropole, outcomes are determined by the local indigenous population as well as by colonial authority. In British colonies a good example of this dynamic was the "export" of the Garden City movement. Developed from the ideas of British urban reformer and activist Ebenezer Howard's 1902 book, *Garden Cities of Tomorrow*, the central premises were promoting "health, light and air." Practices emphasizing orderly development, low density, physically separate dwellings and extensive greenery were offered as "the peaceful path to reform." These were as much a political as a social message. From the early twentieth century onwards, Garden City ideas were promoted not only for the new capital of Delhi in India but throughout the empire (1996). When exported to the colonies, these ideas were for the white expatriate settlement only, not for the indigenous population. Other imperial powers, such as the French in Senegal, the Belgians in the Congo Free State and the Dutch in Mentem, Jakarta, loosely adopted Garden City principles which translated into low-density settlements, with villas in high-walled compounds and tree-lined avenues. By buttressing white privilege, these configurations reinforced the consciousness of ethnic and racial divisions in colonial society. Each group had its own space without disturbing the overall distribution of power.

PLANNING AND HEALTH

At the root of all town planning practices in metropole and colony was the aim of maintaining public health. This was underpinned for decades by the erroneous belief that diseases like malaria were caused by impure air produced by noxious environmental conditions. Whether in metropole or colony, this notion of impurity determined the location, site, and layout of settlements and buildings. The constant factor in such planning was the idea that the indigenous settlement was the potential source of disease. The same assumptions determined the relatively low density of buildings and the generous allocation of space between buildings, especially in wealthy districts. Even after the discovery (in the 1850s) of the waterborne nature of disease, and more importantly, Robert Ross's discovery of the link between malaria and the anopheles mosquito in 1897, the space standards of European residential settlements in the colonies remained significantly higher in comparison with those of the metropole (Chang and King 2011). Similar reasoning explained why colonial buildings were often located on the highest ground although members of the indigenous population also read this (not incorrectly) as a sign of colonial prestige and power (Njoh 2009). Following independence, the excessive space standards in and around colonial residential buildings, frequently occupied by the indigenous administrative elite, continued to reflect the gross inequalities of space and housing.

What Swanson (1970), writing about South Africa, has termed the "sanitation syndrome" became the driving force in colonial planning, defined according to the political and cultural criteria of the metropolitan power. Colonial powers classified vernacular forms of shelter as unhealthy and overcrowded. Hence, whether in Belgian, British, or Italian colonies (Home 1996; King 1990b; Fuller 2007), minimal "detached" housing units emerged as replacements. These structures, typically surrounded by "light and air" and "open space," totally disregarded local senses of religious, social, symbolic, and historical meaning. This quest for sanitation frequently became a cover for building demolition, exclusion, segregation, and "police sweeps" through the "native quarter." In many instances, these practices have continued long after political independence.

Where large numbers of expatriates or "settlers" were involved, as in Lusaka, Nairobi, or Delhi, very low-density residential environments were built to suit their convenience and cultural habits. These included large housing plots, lavish recreational facilities (all based on the assumption of motorized transport), telephones, and plentiful domestic labor drawn from the indigenous population.

Ambe Njoh wrote extensively on town planning in British and French colonies. He claims that while the European colonial planners may have had laudable goals, their main motivations were to benefit the white population through the cultural assimilation of the indigenous population, at the same time increasing social control, political domination, territorial conquest, and perpetuating colonial rule (Njoh 2007). In his view, a large portion of sub-Saharan Africa's socio-economic problems can be explained by the physical and spatial planning schemes of colonial officials, schemes which the indigenous authorities vigorously continue to maintain (Njoh 1999).

URBAN DEVELOPMENT: INDIA AND AFRICA

Though sharing certain similar features, colonial settlements in Asia and Africa differed. These differences were dependent on both the national origin of the settlers as well as the geography and history of the settlement. In India, prior to the twentieth century, the most widespread example of conscious urban planning was the location and layout of army cantonments, typically linked by rail to other indigenous centers. Alongside the cantonment, the "civil station" accommodated the political-administrative "managers" of the colony. These two units were located close to, but at a distance from, the main Indian settlements. The major cities resulting from initial colonial contact (Madras, Bombay, Calcutta) developed from small extant indigenous settlements but were primarily commercial, entrepôt ports. Over time they became oriented to the metropolitan economy. While developed in conjunction with the Indian commercial and business elite (Chopra 2011), these entrepôt ports had more in common in spatial and built environment terms with the metropole than with indigenous settlements. To optimize their functioning, they were also subject to colonial planning legislation (Dossal 1991).

The major planning exercise during two hundred years of informal and formal colonial control in India was the planning and construction of New Delhi (1911–1940), the colony's new capital. New Delhi was entirely devoted to administrative, political, and social functions with virtually no attempt made to plan for industrial development (King 1976).

In Africa, Europeans tended to stay in towns and cities. While rural areas were for Africans and their movements were restricted, a significant minority lived in urban areas as well. This was true of cities like Salisbury and in colonies that practiced some assimilation. In colonies where Africans were not allowed in urban areas, it followed that no African should be in town except to provide labor and when required by a European employer, and this was also determined by gender. "Only men were required. Urban housing was therefore rudimentary" (Collins 1980: 232). Feminist scholarship has challenged this generalization (Barnes 1999; Schmidt 1992). Using various legal and illegal mechanisms, women lived in towns against the grain of the Master Plan implemented by the European colonizers.

The functions of newly established colonial centers in Africa were political, administrative, and commercial; buildings housing the key institutions of colonial government included the government house, council or assembly buildings, army cantonment, police lines, hospital, jail, government offices, and housing and recreational space for expatriate European colonial officials. Occasionally, housing was provided for local black government employees. In Lagos, colonial capital of Nigeria, the place typically occupied by the central business district was occupied by a race course (King 1990a: 49–51).

Geographer Gareth Myers (2011), from whom the next two paragraphs are drawn, in summarizing recent research on sub-Saharan Africa by African scholars, highlights many of the key problems which colonialism bequeathed to African cities. However, given that Britain, France, Spain, Italy, Germany, Belgium, Netherlands, and Portugal all held colonies in Africa and for very different periods (Italy, for less than a decade but Portugal, for almost five centuries), his first comment highlights the important question of diversity. Making generalizations, though possible, can also be problematic.

Significant urbanization on the continent began following the Berlin Conference of 1884–1885, a meeting where European powers divided Africa into different spheres of influence. The conference also compelled imperial powers to occupy their spheres of influence or potentially lose the colonies. This "partition" and occupation of the continent coincided with the "explosion of industrial capitalism" in Europe, which increased competition between colonial powers for African markets and raw materials. Urbanization tended to occur along the coast and adjacent to the places of resource extraction. African cities became warehouse towns, bureaucratic capitals or both, and not, as in Europe, industrial hubs. Pre-colonial cities, such as Mombasa and others in North Africa, grew even larger superseding older urban centers and upsetting a more logical development of the urban space economy. Most African countries tend to have a poorly developed and unbalanced urban hierarchy, with large primary cities and very few secondary cities. As in India, capitals and especially port cities grew functionally more in connection with the development of the metropole than with other cities in the colony. Cities grew, not from industrialization or economic growth, but from inward migration (despite restrictions on movement), encouraged by rural unemployment, poverty, increasing landlessness and herdlessness.

Colonialism led to an African urban landscape characterized by segregation and high levels of inequality. But as Myers argues, underpinning this urban reality were ideologies of race, class, and culture which spelled out who should be in the city and where they could reside. This system of segregation reinforced social and racial hierarchy, enabling the colonial regime, the metropolitan power, and elites to accumulate resources (Myers 2011). Planning policies and legislation were used as "instruments of power and domination" (Njoh 2009). Legislation regarding the planning of towns, whether in French or British colonies, was introduced from the metropole, without amendment. For example, in regard to the establishment of "garden cities" or to landholding in French colonies, the Napoleonic property doctrine of 1810 stressed individual, as opposed to corporate or communal, property rights (Njoh 2009: 306).

Various European powers administered their colonies differently. A common distinction is often made between regimes, such as the French, which exercised power through "direct" rule, or the British, which favored "indirect" rule, making use of existing indigenous institutions in governing their colonies. In the former case, colonialism impacted the urban landscape through the institutions it introduced, that is, forms of governance and administration, military power, and other mechanisms. This meant urban control was asserted from municipal buildings, systems of law from the courts and police, and military security from police posts, prisons, and barracks. Economic regulation took place through custom houses and attempts at cultural transformation through educational and religious buildings—churches, convents and schools.

URBANISM, CONTROL, AND IDENTITY

In addition to use as accommodation, buildings, their size, scale and architectural form, expressed colonial power as well as cultural and social difference. Centrally located and often on commanding heights, colonial buildings and architecture became part of the apparatus of control. All, however, depended on control of the land. This was the function of planning, which operated, above all, through the Master Plan, a document that guided urban development for two to three decades into the future. The architecture and urban design of the imperial power, sometimes incidental, at other times (as in New Delhi) deliberately dominant, was conceived to convey not just imperial authority but also a Eurocentric identity. In major port cities such as Calcutta or Madras, Europeans adopted neo-classical architectural forms similar to those prevailing in the metropole.

Independence, therefore, posed major challenges to nations searching for ways to represent their own culture and identity; yet removing the signs of a colonial past, whether street names (Bignor 2008), imperial statues and monuments, or the palatial residences of governor-generals, has done little to convey a genuine representation of a national self. Major planning and architectural transformations in post-colonial capital cities can take decades before being implemented. For example, the palatial Japanese Government-General headquarters in Seoul, built in 1912, stood for fifty years after Korean independence before being demolished in 1995 (Kal 2011). In some countries, investing a newly independent city with a new cultural and architectural identity has often waited for decades until more important issues—land redistribution or the improvement of squatter housing—have been addressed (Vale 1992).

COLONIALISM AND HOUSING

One of the most fundamental transformations in the spatial form of urbanism brought by European colonialism was the prevalent type of shelter. The shape of housing and the shape of the city are inextricably connected. In many parts of the world influenced by European colonialism—the Americas, Middle East, Africa, and parts of Asia—typical traditional housing forms were based on the courtyard principle (Rabbat 2010). Though varying according to geography, culture, religion, household structure and other factors, the basic house form was a series of rooms built around an inner courtyard. Such structures often catered for the extended family with particular space usage determined according to age, generation, or gender. External walls linked up adjoining houses, creating a close-knit cluster, with narrow alleyways providing mobility through the city or village. In some instances, perimeter walls encircled the city or village. The courtyard house influenced forms of privacy, especially for females, and provided security for different ages and generations. In addition, this housing form played an important role in controlling temperatures and was an important organizing space for multiple household functions. The close integration of numerous courtyard forms, separated by narrow thoroughfares, clearly influenced the shape of the urban fabric. Over the last century, however, in many places, such traditional house forms have been transformed from residential to commercial uses. Not only has housing shifted outside the original settlement but traditional courtyard forms have been replaced by free-standing, "bungalow compounds" (King 1995). The growing numbers of villa-style houses, each in its own compound, provided the basis for the development of a typical Western-style suburb in colonial cities. Representative of this process is the case of the Kabuli courtyard house in Afghanistan, where the first "Western-style" villa was constructed in 1894 by the local Emir, a self-styled architect. The new forms provided space for myriad cultural, social, and behavioral changes. The house became a symbol of status, a sign of modernity, a site for investment, a space for commodities, and a storage place for personalized transport.

The introduction of Western-style dwellings and architecture also brought with it new models and materials of construction. For centuries, indigenous peoples had used natural materials to build their dwellings—mud, or mud brick, stone, timber, plus cane and grasses for thatched or woven roofs. From the mid- to late nineteenth century onwards, colonial influences brought in pre-fabricated ironwork, glass, and later cement, undermining local building practices and building labor, and transforming traditional design (Oliver 1987). Having adopted these materials and also Western-style expectations and housing models, middle-class decision-makers now looked upon self-built settlements on the city's fringe, as those of social pariahs. The makeshift houses in these areas were condemned as "squatter settlements," "jhuggies," and "slums," the enormous contribution of their residents to the city's economy and daily life ignored.

COLONIALISM, NEO-COLONIALISM, AND GLOBALIZATION

Definitions of globalization abound. For example, Held et al. (1999) describes it as a "speeding up" in worldwide connectedness in all aspects of contemporary social life. Sociologist Roland Robertson (1992) alternatively depicts globalization as a process by which "the world becomes a single place" and, whether virtually or in reality, that place has become increasingly urban. Regardless of definitions, globalization stands as a multifaceted phenomenon—economic, social, technological, cultural, visual, and spatial.

Globalization is an historical as well as contemporary phenomenon (Hopkins 2002). Imperialism and colonialism have been instrumental in laying the foundations of our present world (Darwin 2007). Colonial empires, from Roman to the British, French to the Spanish, established the infrastructure on which contemporary globalization has been built. Part of this infrastructure is physical and spatial, such as interlinked transportation and communications systems, the planned spaces and built environments of networked cities. Other parts are cultural and social: language systems; social, political, and religious institutions; and legal and educational frameworks. Since the mid-1980s, the concept and also the reality of the "world" and "global city" have been acknowledged as a pivotal focus for understanding the processes of globalization (Friedmann and Wolff 1982; King 1990b; Sassen 1991). World cities are seen as the centers of command and control in the neo-liberal globalized economy, and are the bases for transnational companies, international banks, producer and financial services. Competition between such cities has massively affected the development of the cultural industries, transforming the quality and nature of knowledge, education, tourism, publishing, cinema, the media, and museology. It has spawned spectacular "global" architecture, enhanced heritage preservation, and generated frenzied competition to host ever-increasing numbers of global events. Yet while research on the recent development of world cities has flourished, the study of their historical origins has been ignored, despite the fact that initial research shows that of 130 "acknowledged" or "emerging" world cities, some two thirds of that number were at one time either imperial capitals (generally in Europe) or prominent colonial cities (King 2011). Many such cities, characterized by immense economic and social polarization, are also wracked by homelessness, poverty, and totally inadequate social provision. While the spatial transition from colonial to aspiring world city could be illustrated by reference to many cities in the postcolonial developing world, I return here to Delhi.

India's colonial capital, New Delhi, built between 1912 and 1940 to the south of the old city of Delhi, was laid out on a European Beaux Art plan. Radiating from the center were broad, tree-lined avenues with expansive vistas and routes for ceremonial military parades. Equally broad boulevards accommodated hundreds of spacious bungalows for colonial officials, the size and location of each compound determined according to rank in the carefully structured bureaucratic hierarchy. The city's function, with colonial parliament building and vast Secretariat, was administration and governance.

Following Independence (1947) the ceremonial space was partially filled with government buildings and official housing. Public and private suburban expansion vastly extended the size of the city. Following Partition (the division of India and Pakistan) in 1947 millions of refugees and migrants flowed into the vastly overcrowded Old Delhi, three miles from the new city and the rapidly expanding city suburbs. Yet the original Beaux Art city plan remained in place; as India's capital, the colonial Lutyens/Baker landscape is kept (visibly) clear of squatters.

Since the 1990s, after decades of state control, the built-up landscape has been rapidly transformed by powerful neo-liberal policies. The result has been the multiplication of luxury hotels, business centers, export villages, and high-rise and gated housing for the rapidly growing middle class. Investment from some of the 30 million "Non-Resident Indians" around the world has encouraged consumer-oriented expansion: supermarkets, shopping malls, condominiums, corporate towers, apartments, gallerias—all within the ever-expanding spatial matrix of the colonial plan (King 2004: 151).

By 2007, the aim of Delhi's Master Plan was "to make Delhi a global metropolis and world class city" (Dupont 2011). The city's earlier colonial and postcolonial history was instrumental in shaping this objective. Its vast, middle-class, English-speaking population has myriad connections with developments in the West. "Becoming global" has its visual, especially spatial and architectural consequences. In Gurgaon, Delhi's new corporate town, massive office buildings housing multinational call centers are part of the new "rigid landscape of global architecture, devoid of any response to the local setting and social milieu" (Mehrotra 2011: 95–97).

Hosting the Commonwealth Games in 2010 aimed to raise the global profile of the city by "putting India on the world map" and "enhancing the city's global recognition and image" (Dupont 2011). The obverse of this image is the negative impact of these developments; diverting resources from the massive housing deficit and the continuing policy of "spatial purification," that is, keeping beggars off the streets and the persistent removal of the all-too-visible squatter settlements, home to a quarter (about three million people) of the city's population. Echoing the long, yet often spurious, imperial connection of planning with public health, Dupont writes, "the hygienist oppression of British urbanism has pervaded the justification for contemporary slum demolitions"

(Dupont 2011). In these ways colonial attitudes and practices live on in the world and global city.

REFERENCES

Abu-Lughod, J. (1965) "Tale of Two Cities: The Origins of Modern Cairo," *Comparative Studies in Society and History*, 7: 429–457.

AlSayyad, N. (1992) *Forms of Dominance: On the Architecture and Urbanism of the Colonial Enterprise*, Aldershot: Avebury.

Barnes, T. (1999) *We Women Worked So Hard: Gender, Urbanization and Social Reproduction in Colonial Harare, Zimbabwe, 1930–1956*, Portsmouth: Heinemann.

Basu, D. Ed. (1979) *The Rise and Growth of the Colonial Port Cities in Asia*, Santa Cruz: Center for South Pacific Studies, University of California.

Basu, D. Ed. (1985) "The Rise and Growth of the Colonial Port Cities in Asia," *Conference Proceedings. Monograph Series, Center for South and South-East Asian Studies, University of California*, Lanham: University Press of America.

Bignor, L. (2008) "Names, Norms and Forms: French and Indigenous Toponyms in Early Colonial Dakar, Senegal," *Planning Perspectives*, 23(4): 479–450.

Castells, M. (1977) *The Urban Question*, London: Edward Arnold.

Celik, Z. (2008) *Empire, Architecture and the City*, Seattle: University of Washington Press.

Chakrabarty, D. (1992) "Provincialising Europe: Postcoloniality and the Critique of History," *Cultural Studies*, 16(1): 337–357.

Chang, J. and King, A. D. (2011) "Towards a Genealogy of Tropical Architecture: Power-Knowledge, Built Environment and Climate in British Colonial Territories," *Singapore Journal of Tropical Geography*, 32(3): 283–300.

Chattopadhyay, S. (2005) *Representing Calcutta: Modernity, Nationalism, and the Colonial Uncanny*, London and New York: Routledge.

Chopra, P. (2011) *A Joint Enterprise: Indian Elites and the Making of British Bombay*, Minneapolis: University of Minnesota Press.

Collins, J. (1980) "Lusaka: Urban Planning in a British Colony," in G. E. Cherry, Ed., *Shaping an Urban World*, London: Mansell.

Darwin, J. (2007) *After Tamerlane: The Rise and Fall of Global Empires: 1400–2000*, Oxford: Oxford University Press.

Demisse, F. Ed. (2007) *Postcolonial African Cities: Imperial Legacies and Postcolonial Predicaments*, London and New York: Routledge.

Dossal, M. (1991) *Imperial Design and Indian Realities: The Planning of Bombay City 1845–1875*, Delhi: Oxford University Press.

Dupont, V. D. N. (2011) "The Dream of Delhi as a Global City," *International Journal of Urban and Regional Studies*, 36(3): 523–554.

Fanon, F. (1968) *The Wretched of the Earth*. New York: Grove reprint. Originally published as *Le Damnes de la Terre*, Francois Maspero, Ed., Paris 1961.

Friedmann, J. (1986) "The World City Hypothesis," *Development and Change*, 17(1): 69–84.

Friedmann, J. and Wolff, G. (1982) "World City Formation: An Agenda for Research and Action," *International Journal of Urban and Regional Research*, 6(3): 309–344.

Fuller, M. (2007) *Moderns Abroad: Architecture, Cities and Italian Imperialism*, London and New York: Routledge.

Glover, W. (2008) *Making Lahore Modern: Constructing and Imagining a Colonial City*, Minneapolis: University of Minnesota Press.

Held, D., McGrew, A., Goldblatt, D., and Perraton, J. (1999) *Global Transformations: Politics, Economics and Culture*, Cambridge: Polity.

Home, R. (1996) *Of Planning and Planting: The Making of British Colonial Cities*, London: Spon.

Hopkins, A. G. Ed. (2002) *Globalisation in World History*, London: Pimlico.

Horvarth, R. (1969) "In Search of a Theory of Urbanisation: Notes on the Colonial City," *East Lakes Geographer*, 5: 68–82.

Hosagrahar, J. (2005) *Indigenous Modernities*, London and New York: Routledge.

Howard, E. (1902) *Garden Cities of To-Morrow*, London: S. Sonnenschein & Co.

Kal, H. (2011) *Aesthetic Constructions of Korean Nationalism: Spectacle, Politics, History*, London and New York: Routledge.

King, A. D. (1976) *Colonial Urban Development. Culture, Social Power and Environment*, London and Boston: Routledge and Kegan Paul.

King, A. D. (1985) "Colonial Cities: Global Pivots of Change," in R. Ross and G. T. Telkamp, Eds., *Colonial Cities: Essays on Urbanism in a Colonial Context*, Dordrecht: Martinus Nijhoff, pp. 7–32.

King, A. D. (1990a) *Urbanism, Colonialism and the World-Economy: Cultural and Spatial Foundations of the World Urban System*, London and New York: Routledge.

King, A. D. (1990b) *Global Cities: Post-Imperialism and the Internationalisation of London*, London and New York: Routledge.

King, A. D. (1995) "Writing Colonial Space, A Review Article," *Comparative Studies in Society and History* 37(3): 541–554.

King, A. D. (2004) *Spaces of Global Cultures: Architecture, Urbanism, Identity,* London and New York: Routledge.

King, A. D. (2011) "Imperialism and World Cities," in B. Derudder, M. Hoylake, P. Taylor, and F. Wilcox, Eds., *International Handbook of Globalization and World Cities,* London: Edward Elgar.

Kusno, A. (2000) *Behind the Postcolonial: Architecture, Urban Space and Political Cultures in Indonesia*, London and New York: Routledge.

Loomba, A. (1998) *Colonialism/Postcolonialism*, London and New York: Routledge.

Mehrotra, R. (2011) "Simultaneous Modernities: Contemporary Architecture in India," in W. Lim and J.-H. Chang, Eds., *Non-West Modernist Pasts,* Singapore: World Scientific Publishing Co., pp. 91–104.

Myers, G. A. (2011) "Moving Beyond Colonialism: Africa's Postcolonial Capitals," in G. Myers, Ed., *African Cities: Alternative Visions of Urban Theory and Practice,* London: Zed Books.

Nasr, J. and Volait, M. Eds. (2003) *Urbanism: Exported or Imported*, London: Wiley.

Njoh, A. (1999) *Urban Planning, Housing and Spatial Structures in Sub-Saharan Africa*, London: Ashgate.

Njoh, A. (2007) *Planning Power: Town Planning and Social Control in Colonial Africa*, London: University College Press.

Njoh, A. (2009) "Urban Planning as a Tool of Power and Social Control in Colonial Africa," *Planning Perspectives*, 24(3): 301–331.

Oliver, P. (1987) *Dwellings: The House Across the World*, Austin: University of Texas Press.

Osterhammel, J. (1997) *Colonialism: A Theoretical Overview*, London: Marcus Wiener.

Pacione, M. (2001) *Urban Geography: A Global Perspective*, London and New York: Routledge.

Rabbat, N. O. Ed. (2010) *The Courtyard House*, Farnham: Ashgate.

Redfield, R. and Singer, M. S. (1954) "The Cultural Role of Cities," *Economic Development and Cultural Change*, 3: 53–73.

Robertson, R. (1992) *Globalization: Social Theory and Global Culture*, London, Newbury Park, New Delhi: Sage.

Sassen, S. (1991) *The Global City: New York, London, Tokyo*, Princeton, NJ: Princeton University Press.

Schmidt, E. (1992) *Peasants, Traders, & Wives: Shona Women in the History of Zimbabwe, 1870–1930*, London: Heinemann.

Swanson, M. W. (1970) "Reflections on the Urban History of South Africa," in H. L. Watts, Ed., *Focus on Cities*, Durban: Institute of Social Research, University of Natal.

Taylor, P. J. (1985) *Political Geography: World-Economy, Nation-State and Locality*, London: Longman.

United Nations (1971) *Climate and House Design,* New York: United Nations.

Vale, L. J. (1992) *Architecture, Power and National Identity,* Newhaven, CT: Yale University Press.

Yeoh, B. (1996) *Contesting Space: Power Relations and the Urban Built Environment in Colonial Singapore*, Oxford: Oxford University Press.

Young, R. (1990) *White Mythologies: Writing History and the West*, London, New York: Routledge.

"Cities Interlinked"

City Worlds (1999)

Doreen Massey

THE SQUARE OF THE THREE CULTURES

Not far from the center of today's Mexico City there is a square called *La Plaza de las Tres Culturas*—the square of the three cultures. It is really just a part of the city in which are exposed to view in immediate proximity elements of the three major cultures which have gone in to making this place. Three cultures juxtaposed. There are the excavated ruins of an enormous Aztec pyramid. There is a seventeenth-century baroque Roman Catholic Church. And there is a complex of buildings in the international style: the pre-Columbian, the Hispanic-colonial, and the modern.

La Plaza is a moving and impressive monumental space. The pyramid, now just a vestige of its previous

Figure 17 *La Plaza de las Tres Culturas.*
Source: © Routledge in association with the Open University.

self, somehow silent and unreachable; the church (built, in a defiant gesture, right up against the razed pyramid), still used but now showing its age; and the (relative) newness and shininess of the more recent constructions. Together they constitute a monument which makes you think. How is it to be understood, what is it trying to say?

On the one hand, it is possible to read the monument as a reminder of a lost past. The devastation of the pyramid and the physical assertiveness of the church in its positioning are certainly emblematic of the destruction this city underwent on the arrival of the Spanish. Physically and culturally much was lost. Perhaps, then, the monument can be read as a burrowing back through the layers of time in search of the city's roots.

Or one could interpret it another way. Perhaps the monument is rather a bringing together of all the elements which still today (and however unequally) make up this city. For while the buildings certainly succeeded each other, the social forces and cultures which they represent each still have a presence here. Not, then, so much a search for roots as a celebration of the mixture which is this place. It is this second interpretation which coincides more closely with the intentions of those who created this project in the 1960s.

[. . .]

These buildings in *La Plaza de las Tres Culturas* bring together periods of the city's history just as does the mixture of buildings in Sao Paulo's Anhangabau. They reveal elements of the city's "different histories." But this square is not just a jumble of city buildings from different periods; it is preserved and designed as a monumental space. What was it Lefebvre argued about the function of monuments?

Lefebvre argued that monuments have a function of establishing membership. His view was that "monumental space" offers "each member of society an image of that membership," that it constitutes "a collective mirror" (Lefebvre 1991: 220). [. . .] what is at issue is *recognition*.

When *La Plaza de las Tres Culturas* was constructed (that is, when the modern buildings were built and the unity of the three periods/cultures was bound into the design) it was precisely with this kind of message in mind. On the one hand it was a physical recognition of the contribution of indigenous and Hispanic cultures to the Mexico of today. And on the other hand it was an invitation to all Mexicans—indigenous, mestizo, white—to recognize themselves in, as part of, the modern city. This, then, is a monument which recognizes the city of today as a mixture evolved over time.

It is, moreover, a mixture in a geographical sense as well. The monument does not speak of this explicitly, yet it is there in the very structures which form it. Thus, the modern buildings share much with such buildings all around the world. You might see their like in Africa or in Asia. The church is reminiscent in its architectural style of Spanish Catholicism in Europe. Neither architecture is simply and authentically "of this place." Both, in their physical construction, express *both* the specificity of this place *and* links with the world beyond. This composite nature (of local and global, one might say), moreover, can be detected in the pyramid too. Without doubt a temple of the local Aztecs, the architecture reflects also the traditions of the wider region of Mesoamerica. Each of the "elements," the societies and cultures, which has gone into making Mexico City what it is today, was itself tied into a wider geographical set of interconnections. Each brought together elements from near and far to create its own specificity.

The city is, then, not only a developing mixture through history, but also in each moment of that history, *the focus of a wider geography,* bringing together differences in space. Cities are foci of changing patterns of interconnections.

The aim of this chapter is to consider cities in the context of these wider geographies. It will draw back a little from the exploration of the internal intensities of individual cities (though we shall keep an eye on these too) in order to develop an approach to cities within their wider sets of interrelations.

Even this brief consideration of *La Plaza de las Tres Culturas* has enabled us to understand Mexico City as both

- a developing mixture through history, and also
- in each moment of that history the focus of a wider geography, bringing together differences in space.

If you want to put it at its most abstract, what I am proposing is the city as an intense focal point or a node of social relations in time and space. And this is how we shall interpret cities in this chapter.

A BIT OF HISTORY

When the Spanish arrived in this city (in their year of 1521), it really was a meeting of two worlds. Indeed, in these supposedly more enlightened times it has been re-named: from *la conquista*—the conquest—to *el encuentro*—the meeting. It was the coming-together of two cultural histories which up until that point had had no knowledge of each other. They had been separated by an ocean across which ran no current social or cultural connections. Geography had kept the two histories apart, their stories had remained separate. But now they were to meet.

Each, indeed, began with massive misconceptions about the other. The Spanish were part of that European exploration westwards led by a Columbus who firmly believed he was on the way to India. He and those first explorers had no notion that a whole other continent lay across their path. And for their part, the Aztecs completely misinterpreted the arrival of these fair-skinned men on strange animals. For them, this was not the mid-sixteenth century when the Spanish had first landed on the Yucatan coast far to the east of the city (in Spanish year 1519). It was to the Aztecs the year One Reed in their 52-year cycle. And that was a year of significance: it was the year in which the deified king Quetzalcoatl had disappeared towards the east (was he therefore about to return? was the news coming in of these strange landings a warning of his imminent arrival?) and it was also a year of special significance for the cosmological sign of the plumed serpent—of Quetzalcoatl himself. This too was ominous. There has been much academic debate about the significance of these temporal coincidences for the Aztecs. But there seems to be agreement that, as Moctezuma [Ed. note: the king at the time] puzzled over the news coming in from the east, it all added to his uncertainty and sense of foreboding.

As they rode down into the wide and spreading Valley of Mexico in which lay the Aztec city of Tenochititlán, the Spanish were stunned (see Figure 18). They wrote afterwards about their complete amazement. For here was a city equal to anything in Europe: in size, in social complexity, and in built form. The central square was a huge complex of massive stone buildings: pyramids, ball-courts, temples, palaces. The market to the north of the center, at Tlatelolco where *La Plaza de las Tres Culturas* now stands, attracted some 25 000 people every day, and between 40 000 and 60 000 on the days of special markets (once every fifth day) (Townsend 1993, pp. 173–4). There was a complexity of

Figure 18 *The island capital of Tenochititlán* (as imagined and painted by Miguel Covarrubias, 1904–1957).
Source: © Routledge in association with the Open University.

agriculture, craftwork, trade, and military activity. This was an accomplished imperial capital.

Many of the local products were completely unknown to the Spanish, while the Aztecs had never seen horses before. The two cultures had different concepts of space and time. They valued different things. In material terms the Aztecs valued jade and turquoise and this is what they offered the Spanish when the latter demanded precious goods. As Vaillant (1944/1950, p. 133) puts it, "Such misguided compliance was highly irritating to Cortes and his men!" What the Spanish were after—what *they* valued—was gold. But gold was only valuable to the Aztecs for the ornaments which could be made of it; for them it was not the foundation of an exchange and currency system, and they remained puzzled.

This, then, really was the beginning of the meeting of two worlds. Two separate histories, up to now operating in completely separate spaces, met up with each other in the first half of the sixteenth century.

But, even if we now call it a "meeting," this was a coming-together of two highly unequal powers. The Aztec city was virtually razed to the ground, and a new built environment was created. The baroque church was built where once stood the pyramid at Tlatelolco, and the present Spanish main square, with cathedral and national palace, was erected where once had been pyramids and ball-courts and temples. Indeed, they were not only built "in place" of the Aztec city, but—it is thought—they probably used the same materials: the masonry of the pyramid, or temple, was taken to build the church. One city center almost obliterated the other. Only the occasional Aztec building remained among the devastation. The physical city was easier to destroy than the social and the cultural. But the Spanish—or some of them, and at first—did their best. Records were destroyed, cultural habits disrupted, even concepts of space and time were obliterated by the imposition of the new Spanish ways. And almost as completely as the material city, something else was destroyed: the city as an Aztec center of a wide and complex geography of cultural interconnections, trade, and military power.

If one of the things which is crucial to a city's creativity and power is its position—as a focus of social relations—then the Aztec city in this sense was also destroyed at this time. Tenochititlán was a focus both of trade and of empire. Its network of trade routes spread over all of what is now Mexico, and beyond, and the empire extended from the Pacific Ocean to the Gulf of Mexico. As part of the acknowledgement of imperial relations, surrounding areas sent in to Tenochititlán tributes and levies of foodstuff, raw materials, and manufactured goods such as clothes and pottery. The empire and imperial relations were destroyed by the Spanish conquest. And within only five years of that conquest the long-distance trade network had also disappeared, a fact explained by Townsend (1993, p. 186) as due to the network's dealing in importing luxury items—the feathers of tropical birds, greenstones, and exotic animal hides—which had high value for the Aztecs but not for the Spanish. From being the dominant focus of trade and imperial connections across Mesoamerica, the city turned towards Madrid. Mexico City was installed as the capital of what now became New Spain (a part of the wider Spanish colonies in the Americas), but its local dominance was now in turn subordinated to an even greater power, a new imperial capital across the Atlantic. Compare maps in Figure 20(a) and 20(b). In terms of its positioning within wider networks of social relations, the meeting of two worlds in this city—the passage from Tenochititlán to Mexico City—meant a complete reorientation.

Yet out of all this something new was created, something once again unique. The contribution of the Aztec was not completely erased. If you go today as a tourist to Mexico City, you will probably be most immediately conscious of the Aztec presence through the heritage sites of the brochures: recent major excavation works have exposed the foundations of the Templo Mayor which once stood in the central precinct of Tenochititlán (today it is on the corner between the cathedral and the national palace); there is the immense Museum of Anthropology in Chapultepec Park, with exhibits not only of Aztec culture but of the enormous variety of other societies which lived in what we now know as Mexico before the Spanish arrived; on the southern fringes of the city there are the famous "floating gardens" where (according to a tourist brochure which I have at this moment to hand) "flower-bedecked boats glide among aquatic plants to the music of *mariachi* orchestras" but which are also the remains of the *chinampas* where, before the Spanish arrived, people toiled in the sun at their water-based agriculture.

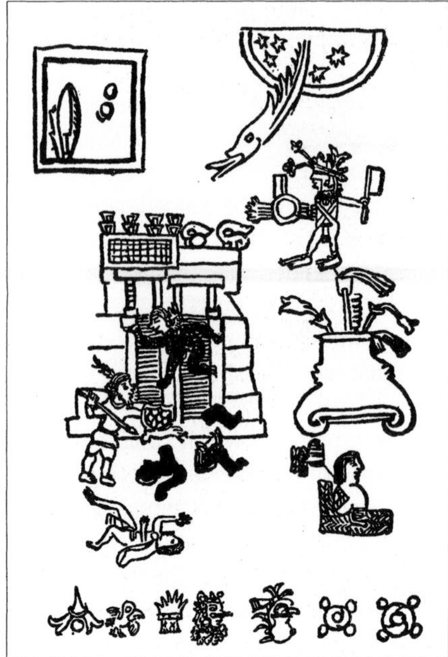

Figure 19 The story of the conquest of Mexico in indigenous symbols.
Source: © Routledge in association with the Open University.

Such tourist sites celebrating the Aztec "presence," however, locate that presence in the past. In fact it is more like an absence. There are other signs about the city which bring its contribution closer. There are the place names—Azcapotzalco, Tequesquinahuac, Tlalpantla, Tenayuca, Nezahualcoyotl; there is the food—the smell of cornbread in the morning (along with pollution) pervades the city; there are certain understandings about the way of living everyday life. All this is set in a city where the language is Spanish, the main religion Roman Catholic, and the dominant cultural references are to Europe and the USA. Out of this long history—indigenous and Spanish (and now independent modern Mexico)—has emerged, gradually and eventually, something new—something different from either of the two components. Mexico calls itself the mestizo state: the hybrid state. Mexico City is a mestizo city.

Yet to walk the streets of Mexico City is also to begin to understand the terms of this historical and geographical mixing. (And here we see that the indigenous "presence" is not just in remnants from the past—it is an active part of the city today.) The three cultures of the square are here together. It is an unequal and potentially and occasionally conflictual mix. And this inequality is expressed in the spaces and the times of the city. The people who trip lightly in and out of the gleaming office buildings, perhaps rushing between "urgent" meetings, have paler skins (look more "European") than those who sit all day on the pavement outside, begging or selling embroidery. And the homes each retreat to at night, and the parts of the city where these homes are found, will be as different as it is possible to be. The dominant structures and images of Mexico City today revolve around those with paler skins, and are in the more "modern"/ international mode. Indigenous lives receive more "recognition" in the monument than in the economic and social structures of the modern city.

[. . .]

REFLECTIONS

So what can be learned from this bit of a city's history? The most important point concerns a basic way of understanding cities in the world. This way of reading the history of Tenochititlán-Mexico City emphasizes the city as a focus of wider networks of social connections. A city is a place where

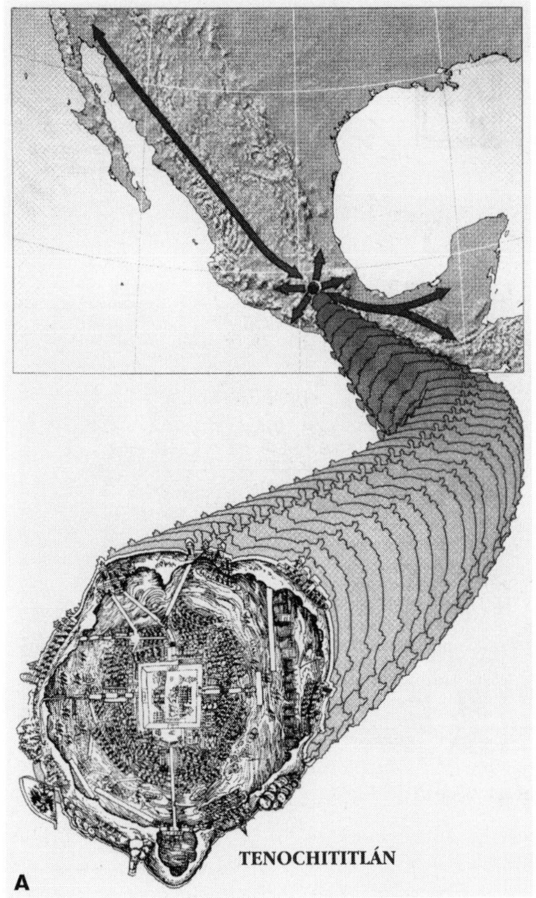

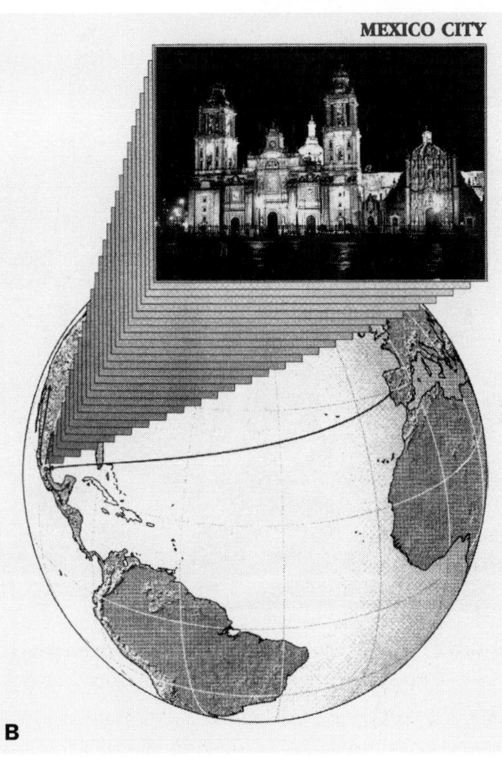

Figure 20 (a) Map showing Tenochititlán; (b) Map showing Mexico City.
Source: © Routledge in association with the Open University.

meetings happen: the meeting of the migrant with the already established resident; the meeting of traders in markets; the meetings of the powerful in dignified offices.

- Cities are set in wider geographical spaces within which they are in some sense a focus. That is the basic point.

But from that, other points can be drawn.

First, in general, cities are places of mixing. Precisely because of their role as foci *they bring together different histories*. That is what happened on a grand scale when the Spanish rode in to the Aztec city. But it happens, constantly and less dramatically, on a smaller scale all the time and in most cities. New people arrive, new stories are added to the ones already there. This process, however, is not a simple one and we can already note three things about it. To begin with, the different peoples, or histories, which a city brings together are themselves already mixtures. This may seem a strange point to make, but it is important. We tend very often to think of societies as having their places, of different cultures being based in particular areas. And we then interpret the historical increase in travel and communication, and the current frenzy of globalization, as linking together what were previously separate cultures and—on what is seen as the "pessimistic" scenario—as breaking down all their authenticity and particularity and leading to a world in which "everything looks the same". This is not entirely untrue. Historically the amount and intensity of geographical movement and exchange has indeed increased enormously. But what is becoming much clearer from research in recent years is that the picture of once

Figure 21 *Moctezuma meets Cortes* (painted screen panel by Roberto Cueva del Rio).
Source: © Routledge in association with the Open University.

isolated societies gradually beginning to interact, or being invaded by the West, is a gross simplification. Thus Eric Wolf in his book *Europe and the People without History* (1982) has pointed to the enormous degree of interconnection between societies around the world in the centuries before the Europeans set out to explore. There were, he argues, no "primitive isolates," no simply isolated cultures. And Janet Abu-Lughod in her book *Before European Hegemony* (1989) analyzed what she calls "the world system between 1250 and 1350" and found in fact an intricate and active system of interconnections which formed a network of links (with cities as nodes); the "world" which stretched from a fairly primitive Western Europe, through the Mediterranean, to India and China. When "the Spanish" met "the Aztecs" both were already complex products of hybrid histories.

Moreover, if cities are places of cultural mixing, then the social terms of that mixing will vary both historically and geographically. The different communities and cultures may remain relatively self-enclosed, inward-looking, or they may meld together, mix comfortably or irritably, or remain in conflict. There may be the "ethnic quarters" of old Chicago, or the bounded and purified spaces described and deplored by Richard Sennett [. . .], there may be the class and other separations produced by money and market, or there may be the glorious "mixity" evoked by the likes of Jane Jacobs. And even this last will have its terms, its power relations. While there may on occasions be egalitarian jostling, recent more skeptical writing has pointed to middle-class invasions of inner-city areas as attempts to enjoy different peoples as exotica, not really to engage with difference but to treat it, rather, as "local color." (See, for example, May 1996.) The social terms of mixing in Mexico City—as we saw—are unequal and occasionally contested.

Finally on this point, whatever the terms of this mixing, something new will be produced. The coming together of histories in space will produce new stories. The long and painful difficulties of the Aztec/Spanish gave birth to the mestizo city. Putting this more generally, new "geographical juxtapositions" produce new histories.

[. . .]

Second, geography (or "space") may also be of fundamental importance in the expression and organization of this urban "mixity." As we have seen, in some cases there may be spatial integration (though the situations where this happens thoroughly are rare), but more often cultural variation will express itself, to one degree or another, in geographical terms: the enforced segregation of South African apartheid, the formalized enclaves for different nationalities in the treaty ports of China, the dual geographies of North African colonial cities, the tree lined spaciousness of the French area against the tight huddle of the Arab medina; the sorting of cities by income group between suburbia and slum; the gay areas of San Francisco and Manchester. The word "ghetto" originated in renaissance Venice where it described the areas in which Jewish people were forced to live: let out at dawn to do their business in the city, at nightfall they were obliged to return, the "gates were locked, the shutters of its houses that looked outward closed; police patrolled the exterior" (Sennett 1994, p. 215). The important point here is that these divisions within the city are not just the result of mapping already existing, different communities on to distinct spaces. It is also that the spatial organization itself—the geography—is important in maintaining, maybe even in establishing, the difference itself. [. . .] But once again the power relations vary. The Venetian ghetto was an expression of power relations and a hatred (and fear?) of people who were different. Here it was the dominated people who were walled in. In the gated communities of North American cities it is the relatively powerful, in an expression of their vulnerability, who wall themselves in. In San Francisco and Manchester the gay streets and villages are part of the process of establishing that community's existence—and right to existence. Geography and community, and difference in the city are inextricably interrelated. One of the most troubling and continuing questions about urban areas is this: how is it best to organize spatially the different elements which go to make the variety which is so characteristic of so many cities?

Both of those first two points in a sense follow from the *third*: the fact, which we have emphasized (throughout), that cities are essentially open, that they are places of wider interconnection. Moreover, and this is the really important point, these connections have to be constructed and actively maintained. When the Aztecs first arrived in the Valley of Mexico (they called themselves the Mexica then) they were considered by those already living there as a somewhat unsophisticated intrusion. It took years of negotiations and treaties, and defeats, before the city was even founded, and many years more before what began as a tentative settlement in the most unpropitious of locations had managed to establish itself as the center of a network of trade and imperial tribute. The later connection to Spain was even more fiercely won: the weeks at sea, long miles across mountain and plain from the coast, to pull the orientation of the city towards Europe, towards Spain, to establish it as a focus in a new and different network of communications.

That process of constructing and re-constructing, negotiating and re-negotiating, their place in the world's geography is, as we shall see, crucial to the survival and success of cities. To drop out of networks can mean decline; to renegotiate your place within them can imply a change in trajectory, a different future. By the early nineteenth century the new Mexicans were beginning to find intolerable the imposed domination of their networks of contact, especially of trade, by the colonial power in Madrid. The battle for independence had as an important element the desire to re-orient Mexico yet again in the world, to re-balance and renegotiate the geography of its networks of interconnection. And with national independence in 1821, the nature of Mexico City's placing in the world geography of interconnections changed once more.

[. . .]

REFERENCES

Abu-Lughod, J. L. (1989). *Before European Hegemony: The World System AD 1250–1350*, Oxford: Oxford University Press.

Lefebvre, H. (1991). *The Production of Space*, (tr. David Nicholson-Smith), Oxford: Blackwell. (First published in French in 1974.)

Sennett, R. (1994). *Flesh and Stone: The Body and the City in Western Civilization*, London: Faber and Faber.

Townsend, R. F. (1993). *The Aztecs*, London: Thames and Hudson.

Vaillant, G. C. (1944/1950). *The Aztecs of Mexico*, 1944: New York: Doubleday and Doran; 1950: New York: Pelican Books.

Wolf, E. (1982). *Europe and the People without History*, Berkeley: University of California Press.

TWO

SECTION 2
Development and Urbanization

Editors' Introduction

Development and urbanization are closely articulated fields of study and practice. While urbanization and cities are not an invention of the development era or the industrialized world, the discourse of development has helped shape the discourses of contemporary urbanization. How development is defined, how it is measured, whose development experience is counted and recognized, contemporary all have been the subject of debate and critique with significant implications for urban policies in the global South. The overall goal of this introduction is to highlight the shifting terrain of discourses, institutions, and actors of development and urbanization and their impact. "*Whose development?*" and "*whose cities?*" are two questions looming large in these debates, around which we organize this brief introduction.

KEY ISSUES

Discursive shifts: whose development?

Since World War II, the record of the development enterprise and its glaring failure to bring about a dignified livelihood for the majority in the global South has invoked a range of important critiques from various corners. In the 1960s and 1970s, against the backdrop of policies that understood development as national economic growth, scholars looking at economic growth and poverty in growing cities declared that development was not benefiting the poor. Some called for a kind of development that addressed the basic needs of people and advocated "growth with equity" (Streeten 1995; Burkey 1996). Others advocated a self-help movement in housing that learned from the poor and their informal strategies (Turner 1977). These critics were joined by feminist scholars and activists who in the 1970s had scrutinized agricultural modernization from a gendered perspective (Boserup 1970). In the 1980s they demonstrated that economic development as promoted through modernization and industrialization policies was also not beneficial for poor women in urban areas. They argued that development diminished the socio-economic status of women and their power within the household, even as it increased their domestic burden (Brydon and Chant 1989; Potts 1999; Mies et al. 1988; Crewe and Harrison 1998). In the same period, environmental movements demonstrated that decades of implementing development policies and projects for economic growth had increased environmental problems with devastating consequences for the poor, particularly for indigenous communities. They declared development as not only excluding the poor but also damaging to the earth (Esteva and Prasak 1997; Peet and Watts 1993; Hecht 2004; Mies and Shiva 1993). By the 1990s, post-developmentalist scholars and activists pushed these oppositional voices further, arguing that development was not good for humanity, period. They declared it was the success of development, not its failure, that we should fear (Sachs 1992; Ferguson 1994; Escobar 1995).

In their own way these critiques influenced the formulation of urban development policies and emerged from the larger global and regional political economic restructurings of their time. In the 1960s and 1970s for example, with the anxieties of the Cold War era and the fear of political radicalization of the poor in rapidly growing cities, the bilateral and multilateral development institutions and national governmental agencies paid greater attention to self-help advocates and critics who called for development with a human face to address people's basic needs. By the late 1970s, in regions like Latin America a sharp urban population growth was taking place. This shift translated to a greater urban focus in development, especially site and services projects, which adopted the incremental self-help strategies of the urban poor to access basic shelter.

With a global restructuring of capitalism in the 1980s that mandated a "leaner" central state, mainstream institutions began to pay greater attention to non-economic factors as well as local processes and actors for development issues for which feminists, along with advocates of "development with a human face" and the "basic needs approach," had long argued. For example, while prior to the late 1980s mainstream development institutions assessed development solely on national level economic performance, by 1990 the United Nations had adopted the Human Development Index (HDI), a composite measure that went beyond GNP, and GDP. Spearheaded by economists Mahbub ul Haq and Amartya Sen, HDI broadened development's definition to include gender gaps in respect to life, health, education, and income. Most importantly for purposes of our discussion, these measures included urban-rural gaps with respect to infrastructure and sanitation (Haq 1995).

An important force prompting this "local and humanist turn" among mainstream development institutions and their critics was the collapse of the Soviet bloc in 1990. This unleashed the forces of free market capitalism, diminishing the role of the developmental state in the global South and the welfare state in the global North. The post-Cold War era clearly marked a severe crisis in the discourse and practice of development. While some declared development as "dead," others set out to search for alternatives (Fine 1999). Scholars of urbanization in the global South turned to the grassroots as a source of inspiration. Building on what self-help advocates of the 1970s and feminist activists of the 1980s had promoted, they emphasized the role played by a range of actors, including women who served as city builders and urbanized much of the global South through everyday acts of community building (Moser and Peake 1987). Focusing on the urban poor and how they develop their neighborhoods, Friedmann theorized alternative development with the grassroots as the new active agents. In his influential work, *Empowerment: The Politics of Alternative Development* (1992), Friedmann, for example, reflected on the experience of cities in Latin America and made a case for a development paradigm shift, from national level economic development plans to household and the neighborhood level livelihood strategies–what he called an "empowerment paradigm."

While some scholars searched for an alternative to development, Bretton Woods institutions such as the World Bank turned to social capitalists to offer their version of an alternative. For example, the World Bank adopted strategies of grassroots and poverty alleviation for development resuscitation (see Hart 2001). The UN took a slightly different path, declaring alleviation from abject poverty, access to basic infrastructure, and improved sanitation in cities as millennial development goals or MDGs. Gillian Hart captured the discursive tensions in the development debates taking place at this historical moment as those between the big and small D developments: the former relying on the large scale national state developmental projects, the latter relying on grassroots and civic associations for a supposedly more humane development (Hart 2001).

In search of alternative development focused on the poor and the marginalized, urban scholars and multilateral and bilateral development agencies seem to have converged on a focus on grassroots urbanization. This convergence, however, does not mean development with a capital D was abandoned. Capital D development remained alive and well but functioned through more complex institutional and regional dynamics. In the context of a rollback in government expenditure as per structural adjustment policy prescriptions, actors of development conveniently broadened beyond national governments to include international organizations and their subsidiary local NGOs, grassroots and community-based groups, as well as transnational corporations and large development companies. (See section on governance in this *Reader.*) Emerging economies like Brazil, Russia, India, China and South Africa (known as BRICS) have increasingly been able to challenge the post-World War II order of development and its focus on particular populations (Chatterjee 2006). For instance the governments of Brazil, India, and China have gradually taken over the role that multilateral and bilateral development agencies used to play for urbanization and infrastructure development

(McGranahan and Martine 2012). The national governments of the BRICS countries use legal tools to create zones of interest and the best environment for investments domestically and abroad, particularly in the global South. The Chinese government, for example, is currently one of the largest development agencies in Africa involved in urban infrastructure and construction projects (Kragelund and Van Dijk 2009).

In the post-Cold War era, with the global unleashing of market forces, the distinctions among diverse institutional actors have increasingly blurred. Similar processes are taking place in the regional dynamics of large scale development projects. Western (Euro-American) multinational corporations, development agencies, and state agencies do not only pursue development projects on their own national territories. They are going global. Furthermore, the BRICS governments are using their large economies to engage with local, national, and international development processes and debates on their own terms. In light of shifting discursive, institutional, and regional terrains of the development enterprise, the question, "whose development?" only becomes more difficult to parse.

Urbanization shifts: whose cities?

In the post-World War II era, strategy for economic growth in many regions of the global South focused on industrialization to substitute imports. As discussed by Goldman in the first article selection for this section, the urban bias of this development strategy created ground-shifting conditions for rapid migration (some call it displacement) of rural populations to cities. In Latin America and the Caribbean, for example, during the period of 1950 to 1975 the distribution of population changed from majority rural to majority urban–from 59 percent rural in 1950 to 61 percent urban by 1975 (Latin America United Nations Department of Economic and Social Affairs 2008). But this urbanization trend in Latin America differs from what happened in other regions at the time and what happens in the region today. While rural–urban migration has stagnated in Latin America in the last decade, this trend has increased in a number of other regions. In Africa, for example, where urbanization progressed slowly until the late 1970s, the population of cities has increased at an exponential rate since. Between the 1970s and 2005 the number of urban dwellers in Africa quadrupled. No other major world region experienced such a fast rate of urban growth. Asia's urban population, for example, tripled in that period (Zlotnick 2006:19). Some analysts link this to the effects of 1980s structural adjustment policies on agricultural livelihoods. As Potts (1999) has noted, these policies were promoted as part of the export-oriented development.

Enabled by technological improvements and decentralized forms of production in the 1980s, export-oriented development brought about distinct patterns of urbanization across regions. In Mexico, export-oriented development encouraged almost overnight *maquila* urbanizations in border areas where US industries had moved. These *maquilas* primarily drew on populations from other cities and towns in the country. East Asian Newly Industrializing Countries (NICs)–Korea, Taiwan, Singapore, Hong Kong later joined by Indonesia, Thailand–followed the example of Japan as an export-oriented model. However, these countries primarily drew on rural migration for their huge increases in urbanization. By suspension of labor rights and increased use of female labor, export zones in East Asian NICs, enterprise zones in the Caribbean, *maquilas* in Mexico, or Special Economic Zones in the case of China have all taken a great toll on local environments and their local labor force (Fernández-Kelly 1983; Cross 2010; Summerfield 1995).

In the case of China, the economic, political, and ideological shifts to an export-oriented model of economic growth have been drastic yet revealing of the development-urbanization connections and their cost on urban inhabitants. When the People's Republic of China was founded in 1949, only about 13 percent of the population lived in urban areas. Mao Zedong considered capitalist cities parasitic centers of consumption, but socialist cities necessary for production. During the Great Leap Forward campaign of 1958–1961, heavy, Soviet-style industrialization was encouraged in China's cities especially in the west, where population grew rapidly with disastrous outcomes especially in terms of the movement's impact on food supply for urban inhabitants. But during the cultural revolution of 1966–1976, many urban youth were forcibly sent back to the countryside away from the "bourgeois" cities. This reduced China's urban population almost to what it had been in 1949.

Since the adoption of an export-oriented industrialization model in the late 1970s, China has urbanized rapidly. Now more than half of its population lives in cities. This time, however, cities are swelled by rural–urban migrants who live illegally in cities despite official residence status or *hukou* in rural areas.

In addition to distinct patterns of urbanization, the global dominance of neoliberal capitalism has also given rise to a range of new city building projects that equally beg the question: "whose cities?" Goldman in the reading that follows introduces speculative cities whose *raison d'être* is creating a use for accumulated financial capital, just as Mitra (in Part III of this volume) notes the ways in which capital is anchored in cities in various ways.

Indeed, capitalism seems to have turned to urban projects to solve the problem of excess accumulation of profits. These projects take many forms–the border towns and enterprise zones of Latin America and the Caribbean, the vast expansion of construction in Chinese and Indian cities of the last decade, or the elitist City Doubles of Africa that Martin Murray (forthcoming) documents. Given the scale and nature of these projects the questions of "whose cities?" and "whose development?" may be more important than ever.

ARTICLE SELECTIONS

There are two selections for this section: a commissioned piece by sociologist Michael Goldman, and an excerpt from an influential book by geographer Jennifer Robinson. The two pieces dovetail in their critique of the global city as an idealized Western model of what cities ought to be–a dangerous idealization that suppresses the imagination of a range of cities by and for their inhabitants. While Robinson focuses on the global city debate, Goldman sets the stage by reviewing more than six decades of development discourse in relation to urban policy interventions in the global South. He shows how ideas of development have a formative influence on urban policies and city building. In his prior research and publications, Goldman has highlighted the hegemonic role that the World Bank plays in development processes through its knowledge production about development (i.e., his 2006 volume on the greening of the World Bank titled *Imperial Nature*). In this review essay he makes a compelling case on the role that influential global financial institutions like the IMF and the World Bank play in respect to Third World urbanization, often as much through their "problem framing" as their role in the production of so called "solutions." Most recently, for example, the trope of the global city has served as a means of speculative urbanization that excludes the majority of inhabitants from an urban livelihood, with the promise of building a globally competitive city.

The second selection is from Jennifer Robinson's book *Ordinary Cities: Between Modernity and Development* (2006). With extensive research and publication on the relationship between power and space in South Africa's geographies of apartheid, Robinson is well known for her pointed critique of global cities. She argues against creating hierarchically positioned categories of cities (global cities, Third World cities, megacities) with an idealized position given to the global cities model. She suggests all cities be understood as ordinary cities that can fulfill their own potential. Her postcolonial critique of emulating certain Western cities as "global" has important implications for urban studies and practice worldwide. Her critique is important because patterns of urbanization and the types of cities people live in vary widely across contexts; but her remarks are particularly timely in an era of marketing global cities as a blueprint for consumption by city planners and city builders across the world at the cost of ordinary inhabitants of cities.

Figure 22 Development diktats.
Source: © Zapiro.

REFERENCES

Boserup, E. (1970) *Woman's Role in Economic Development*. London: Allen & Unwin.

Brydon, L. and Chant, S. (1989) "Introduction." In L. Brydon and S. Chant (Eds.), *Women in the Third World: Gender Issues in Rural and Urban Areas*. New York: Rutgers University Press, pp. 1–46.

Burkey, S. (1996) *People First*. London: Zed Books.

Chatterjee, P. (2006) *The Politics of the Governed: Reflections on Popular Politics in Most of the World*. New York: Columbia University Press.

Crewe, E. and Harrison, E. (1998) *Whose Development? An Ethnography of Aid*. London: Zed Books.

Cross, J. (2010) "Neoliberalism as unexceptional: Economic zones and the everyday precariousness of working life in South India." *Critique of Anthropology*, 30(4): 355–373.

Escobar, A. (1995) *Encountering Development: The Making and Unmaking of the Third World*. Princeton, NJ: Princeton University Press.

Esteva, G. and Prasak, M. (1997) "From global thinking to local thinking." In M. Rahnema and V. Bawtree (Eds.), *The Post-Development Reader*. London: Zed Books, pp. 277–289.

Ferguson, J. (1994) *Anti-Politics Machine: Development, Depoliticization, and Bureaucratic Power in Lesotho*. Minneapolis: University of Minnesota Press.

Fernández-Kelly, M. P. (1983) *For We Are Sold, I and My People: Women and Industry in Mexico's Frontier*. Albany, NY: State University of New York Press.

Fine, B. (1999) "The developmental state is dead–long live social capital?" *Development and Change*, 30(1): 1–19.

Friedmann, J. (1992) *Empowerment: The Politics of Alternative Development*. Oxford: Blackwell.

Goldman, M. (2006) *Imperial Nature: The World Bank and Struggles for Social Justice in the Age of Globalization*. New Haven, CT: Yale University Press.

Haq, M. U. (1995) *Reflections on Human Development*. Oxford: Oxford University Press.

Hart, G. (2001) "Development critiques in the 1990s: Culs-de-sac and promising paths." *Progress in Human Geography*, 25: 649.

Hecht, S. (2004) "Invisible forests: The political ecology of forest resurgence in El Salvador." In R. Peet and M. Watts (Eds.), *Liberation Ecologies: Environment, Development, Social Movements*. London: Routledge, pp. 58–97.

Kragelund, P. and van Dijk, M. P. (2009) "China's investment in Africa." In M. P. van Dijk (Ed.), *The New Presence of China in Africa*. Hilversum: Amsterdam University Press, pp. 83–100.

Latin America United Nations Department of Economic and Social Affairs/Population Division (2008) *World Urbanization Prospects: The 2007 Revision*. New York: United Nations.

McGranahan, G. and Martine, G. (2012) *Urbanization and Development: Policy Lessons from the BRICS Experience*. Human Settlements Discussion Paper: December 2012. London: International Institute for Environment and Development IIED.

Mies, M., Bennholdt-Thomsen, V., and Von Werlhof, C. (1988) *Women: The Last Colony*. London: Zed Books.

Mies, M. and Shiva, V. (1993) *Ecofeminism*. London: Zed Books.

Moser, C. and Peake, L. (Eds.) (1987) *Women, Human Settlements, and Housing*. New York: Tavistock Publications.

Murray, M. J. (forthcoming) "'City-Doubles': Re-Urbanism in Africa." In F. Miraftab, D. Wilson, and K. Salo (Eds.), *Cities and Inequalities in a Global and Neoliberal World*. New York: Routledge.

Peet, R. and Watts, M. (1993) "Introduction: Development theory and environment in an age of market triumphalism." *Economic Geography*, 69(3): 227–253.

Potts, D. (1999) "The impact of structural adjustment on welfare and livelihood: An assessment by people in Harare, Zimbabwe." In Sue Jones and Nici Nelson (Eds.), *Urban Poverty in Africa*. London: ITP, pp. 36–48.

Robinson, J. (2006) *Ordinary Cities: Between Modernity and Development*. London, New York: Routledge.

Sachs, W. (1992) "Introduction." In W. Sachs (Ed.), *The Development Dictionary*. London: Zed Books.

Streeten, P. (1995) *Thinking about Development*. Cambridge: Cambridge University Press.

Summerfield, G. (1995) "The shadow price of labour in export processing zones: A discussion of the social value of employing women in export processing in Mexico and China." *Review of Political Economy*, 7(1): 28–42.

Turner, J. F. C. (1977) *Housing by People: Towards Autonomy in Building Environments*. New York: Pantheon Books.

United Nations Department of Economic and Social Affairs/Population Division (2005). *World Urbanization Prospects: The 2005 Revision*. Available online at http://www.un.org/esa/population/publications/WUP2005/2005wup.htm (accessed July 7, 2013).

Zlotnick, H. (2006) "The dimensions of migration in Africa." In M. Tienda, S. Findley, S. Tollman, and E. Preston-Whyte (Eds.), *Africa on the Move: African Migration and Urbanisation in Comparative Perspective*. Johannesburg: Wits University Press, pp. 15–37.

"Development and the City"

Essay written for *Cities of the Global South Reader* (2015)

Michael Goldman

> I have understood the population explosion intellectually for a long time. I came to understand it emotionally one stinking hot night in Delhi a few years ago. The streets seemed alive with people. People eating, people washing, people sleeping. People visiting, arguing, and screaming. . . . As we moved slowly through the mob, hand horn squawking, the dust, noise, heat and cooking fires gave the scene a hellish aspect. Would we ever get to our hotel? All three of us were, frankly, frightened . . . since that night I've known the feel of overpopulation. (Ehrlich 1968)

REVILED AND DESIRED: SCHIZOPHRENIA OVER THE THIRD WORLD CITY

In the once-rocky fields southwest of Delhi sits triumphantly the Indian metropolis Gurgaon, built by a real estate mogul and now a magnet for Fortune Global 500 firms. Along with the fastest-growing cities in the world—Toluca in Mexico, Palembang in Indonesia, Chittagong in Bangladesh, Beihai in northern China—Gurgaon may be the next Wall Street pick for spectacular rates of financial return. Meanwhile Shanghai, home to more than 20 million people, remains the envy of and model for World Bank urban planners. Five decades after the Ehrlichs were chased from "frightening" Third World cities, how would they interpret this amazing urban turn? City planners now desire to *populate* their cities—some cities of the North are suffering from *de*-population—while China's planners expect to transform 100 towns into state-of-the-art global cities within 20 years. Too many people fighting over too few resources? On the contrary, the only scarce resource global-city planners' lament is the lack of capital investment, inadequate to keep up with the torrent pace of urban expansion in the global South.

As one British fund manager explained to me at an investors meeting in New Delhi in 2010, it is not a question of a world divided between "developed" (i.e., the North) and "emerging" (i.e., global South) markets, in which the latter is too risky for investors. US and European cities are now seen by Wall Street as "declining" and global South cities are the new "growth" markets overflowing with opportunities for high returns. "Who is still willing to invest in the dying markets of the North?" he asked with a hint of irony. Large pension funds from California, Texas, and Ottawa, and sovereign wealth funds from Abu Dhabi and China have plenty of capital and are eager to invest *if* urban governments are willing to offset the huge risks of world-city projects with guarantees of attractive financial returns.

Along with PricewaterhouseCoopers, UN Habitat, and European Chambers of Commerce, the World Bank has remapped the world as a "system of cities," emphasizing *not* Tokyo, London, and New York, but Dubai, Shanghai, and Sao Paolo as key nodes of urbanization (World Bank 2011). Rapid city growth is seen as a key factor in reversing disturbing poverty trends in Sub-Saharan Africa and throughout Asia (PricewaterhouseCoopers 2010). In 2009, the OECD Director declared in Copenhagen at the UN Climate Change Conference that one of the most important contributions to addressing the global warming problem is to invest much more capital into our expanding cities, not to reduce their size as environmentalists once demanded (OECD 2009; World Wildlife Fund 2010). In other words, to many policy experts, cities have now become the key to environmentally sustainable development, global economic growth, and poverty alleviation.

How did this happen? Why did this remarkable *global urban turn* in Northern expertise on development occur? Once disparagingly represented as "megacities" with intractable mega-problems, global South cities are now positively portrayed as "global cities," beacons of a new regime of urbanism and planetary sustainability. What has been the effect on the poor of this push to transform cities into global cities? What impact has the new urban turn had on other cities that have no appeal to global investors? How has this priority shift affected the majority urban populations that struggle daily to gain access to basic urban goods and services?

The goal of this chapter is to review the scholarly literature on development and the city, highlight the shifts in the development enterprise's approach to the city (e.g., how the World Bank and its partner organizations understand cities of the Third World), and delineate some of the most pressing issues for inhabitants of these rapidly changing urban landscapes. The chapter emphasizes the making of global cities because that is the trend today in global urban planning, a trend worthy of scrutiny. I question the policy to shift resources to these projects and away from what Jennifer Robinson (see next chapter) calls "ordinary cities." These more established cities are now off the radar of investment firms, business schools, and development banks, but remain home to many urban denizens who also need secure and ample access to the public goods and services, living wages, and social support systems that cities can provide (Robinson 2006; Watson 2009). The chapter concludes by highlighting themes and questions from scholars interested in creating an analytic framework and policy approach that more accurately captures the diverse lived experiences and histories of urban dwellers. I review perspectives that take seriously the past effects of colonial exploitation on city life, as well as current global capitalist practices, in order to appreciate why social inequalities seem to increase over time even with development policy infusions. Finally, I consider scholarship on more localized forms of power that have constrained and enabled non-elite majorities in efforts to make their cities livable based on their own, and not global, standards of city life.

DEVELOPMENT AFTER WORLD WAR II, AND ITS INFLUENCE ON CITY MAKING

The twentieth century was full of nationalist hope and ambition. In the post-World War II era, declarations of fulfilling "basic needs" of providing food, home, and security to the majority, by developing with a strong state and an economy geared toward producing for the nation became the goal. These goals included the desire to nationalize resources and industries controlled by ex-imperial powers abroad. As countries fought for national independence, the political question for nationalist movements was how to keep national wealth circulating *within* the country and *across* social classes so that the uneven and exploitative circuits of the colonial past could be broken and new social relations around production could be created, nationally and globally.

Some Southern countries experimented with a political-economic regime called "import substitution industrialization" (ISI) between the 1930s and 1970s, which countered the demands of Europe and the US for raw materials-producing countries to make these resources cheaply available. With the worldwide depression of the 1930s, the new strategy—among newly independent nations as well as many Latin American countries—was to protect national industries and populations from the extreme volatilities and uncertainties that triggered crisis for the global economy. Bryan Roberts explains how ISI worked in Latin America: because of the relative weakness of local capital, the Latin American state became an active agent in urban development (Roberts 1999). State employment was high and concentrated in cities; cities became "places of refuge" for a rural and small town population "for whom the agrarian structure no longer provides sufficient income for household subsistence" (Roberts 1999: 674). Government-led industrialization created internal demand for locally produced goods, created jobs, set rent controls for affordable housing (government built), offered food subsidies, and supported relatively strong trade union-backed rights.

Across the globe in Bandung, Indonesia, 29 independence leaders representing more than 1.5 billion people met in a conference hall in April 1955 to discuss African and Asian solidarity, opposition to colonialism and the vitriolic cold war politics of the US and Soviet Union, as well as search for ways to engage each other culturally and economically (now known as the "Bandung conference"). Subsequently, progressive national leaders including Nyerere of Tanzania, Senghor of Senegal, Nkrumah of Ghana, and Touré of Guinea initiated a policy of "African socialism" (a traditional form distinct from Europe's) in which economic development was guided by a large, job-creating public sector. In 1967, Nyerere released the Arusha Declaration that promoted an African model of development based on what he called *Ujamaa*, a Swahili word meaning "extended family"—a moral compass for a national plan based on the notion of shared development.

All of these development models stood in stark contrast to those coming from European and US governments and from those Northern institutions they led, such as the World Bank and the International Monetary Fund (IMF). These twin agencies were created at a meeting in Bretton Woods, New Hampshire, in July 1944, as World War II was winding down and the global economy was in collapse. In response, US and British economic advisors John Maynard Keynes and Henry Morgenthau called together allied world leaders to figure out how to rebuild the global economy, secure devastated currencies, and get production back on track. The most influential development organization to emerge was the International Bank for Reconstruction and Development, or the World Bank (McMichael 2009; Goldman 2005). Over time, the bank's Board of Directors and upper management comprised officials selected mainly from the UK, Germany, France, Italy, US, Canada, and Japan. Most of its earliest projects went to build up the infrastructure that fed the global economy: rebuilding the mines, power plants, railways, and ports that would expeditiously extract minerals from the mines for the industrial factories in the global North. The IMF's job was to harmonize national currencies and encourage rules to stabilize wild fluctuations in currency values and activities, and promote the unimpeded movements of capital and goods, so as to end what Morgenthau called "economic nationalism" (McMichael 2009).

From the 1950s through the 1970s, development policy makers and project lenders thus found their mission *outside* the cities, in the rural-based sectors of energy, mining, industry, agriculture, and transport. From an urban perspective, rural-based development could stop the flow of the rural poor migrating into the city in search of remunerative work; and, from the perspective of Robert McNamara, World Bank president from 1968 through 1981, such capital infusions could also stem the tide of rebellions and revolutions in the countryside of ex-colonies such as Vietnam, Indonesia, and Cambodia.

While the World Bank President justified these rural investments to his staff and clients as contributing to overall economic growth, and perhaps reducing the stress of in-migration on cities, he had a difficult time selling the idea of lending directly to large cities. Since the bank was dominated by neoclassical economists with a Wall Street business sense (all of its presidents have been Americans, and most came from Wall Street), the prickly question became: on what grounds could it justify lending money to cities? The World Bank needed to find the urban equivalent of the rural peasant, "a targetable population that could be the recipient or direct beneficiary of productive investments, not simply welfare transfers," as political scientist Edward Ramsamy put it (Ramsamy 2006: 83). In the mid-1970s when World Bank President McNamara introduced an urban department into the institution's growing bureaucracy, he was met with extreme wariness. The whole idea of lending for "the urban" sounded unfamiliar to most World Bank economists, quite different from lending for energy, farming, or transport. "Some people in the Bank were making jokes that next we are going to have suburban development, or an outer-space development program" (Ramsamy 2006: 83).

Under the rubric of "basic needs" programming, the World Bank did start to finance very small schemes for "slum upgrading" and helped in the creation of UN-Habitat in 1978, a marginally funded UN agency focused on city development to counterbalance the many UN agencies working on rural development. The belief was that if housing and access to water and sewerage facilities were improved, the poor could have their basic needs met and could be better participants in the economy.

Urban development is a complex issue, as Matthew Gandy demonstrates in his work on Lagos, Nigeria (Gandy 2006). Lagos is the largest, most prominent West African metropolis and a city in which development planners have taken a great interest. In the late 1970s, according to Gandy, development experts believed that if modern infrastructure systems were delivered to megacities, the foundation would be laid for major urban improvements. For example, urban planners confidently predicted that within 20 years Lagos would be fully connected with a water system supplying water to all its residents. All the most potent organizations worked toward this accomplishment, and yet by 2003, fewer than 5 percent of Lagos' households had a direct connection, leaving most of its 15 million denizens to buy from private tankers or vendors, or expensively dig their own bore wells (Gandy 2006).

Part of the explanation for what happened in Lagos rests with the political and developmental shift for many global South cities in the 1970s and 1980s, as the ideal of the socially planned city deteriorated due to uneven forms of capitalist urbanization in which corporate investors primarily built hotels, hospitals, and infrastructure to benefit the elite while basic infrastructure for the non-elite majority was left in disrepair. Development policies were rarely written to mitigate these inequalities, in large part because the World Bank and others encouraged investment in private property and wealth accumulation in cities, imagining a "trickle-down" effect to the poor from such private forms of accumulation. For example, a common World Bank loan would be for road construction with the intent of relieving traffic, but these loans are both expensive and only beneficial for the few who own cars. Moreover, the majority who do not own cars can often lose access to a public good such as the city road, as their modes of transport—walking, biking, moving by ox-carts, public buses—are neglected or pushed aside.

Many development scholars and practitioners diagnosed the impoverishment of the postcolonial South using an analytical framework called *modernization theory*, premised on three assumptions. First, according to modernization theory, societies go through evolutionary stages of development, from traditional to modern, from agrarian to urban. Second, technology more than social organization and structure is considered a prime engine and defining feature of social change, and third, industrialism rather than capitalism is believed to be the major force behind such "progress." In the early 1960s, a national security advisor for President Johnson during the Vietnam War, Walt Rostow, promoted his book, *Stages of Economic Growth*, and his beliefs on how countries develop economically, with a US-centric understanding of progress and development (Rostow 1960). Despite its profound shortcomings (e.g., ignoring the structural roles of European-derived colonial and capitalist relations that *underdeveloped* parts of the global South), it caught the imagination of Northern policy experts and economists. Modernization has since become the foundational framework for understanding global North–South relations, as well as the justification for the role and logic of development institutions such as the World Bank.

Another perspective emerged during the 1970s, from development consultant and scholar Michael Lipton who detected an "urban bias" within development institutions, as well as in national politics, favoring the needs of the urban elite. Lipton interpreted the large accumulation of wealth in the city as built on the backs of rural areas (Lipton 1977). He argued that resources were often sucked out of the countryside and funneled into the cities, through political pressures to keep crop prices low so that urban food prices would be kept low. This bias toward the city benefited urban workers as well as their employers who could pay them less as a result.

Other debates run through the scholarly literature around the questions of how cities grow, how they generate wealth, and how they allocate resources, public space, and social goods such as open public areas for markets and hawking. Many large Southern cities were (and some still are) important manufacturing sites and hence crucial to the national economy. By the 1980s, for example, the Bangkok metro area contributed almost 90 percent of Thailand's gross national product (GNP) in services and 75 percent in manufacturing, while having only 10 percent of the nation's population (Kasarda and Crenshaw 1991). Similarly, Lagos produced half of Nigeria's manufacturing value and Mexico City alone generated 30 percent of Mexico's GNP, as Sao Paolo did for Brazil. Yet access to basic public goods such as health care and clean water, and especially a living wage, was not part of the lives of the working majority.

By contrast, and as a strong critique, *dependency* and *world-systems* theorists challenged the dominant perspective of modernization, arguing that capitalism is a historically unique and powerful social formation (emerging from colonialism) that creates unequal structures of exploitation and expropriation (Frank 1966; Cardoso and Faletto 1979; Wallerstein 1974; Amin 1976). The beneficiaries are mostly elite classes in the major cities of the North (called "the core"), and secondarily in the urban South (or "periphery"). Some called this process "underdevelopment," whereby the wealthier classes and governments in the North *underdeveloped* governments and populations in the South (Amin 1976; Frank 1966). Such scholars argued for a historical analysis that revealed the dual processes of urbanization that inextricably linked the wealth accumulation and cosmopolitan living of New York, Paris, and London with the generation of poverty and inequities of Mexico City, Dakar, and Dhaka. Whereas modernization scholars thought of countries as discrete objects with their own internal plans and successes/failures, "developing" *sequentially*, these critical social theorists understood history in *relational* terms. That is, England became wealthy and powerful *because* it exploited India's wealth and possibilities, thereby "underdeveloping" it (McMichael 2010). In their view, as capitalism evolved after colonialism, new hierarchies and social relations emerged, such that cities like Singapore, Shanghai, and more recently Dubai contributed to the reshaping of a multi-polar system of capitalism. In either case, these scholars argued that one should not diagnose and solve developmental problems as if they exist in a vacuum, or are due to purely localized problems of "lacks" or deficiencies.

Manuel Castells added an urban dimension to the dependency school approach. Dependency, he argued, was not *only* an external condition and imposition, but also an internal set of practices, reproducing localized forms of inequality, as one can see in the social tensions within Latin American cities where he studied (Castells 1977). Therefore, he contended that one should study the interaction between global and national structures of inequality as *interdependent* processes. Simply put, local political and economic elites also have agendas that can lead to greater class disparities and social injustices, and the effects can be seen in the rise of the gated enclaves in and around cities and the vast areas of *favelas* (slums), where the poor, who do so much of the daily work for the city, live. The majority of these workers produce the wealth for the city but the city's public services provide little in terms of safe and adequate housing, drinking water, health care, waste removal, schools, and most importantly, living wages (i.e., enough income to work oneself out of poverty).

URBAN SHOCK THERAPY: STRUCTURAL ADJUSTMENT AND NEOLIBERALISM

By the 1970s in Chile and Argentina, and almost everywhere else in the 1980s (except, notably, China), the state-centered model of development ended and a new approach, which privileged the private sector and "market mechanisms," was embraced. This development shift paralleled a global political project emphasized by US President Ronald Reagan and British Prime Minister Margaret Thatcher in the early 1980s, which critical scholars call *neoliberalism*. This ideology reflected the belief that market actors, such as corporations, should direct national economies rather than governments (or trade unions), which are supposedly dominated by political interests. In the development industry, this political ideology drew strong support from the World Bank and the IMF, both of which pushed the idea that less regulation and more freedom of market actors such as large firms would reduce corruption, politics, and market inefficiencies. Beginning in the 1970s, the World Bank imposed *structural adjustment programs (SAPs)* on countries of the global South, with the goal of reducing the role of the state. They insisted that their borrowers sell off important national assets such as steel factories, coal mines, railways, and telecommunications companies to local or foreign firms. They also insisted that governments charge "user fees" for public services (e.g., education and health care) that were once provided for free or very cheaply to the poor majority.

This new development approach was based on the belief that firms, especially more "experienced" Northern firms, knew best how to run a business and that cities basically comprised numerous government enterprises that functioned poorly. They believed that converting these once-public services to a private, fee-for-service model would both lead to greater efficiency in their use and convert urban citizens into reliable and responsible customers. One of the sectors in which these ideas were put to the greatest test—and failed—was water (see box). Furthermore, the World Bank and the IMF required that governments devalue their currencies, cut tariffs on imports, and re-orient the output of their farms and factories toward those who could pay, namely, foreign rather than local customers. Since the Bank and the IMF held the purse strings, cash-starved countries reluctantly followed these controversial prescriptions, despite the fact that they were never promoted in Northern countries in such an extreme fashion.

The net effect was disastrous for the poorest populations. Indicators for national-level health, life expectancy, income equality, and per-capita GDP

WATER WARS

One of the most heavily criticized development policies during the 1990s was the attempt to privatize large-city water supplies and services. The World Bank's idea of turning over public provisioning of water to European and US firms spread quickly, sweetened with the green and humanitarian promise of *finally* delivering safe and efficiently distributed water to the world's poor urban population. With governments broke and old colonial water systems woefully inadequate, the Bank sold the idea that well-capitalized international firms were better positioned than politicized governments to provide water. The free market was to be the savior. The World Bank and the IMF offered guaranteed access to much-needed capital and debt relief in exchange for government's implementing water privatization. The pressure was substantial: by 2001, all 11 of the World Bank's water and sanitation loans carried conditionalities that required borrowing governments to either privatize these services or dramatically increase the price to consumers. In that same year, IMF "poverty reduction" loans to highly indebted countries had water privatization as a key conditionality. Highly indebted countries could not receive relief without privatizing their municipal water systems.

But as water bills rose to unwieldy heights, people took to the streets: in Guinea, water prices rose more than fivefold for the majority poor; in Johannesburg the price doubled and many of the poorest were completely cut off for not being able to pay their bills. In the Bolivian city of Cochabamba, the price spiked 200 percent, and some users were fined for harvesting rain water without paying Bechtel as the contract drafted by the Bank required. Cities erupted in protest as the cost of water became prohibitive, promised pipes were never run to the poor neighborhoods, and international firms held city governments hostage to contracts that guaranteed rates of returns to the firms with no guarantees to the city that all would receive a fair share of the city's water. By the early 2000s, 80 percent of these World Bank water contracts were nullified by angry citizens, vulnerable politicians, and/or frustrated firm shareholders demanding their companies withdraw from money-losing ventures.

From Ecuador to Bolivia and Paraguay, these "water wars" fed into urban movements against resource and service privatization, which forced out neoliberal governments and replaced them with anti-World Bank, anti-privatization governments. The *urban revolution* was back on the political landscape.

Source: Goldman 2007.

growth, all fell dramatically during the neoliberal, structural adjustment period of the 1980s and 1990s (Weisbrot and Ray 2011; Ismi 2004). When development experts insisted that countries introduce user fees for children attending elementary schools and for families using local health clinics, attendance dropped precipitously. Mortality rates increased and life expectancies dropped. Health and education budgets, already low by comparison to the North, reached their lowest point since the colonial era (Ismi 2004). Development scholar Asad Ismi summarizes the effects on the African continent as follows: as a result of structural adjustment policies, "Africa spends four times more on debt interest payments than on health care. This combined with cutbacks in social expenditure caused health care spending in the 42 poorest African countries to fall by 50% during the 1980s" (Ismi 2004:12).

This draconian development policy reversed most of the gains that countries had enjoyed during the post-independence, nation-centered development period. The worst hit countries were the poorest, those that had little choice but to follow World Bank and IMF prescriptions if they did not want to lose access to international capital and loans. Middle-income countries and China were notable exceptions, as they pursued decidedly state-led, non-neoliberal plans.

Neoliberal policies undertaken by many countries in the global South reduced employment in major cities where government offices were located; they also removed employment protections and important subsidies for public goods and services. By and large, electricity, water, health services, schooling, and waste removal were priced at levels that most consumers could not afford. With rising water and electricity prices and the loss of subsidies for locally consumed food grains, local diets often became unaffordable. Housing prices were de-subsidized and housing became part of a globally competitive real estate market, which priced many poor city dwellers out of their homes. There is a large literature about the near universal failure of these development policies to improve the poor's lives (Ismi 2004; Naiman and Watkins 1999; SAPRIN 2002; Weisbrot and Ray 2011). Ismi concludes that "after 15 years of following World Bank and IMF-imposed policies, Latin America, by the late 1990s, was going through 'its worst period of social and economic deprivation in half a century.' By 1997, nearly half of the region's 460 million people had become poor—an increase of 60 million in ten years" (Ismi 2004: 9).

But cities have also been the sites of popular protests against draconian development policies. In the 1980s, city streets broke out in "bread riots" and anti-IMF protests, closing down business and governments (Walton and Seddon 1994). Some eventually led to small and large changes, including the toppling of conservative political regimes, which often were the most compliant World Bank clients. In the late 1990s and early 2000s, Latin American governments were voted out in Uruguay, Ecuador, Bolivia, Brazil, Argentina, Chile, Paraguay, and Venezuela. Election campaigns highlighted the desire to kick out the World Bank and IMF and nationalize ownership of natural resources. A new Banco del Sur (Bank of the South) was inaugurated in September 2009 with the objectives of boycotting the major development agencies and pooling capital resources from progressive Latin American governments (starting with $20 billion) to invest in social welfare and development projects in the poorest regions of Latin America, a socially progressive set of priorities that pushed against the grain of global capitalist development ideology and practices.

SPECULATING ON GLOBAL CITIES

In 2009, World Bank President Zoellick announced:

> Infrastructure is a cornerstone of the World Bank Group's recovery strategy for the global economic crisis. These investments can create jobs today and higher productivity and growth tomorrow ... the needs are huge: an estimated 880 million people still live without safe water; 1.5 billion people without electricity; 2.5 billion without sanitation; and more than one billion without access to an all-weather road or telephone service. The world's urban population is expected to increase from 3.3 billion to five billion by 2030—with Africa and Asia doubling their urban populations—creating new infrastructure demands for transport, housing, water, waste collection, and other amenities of modern life ... These infrastructure choices will shape cities and lifestyles for many decades or even a century to come. (Zoellick 2009)

Thus began the current era in which the global South city itself becomes the marketable commodity, under the guidance of global development (De Soto 2000; World Bank 2008; Glaeser 2012). From the perspective of World Bank official Hernando de Soto, the misery of the South is rooted in the unvalorized potential that sits right under the poor and their land—including the public spaces they inhabit such as sidewalks, plazas, and slum areas squatted upon by hard-working but low-income populations. The solution to this problem has now become: expand the railway station into a shopping mall, the sidewalks into private property, slums into central business districts, and charge fees for government services outsourced to international firms. Streamline, upgrade, and globalize resources, services, public spaces, and land, and you can ignite the transformation of the wretched mega-city into a global city, a place where international capital and its employees would desire to invest and settle.

The development enterprise did not fall in love with the idea of globalizing cities on its own; indeed, as is true with its earlier shifts in lending policies, it tends to follow, and support, trends emerging from various other quarters. The 1980s and 1990s witnessed a tremendous consolidation and unleashing of power from what were once fairly discrete organizations of local and international capital: insurance, housing, banking, investing, real estate. This consolidation has created large pools of finance and real estate capital working across national boundaries to develop new types of investment strategies, and new social imaginaries as to how cities can be made globally competitive and more profitable for investors, and, allegedly, more developed and livable.

These global-city strategies were contrived by transnational policy networks, creating a global network of expertise on cities, led by consultants (e.g., McKinsey, PricewaterhouseCoopers), UN agencies (e.g., UN-Habitat, the UN Development Program), international finance institutions (e.g., the World Bank, the Asian Development Bank), and global forums such as the World Cities Summit and the C40 Cities Mayors Summit. These forums are often co-sponsored by real estate and capital goods firms. On one end of the spectrum, the World Bank and bilateral agencies advocate for the "urban turn" in lending and urban planning; at the other end, environmental NGOs and experts have deemed cities as the new site of green innovation for carbon-cutting built environments, living spaces, and transport systems. Whereas once these actors deplored the megacities of the Third World, today, cities are now seen as the fount of innovation and "best practices" for "ending poverty" and for "sustainable lifestyles." By 2012, the World Bank declared that financing city infrastructure was one of its most important endeavors (Rethinking Cities 2012).

The rationale for these global investment practices starts with the assumption that Southern cities have been dying under the weight of mega-city problems, including a culture of ineptitude that has made it hard for any national government or financial market to expect large cities to become drivers of a national economy. Select cities, however, do become economic powerhouses that urban boosters claim *can be* replicated elsewhere: Singapore and its city-state model of capital accumulation (Chua 2011), Shanghai and its massive infrastructural transformation, Dubai and its capital-flush real estate sector, and Bangalore and its phenomenal information technology industry. These models of urban transformation have incited ambitious master plans to convert urban landscapes into global cities with spectacular skylines and splashy must-see infrastructure.

Coupled with these assumptions and beliefs is a new agenda from the world's largest real estate and financial firms, to build up various forms of global cities. For example, in 2007, Dubai's largest developer and real estate firm, Limitless LLC, committed $15 billion to finance and build India's first privately owned small city outside of Bangalore, India's "Knowledge City," the first of five multi-billion dollar private "greenfield" cities on the rapidly urbanizing rural periphery of Bangalore. Before 2007–2008, this same firm had large-scale financial commitments for building luxurious cities adjacent to Moscow, Istanbul, Riyadh, Amman, and in and around Dubai—with ambitious names like "The World" and "The Universe." As of today, most of these projects have been stalled due to lack of capital and demand.

Meanwhile, China is planning to build more than 100 global cities over the next two decades, an extraordinary feat by any standard. It is also building a string of "eco-cities" that are hoped to become prototypes to export abroad. In its interest to acquire much-needed farm land and minerals to continue its city building campaign, China is also purchasing land and mines throughout Africa, and has agreed to build up urban infrastructure in exchange, including lavish state palaces and halls, soccer stadiums, and central business districts, and in oil-rich Angola, a $2.5 billion upscale mini-city of apartment complexes (called Nova Cidade de Kilamba) with condos selling at New York prices in a land where people make an average of $2 per day. In 2013, it sits mostly empty.

Is this the way to build up a city with equitable access to basic services, good employment, and affordable housing and clean water? These speculative, and seemingly failed, urban projects and visions can do harm to more than just their investors. For one, they require large swathes of public or farming land, discounted by governments and often with guarantees of 24/7 water, electricity, and the latest transport systems—resources that nearby communities rarely receive. World Bank loans are oftentimes featured in the lure for investors to come to town, even though the greater population must contribute to the repayment of these loans. Second, this type of speculative urbanism has sparked a new form of urban governance, one vulnerable to the rise and fall of stock markets and the whims of investment capital, whose actors are often based in cities immune to such volatilities, such as Singapore, London, Shanghai, and New York (Goldman 2011). Third, and perhaps most importantly, this type of frenetic city building displaces large populations to accommodate investors. In China alone, it is estimated that tens of millions have been displaced in the past decade from city expansion and reconstruction (Hsing 2010).

The latest trend seems to be a grand distraction from any attempts to build secure and affordable cities for the majority. Urban and rural social movements have risen in protest against what is being called "land grabbing" for these luxury enclaves.

Today, China is home to thousands of protests, especially in its smaller cities and large towns. Elsewhere, city dwellers are challenging these development processes and working to reclaim and remake their city in their own image, rather than the imagined one of global-city boosters. One prominent Africanist scholar argues it is often the Africans marginalized by failed global-city projects who mobilize "the city as a resource for reaching and operating at the level of the world" (Simone 2009). City dwellers, local governments, and even many actors in the global economy cannot cope with such volatility, as the crippling global financial crisis of 2007–2008 demonstrated.

NEW APPROACHES TO THINKING ABOUT CITY LIFE AND LIVELIHOODS

As Achille Mbembe (2004) and other African scholars note, there is nothing new to the idea that African cities have been re-made due to transnational forces, connections, and imaginaries (see Miraftab 2012). To complement and counterbalance the premise that external forces have often dictated local realities, these scholars start from the place where the majority of city dwellers live, where they work and struggle, and create, and then assess the influences of external forces *in situ*. AbdouMaliq Simone observes that most African denizens thrive at a distance from the formal center of the city, far from where official practices and structures dominate (Simone 2004a, 2004b). That is, urban life for the majority is not solely defined by the impact of a mega-project or a development policy; rather, the African urban is constituted in large part by the multiple ways that people generate their own provisions and services, make their own markets and social goods, and feel and express their own sentiments and sensations, or what he calls "people as infrastructures." In a similar vein, Mbembe stresses "ways of seeing" that overcome the myopic outsider's view only of the market, either macro/economic structures of dependency or the neoliberal emphasis on markets as saviors, with the "African variation" perceived as backwards and failed. The ironclad hold of developmental thinking on the telling of Africa forecloses the possibility of understanding complex power relations beyond the narrow Western social science categories of state, market, and civil society, which always look inadequate under the Western gaze.

Instead, Africanist scholars argue that a socio-spatial approach based on ethnographic and historical readings of the African city reveal a quite different urban experience. From this perspective, the state does not have a monopoly on power, as people have successfully self-organized across numerous lines of connectivity, locally and globally, based on a wide range of needs, from food to spirituality, in efforts to fulfill their needs and generate alliances (Mbembe 2004; Simone 2004a; Watson 2009). They argue that "African cosmopolitanism" surfaces and circulates in the shadows of, and in spite of, official forms of urban power. They are calling for a more ground-up analysis of how the non-elite majority creatively constitutes the city and persists amidst injustices and scarcities (Ferguson 2006). If we can better understand how urban poverty and wealth are co-constituted and legitimated, through forms of capitalist development, we can better understand the alternatives. This approach reveals that which the developmentalist worldview often masks or erases; from this perspective, one can begin to see global South cities in their complexities and possibilities, not in caricature or in comparison to the North and its standards and misperceptions.

From this vantage point, we can also see the multiple modernities of the urban, to better understand global South peoples as *actors* making cities and histories, as *subjects* constituted in times of crises, resisting authorities and creating aspirations and livelihoods (Mbembe and Roitman 2003; Mignolo 2000; Roy 2009a). This is a perspective that does not assume a half-formed subject or a malignant modern, but it also is not a romanticized reading of reality. Instead, it starts from the premise that the same powers of creativity, expertise, and infinite possibilities that have been associated with the cosmopolitan urbanite of the global North be granted their counterpart in the Southern city. In post-apartheid South Africa, one of the most vibrant urban social movements is *Abahlali baseMjondolo*, shackdwellers pushed to the limits by the police who have had the mandate to aggressively destroy slums, especially in the brutal "clean up" prior to the 2010 FIFA soccer World Cup event. They have no party affiliation; instead they have conjured up their own definition of politics in which they demand the right to their homes and workshops and access to urban public space—the "right to the city" which so many South Africans have lost. Their ambition is to take back the urban commons that have been privatized or made accessible only to elites.

Elsewhere, urban protests against the privatization of city water services sparked a broad social justice agenda. It started in Cochabamba, Bolivia, in the early 2000s and spread across Latin American cities, where people from different backgrounds came together to roll back profitable schemes for elite politicians and international firms, whether it be selling off water, electricity, health care, education, or urban public space (or the "urban commons") such as open-air markets and the land in and surrounding slum dwellings deemed more valuable as commodified "real estate."

In sum, to more accurately understand the relationship between the development enterprise and the global South city, we need to "decenter" global urban theories in scholarship and development practice that stick to master tropes about slums, poverty, and arrested development, and instead consider the significance and worldliness of analytic perspectives that emanate from these communities and their cities (McFarlane 2008; Roy 2009b; Miraftab 2009).

CONCLUSION: CITIES OF INFINITE POSSIBILITY?

Perhaps as momentous as in 1871 (insurrection of the Commune of Paris), 1917 (Russian Revolution), 1968 (urban revolts from Mexico City to Paris), and 1989 (Eastern European revolutions), the year 2011 erupted into a series of unexpected urban revolutions. From Tahrir Square in Cairo, Egypt, to Pearl Square in Manama, Bahrain, a diverse swathe of national populations barricaded the streets of major and minor cities and declared that a new world was possible, one free of autocratic rule, free of austerity and speculation, free of multiple forms of oppression. How shall we theorize the urban nature of these expressions of people's power and social change, as they unfold over the next decade?

No longer should cities such as Tunis and Tripoli be portrayed as provincial outposts far from the modern, and disconnected from the happenings of the world. History teaches us that there would be no affluent London without the work and resources of impoverished Calcutta and Kingston; no Paris without Abidjan and Saigon. Under the most exploitative conditions, global South cities have contributed enormously to our rich-and-poor modern world. How can they also be sites for a different sort of expectation and experimentation, as they are for millions of migrants entering the big city every day with hope and ambition? How can they become cities of justice, peace, and solidarity?

South and east of the tumultuous Arab Spring, one sees another sort of urban prospecting: by introducing new trade relations with resource-rich countries in Africa, China hopes to receive privileged access to the raw materials necessary to construct its newest global cities. In exchange, China is building skyscrapers, government palaces, soccer stadiums, and even Africa's largest university, transforming urban Africa. Challenging the ethics of these closed-door political deals, protest movements have emerged in Ghana, Zambia, and Tanzania. Many worry that the aspirations of Asia's urban century hinge on these highly speculative and risky activities in Africa. The extent to which these desires and practices of global accumulation stick in our cities may influence the next round of urban revolutions, the kind coming from Washington, DC and Beijing, as well as the types arising from the plazas, neighborhoods, and workshops of our lesser known cities worldwide.

REFERENCES

Amin, S. (1976) *Unequal Development: An Essay on the Social Formations of Peripheral Capitalism*, New York: Monthly Review Press.

Amin, S. (2000) "Urban land transformation for pro-poor economies." *Geoforum* 35: 277–288.

Cardoso, F. H. and Faletto, E. (1979) *Dependency and Development in Latin America*, Berkeley: University of California Press.

Castells, M. (1977) *The Urban Question. A Marxist Approach*, London: Edward Arnold.

Chua, B. H. (2011) "Singapore as model: Planning innovations, knowledge experts," in A. Roy and A. Ong (Eds.) *Worlding Cities: Asian Experiments and the Art of Being Global*, London: Wiley Blackwell.

De Soto, H. (2000) *The Mystery of Capital: Why Capitalism Triumphs in the West and Fails Everywhere Else*, New York: Basic Books.

Economic Commission for Latin America (ECLA). (1951) *Theoretical and Practical Problems of Economic Growth*, Santiago, Chile: ECLA, United Nations.

Ehrlich, P. (1968) *The Population Bomb*, New York: Ballantine Books.

Ferguson, J. (1994) *Anti-Politics Machine: Development, Depoliticization, and Bureaucratic Power in Lesotho*, Minneapolis, MN: University of Minnesota Press.

Ferguson, J. (2006) *Global Shadows: Africa in the Neoliberal World Order*, Durham, NC: Duke University Press.

Frank, A. G. (1966) *The Development of Underdevelopment*, New York: Monthly Review Press.

Gandy, M. (2006) "Planning, anti-planning and the infrastructure crisis facing metropolitan Lagos," *Urban Studies* 43(2): 371–396.

Glaeser, E. (2012) *Triumph of the City*, New York: Penguin Books.

Goldman, M. (2005) *Imperial Nature: The World Bank and the Struggle for Social Justice in the Age of Globalization*, New Haven, CT: Yale University Press.

Goldman, M. (2007) "How 'Water for All!' policy B became hegemonic: The power of the World Bank and its transnational policy networks," *GeoForum* 38: 786–800.

Goldman, M. (2011) "Speculating on the next World City," in A. Roy and A. Ong (Eds.) *Worlding Cities: Asian Experiments and the Art of Being Global*, London: Wiley Blackwell.

Hsing, Y. (2010) *The Great Urban Transformation: Politics of Land & Property in China*, New York: Oxford University Press.

Ismi, A. (2004) "Impoverishing a continent: The World Bank and the IMF in Africa," Report for the Halifax Initiative, Halifax.

Kasarda, J. D. and Crenshaw, E. M. (1991) "Third World urbanization: dimensions, theories, and determinants," *Annual Review of Sociology* 17: 467–501.

Lipton, M. (1977) *Why Poor People Stay Poor: Urban Bias in World Development*, Cambridge: Harvard University Press.

Mbembe, A. (2004) "Aesthetics of superfluity," *Public Culture* 16(3): 373–405.

Mbembe, A. and Roitman, J. (2003) "The figure of the subject in the time of crisis," in M. Abdoul (Ed.) *Under Siege: Four African Cities*, Hatje Cantz, Ostfildern, pp. 99–126.

McFarlane, C. (2008) "Urban shadows: materiality, the 'southern city' and urban theory," *Geography* 2(2): 340–358.

McFarlane, C. (2010) "The comparative city: Knowledge, learning, urbanism," *International Journal of Urban and Regional Research* 34(4): 725–742.

McMichael, P. (2009) *Development and Social Change: A Global Perspective*, Thousand Oaks, CA: Sage, 3rd edition.

McMichael, P. (2010) *Development and Social Change: A Global Perspective*, Thousand Oaks, CA: Pine Forge Press.

Mignolo, W. (2000) *Local Histories/Global Designs: Coloniality, Subaltern Knowledges, and Border Thinking*, Durham, NC: Duke University Press.

Miraftab, F. (2009) "Insurgent planning: Situating radical planning in the global South," *Planning Theory* 8(1): 32–50.

Miraftab, F. (2012) "Colonial present: Legacies of the past in contemporary urban practices in Cape Town, South Africa," *Journal of Planning History* 11(4): 283–307.

Naiman, R. and Watkins, N. (1999) "A survey of the impacts of IMF structural adjustment in Africa: Growth, social spending, and debt relief," Preamble Center, Washington, DC.

OECD. (2009) *Competitive Cities and Climate Change*, OECD Working Paper, Paris, France.

PricewaterhouseCoopers. (2010) *Cities of Opportunities*, London.

Ramsamy, Edward. (2006) *The World Bank and Urban Development: From Projects to Policy*, New York and London: Routledge.

Rethinking Cities: Framing the Future. (2012) 6th Urban Research and Knowledge Symposium, Barcelona, October.

Roberts, B. (1999) "Urbanization, migration, and development," *Sociological Forum* 4: 4.

Robinson, J. (2006) *Ordinary Cities: Between Modernity and Development*, London: Routledge.

Rostow, W. (1960) *Stages of Economic Growth: A Non-Communist Manifesto*, Cambridge: Cambridge University Press.

Roy, A. (2009a) "The 21st century metropolis: New geographies of theory," *Regional Studies*, Spring.

Roy, A. (2009b) "Informality and the politics of planning," in J. Hillier and P. Healey (Eds.) *Conceptual Challenges for Spatial Planning*, Surrey: Ashgate, pp. 87–108.

SAPRIN (The Structural Adjustment Participatory Review International Network) (2002) *The Policy Roots of Economic Crisis and Poverty: A Multi-country Participatory Assessment of Structural Adjustment*, Washington, DC.

Simone, A. (2001) "On the worlding of African cities," *African Studies Review* 44(22): 15–41.

Simone, A. (2004a) *For the City Yet to Come*, Durham, NC: Duke University Press.

Simone, A. (2004b) "People as infrastructure: Intersecting fragments in Johannesburg," *Public Culture* 16, 3: 407–429.

Simone, A. (2009) *City Life from Jakarta to Dakar: Movements at the Crossroads*, New York: Routledge.

Wallerstein, I. (1974) *The Modern World-System, Vol. I: Capitalist Agriculture and the Origins of the European World-Economy in the Sixteenth Century*, New York and London: Academic Press.

Walton, J. K. and Seddon, D. (1994) *Free Markets and Food Riots: The Politics of Global Adjustment*, London: Wiley Blackwell.

Watson, V. (2009) "Seeing from the South: Refocusing urban planning on the globe's central urban issues," *Urban Studies* 46(11), 2259–2275.

Weisbrot, M. and Ray, R. (2011) "The scorecard on development, 1960–2010: Closing the gap?" UN-DESA, Washington, DC.

World Bank. (2008) "World Development Report 2009: Reshaping economic geography." Washington, DC: World Bank Press.

World Bank. (2010) "Eco2Cities," Washington, DC: World Bank Press.

World Bank. (2011) "System of cities: Harnessing urbanization for growth and poverty alleviation," Washington, DC: World Bank Press.

World Wildlife Fund. (2010) "Reinventing the city," WWF International, Gland: Switzerland.

Zoellick, R. (2009) "Keynote address," The World Bank-Singapore Urban Hub Inauguration Conference, Singapore, June.

"World Cities, or a World of Ordinary Cities?"

Ordinary Cities: Between Modernity and Development (2006)

Jennifer Robinson

INTRODUCTION

Globalization has transformed urban studies. Cities are now routinely viewed as sites for much wider social and economic processes, and the focus for understanding urban processes has shifted to emphasize flows and networks that pass through cities rather than the territory of the city itself (Friedmann 1995a, Friedmann 1995b; Sassen 1991; Smith 2001). The study of cities now commonly encompasses the flows of global finance capital, the footloose wanderings of transnational manufacturing firms, and the diverse mobilities of the world's elite alongside diasporic and migrant communities from poorer countries. In many ways, urban studies has become much more cosmopolitan in its outlook. Globalizing features common to many cities around the world encourage more writers to consider cities from different regions, as well as wealthier and poorer cities, within the same field of analysis (Marcuse and van Kempen 2000; Scott 2001; Marvin and Graham 2001). This is certainly good news for a post-colonial urban studies, eager to bring different kinds of cities together in thinking about contemporary urban experiences.

The situation appears more propitious than ever, then, for an integration of urban studies across long-standing divisions of scholarship, especially between Western and other cities, including "Third World" and former socialist cities. An analytical focus on the transnational global economy could ensure that such categorizations of cities will no longer be of any relevance. Indeed, this is a claim made by the key advocates of these approaches (Sassen 1994; Taylor 2001). Does this mean that urban studies has come to be sensitive to the diversity of urban experiences, to the wide range of cities across the world? Could this be the basis for a post-colonial urban theory that refuses to privilege the experiences of some cities over those of others?

Many studies of globalization and cities have drawn on the idea of "'world" cities to understand the role of cities in the wider networks and circulations associated with globalization. Some cities outside the usual purview of Western urban theory—"Third World cities"—have been incorporated into these studies insofar as they are involved in those globalizing processes considered relevant to the definition of world cities. This is definitely a positive development in terms of ambitions to post-colonialize urban studies, to overcome the entrenched divisions between studies of "Western" and "Third World" cities. But many cities around the world remain "off the map" of this version of urban theory (Robinson 2002b). And despite the relative inclusiveness of the focus on globalization processes, developmentalism continues to pervade global- and world-city narratives, consigning poorer cities to a different theoretical world dominated by the concerns of development. Although the older categories of First and Third World may have less purchase, world cities approaches have a strong interest in hierarchies, and have invented new kinds of categories to divide up the world of cities. Perhaps most worrying for a post-colonial urban studies, world-cities approaches, by placing cities in hierarchical relation to one another, implicitly establish some cities as exemplars and others as imitators. In policy-related versions of these accounts cities either off the world-cities map or low down the supposed hierarchy have an implicit injunction to become more like those at the top of the hierarchy of cities: they need to climb up the hierarchy to get a piece of the (global) action. Being one of the top-rank global cities can be equally burdensome, though, encouraging a policy emphasis on only small, successful and globalizing segments of the economy and neglecting the diversity of urban life and urban economies in these places.

So while there is much to learn from global- and world-cities approaches, this chapter suggests that there is still considerable work to be done to produce a post-colonial-form of urban studies relevant to a world of cities, rather than simply for selected "world cities." Noting especially the adverse political consequences of analyzes that emphasize hierarchies and categories and that still divide the field of urban studies along developmentalist lines, the chapter presses the importance of letting all cities be ordinary. World cities approaches, it will be suggested, operate to limit imaginations of possible urban futures, especially in relation to poorer cities, and the situation of poor and marginalized people in cities around the world. A post-colonial urban studies needs to move beyond categories and hierarchies and to abandon claims to represent some cities as exemplars for others. It needs to be able to be attentive to the diverse experiences of a world of cities. While global and world-cities approaches have much to offer, ultimately they leave these challenges unmet. Instead, this chapter makes the case that all cities should be viewed as ordinary, both distinctive and part of an interconnected world of cities. The last section spells out what this might mean. And, as we will see, an ordinary-city approach is as important for the wealthiest ("global") cities, as for the poorest.

GLOBAL AND WORLD CITIES

World cities are thought to be different from others to the extent that they play an important role in articulating regional, national and international economies into a global economy. With the rise since the 1970s of national investment flows and the emergence of a new international division of labor based on the restructuring of manufacturing production processes, now involving integrated production processes dispersed across the globe, scholars have been drawn to rethink the role of cities in relation to the global economy. It was suggested that some cities were increasingly serving as the organizing nodes of a global economic system, rather than simply being linked into local hinterlands or part of national systems of cities. At the same time, many populations were being excluded from these new spaces of global capitalism, and thus from the field of world cities: to these writers some cities were becoming "economically irrelevant" (Knox 1995: 41).

[...]

In a prominent contribution to the world-cities literature, Saskia Sassen (1991, 1994) coined the term "global cities" to capture what she suggests is a distinctive feature of the current (1980s on) phase of the world economy: the global organization and increasingly transnational structure of elements of the global economy. Her key take-home point is that the spatially dispersed global economy requires locally based and integrated organization and this, she suggests, takes place in global cities. Although many transnational companies no longer keep their headquarters in central areas of these major cities, the specialized firms that they rely on to produce the capabilities and innovations necessary for command and control of their global operations have remained or chosen to establish themselves there, including advanced business and producer services, legal and financial services. Moreover, it is no longer the large transnational corporations that are the center of these functions, but small parts of a few major cities that play host to and enable the effective functioning through proximity of a growing number of these new producer and business-services firms (Sassen 2001). A similar argument concerning the benefits of co-location for finance and investment firms suggests that these cutting-edge activities are produced in a few major cities. Co-location benefits both these sets of firms as this facilitates face-to-face interaction and the emergence of trust with potential partners, which is crucial in terms of enabling innovation and coping with the risk, complexity, and speculative character of many of these activities (Sassen 1994: 84).

Both global- and world-cities analyses bring into view the wider processes shaping cities in a globalizing world and economic networks amongst cities. However, the emphasis has been on a relatively small range of economic processes with a certain "global" reach. This has limited the applicability of world-cities approaches, excluding many cities from its consideration. Although status within the world-city hierarchy has traditionally been based on a range of criteria, including national standing, location of state and interstate agencies and cultural functions, the primary determination of status in this framework is economic. As Friedmann (1995a) notes, "The economic variable is likely to be decisive for all attempts at explanation" (317). This has become more, not less, apparent in the world-cities literature over time as more recent research has focused on identifying the transnational business connections that define the very top rank of world cities, Sassen's "global cities" (Beaverstock et al. 1999; Sassen 2001).

World-cities approaches have been strongly shaped by an interest in determining the existence of categories of cities and identifying hierarchical relations amongst cities. This led Friedmann to ask, in his review of "World City Research: 10 years on," whether the world-city hypothesis "is a heuristic, a way of asking questions about cities in general, or a statement about a class of particular cities—world cities—set apart from other urban agglomerations by specifiable characteristics?" (1995b: 23). He suggests that it is both, but that the tendency has been to categorize cities into a hierarchy in which world cities are at the top of the tree of influence. This league-table approach has shaped the ways in which cities around the world have been represented—or not represented at all—within the world-cities literature. From the dizzy heights of the diagrammer, certain significant cities are identified, labeled, processed, and placed in a hierarchy, with very little attentiveness to the diverse experiences of that city or even to extant literature about that place. The danger here is that out-of-date, unsuitable or unreliable data (Short et al. 1996; although see Beaverstock et al. 2000) and possibly a lack of familiarity with some of the regions being considered can lead to the production of maps that are simply inaccurate. These images of the world of important cities have been used again and again to illustrate the perspective of world-cities theorists and leave a strong impression on policy-makers, popularizing the idea that moving up the hierarchy of cities is both possible and a good thing. Peter Taylor (2000: 14) notes with disapproval, though, the "widespread reporting of [. . .] a preliminary taxonomy" of world cities. However, revised versions of these taxonomies, based on more substantial research, draw remarkably similar conclusions, and similar maps . . . In contrast to the world-city enthusiasm for categories, the global-city analysis has a strong emphasis on process. It is the locational dynamics of key sectors involved in managing the global economy that give rise to the global-city label. However, the category of global city that is identified through this subtle analysis depends on the experiences of a minor set of economic activities based in only a small part of these cities. They may constitute the more dynamic sectors of these cities' economies, but Sassen's evidence of declining location quotients for these activities in the 1990s (for example, 2001: 134–5) suggests that the concentrated growth spurt in this sector may well be over. And it is important to put the contribution of these sectors to the wider city economy into perspective. In London, for example, where transnational finance and business services are still most dynamic and highly concentrated, the London Development Agency (LDA) suggests that only "about 70 percent of employment is in firms whose main market is national rather than international" (LDA 2000: 18). Even this city, routinely at the top of world-city hierarchies, is poorly served by a reduction of its complex, diverse social and economic life to the phenomenon of globalization, and is certainly poorly described as a "global" city. There have been many criticisms of the empirical basis for claims that global cities are significantly different from other major centers in terms of the composition of economic activities, wage levels or social conditions (Abu-Lughod 1995; Short et al. 1996; Storper 1997; Smith 2001; Buck et al. 2002).

Nonetheless, the global-city hypothesis has had a powerful discursive effect in both academic and policy circles. The pithy identification of the "global city" as a category of cities which, it is claimed, are powerful in terms of the global economy, has had widespread appeal. However, this has depended on continuing, indeed strengthening, the world-city emphasis on a limited range of economic activities with a certain global reach as defining features of the global city. This has the effect of hiding most of the activities within global cities from view, while at the same time also dropping most cities in the world from its vision. The insights of the global-city analysis are very important to understanding the way in which some aspects of some cities are functioning within the global economy. But perhaps it would be more appropriate if these processes were described as an example of an "industrial" district. They could be called new industrial districts of transnational management and control. The core understanding about these novel processes would remain important both theoretically and in policy terms. [. . .] World-cities research, then, has moved on from the time of Friedmann's influential mid-1980s review of the field and, especially in the wake of Saskia Sassen's study, *The Global City*, it has adopted a strong and intensely researched empirical focus on transnational business and finance networks. (See, for example, Beaverstock et al. 1999; Taylor 2004.) In some ways, the focus of attention has narrowed, although there has been a concerted effort to focus on processes and to track connections amongst cities rather than simply to map city attributes (Beaverstock et al. 2000). However, cities still end up categorized in boxes or in diagrammatic maps and assigned a place in relation to a priori analytical hierarchies. A view of the world of cities emerges where some cities come to be seen as the pinnacle of achievement, setting up sometimes impossible ambitions for other cities. This also suggests to the most powerful cities that they need to emphasize those aspects of their cities that conform to the global and world cities account, with sometimes detrimental effects on other kinds of activities and on the wider social life of the city. (See, for example, Markusen and Gwiasda 1994; Sites 2003.) Global and world-cities approaches expose an analytical tension between assessing the characteristics and potential of cities on the basis of the processes that matter from within their diverse dynamic social and economic worlds or on the basis of criteria determined by the external theoretical construct of the world or global economy. (See also Varsanyi 2000.) This is at the heart of how a world-cities approach can limit imaginations about the futures of cities and why I propose instead to think about a world of cities, all quite ordinary.

If global- and world-cities approaches offer only a limited window onto those cities that make it into the league tables, and even those at the top, we should also be concerned about the effect of these hierarchies and league tables on those cities that are quite literally off the maps of the global- and world-cities theorists. Millions of people and hundreds of cities are dropped off the map of much research in urban studies to service the very restricted view that the global and world cities analyses encourage regarding the significance or (ir)relevance of cities in relation to certain rather narrow sections of the global economy. For the purposes of developing a post-colonial urban studies relevant for a world of cities, global and world cities approaches have some serious limitations.

[. . .]

It is hard to disagree that some countries and cities have lost many of the trading and investment links that characterized an earlier era of global economic relations. A country like Zambia, for example, now one of the most heavily indebted nations in the world and certainly one of the poorest, has seen the value of its primary export, copper, plummet on the world market since the 1970s. Its position within an older international division of labor is no longer economically viable, and it has yet to find a successful path for future economic growth. En route it has suffered the consequences of one of the World Bank/IMF's most ruthless Structural Adjustment Programs (Young 1988; Clark 1989). However, Zambia is also one of the most urbanized countries on the African continent, and its capital city Lusaka is a testimony to the modernist dreams of both the former colonial powers and the post-independence government (Hansen 1997). Today, though, with over 70 percent of the population in Lusaka dependent on earnings from the informal sector (government bureaucrats are known to earn less than some street traders: see Moser and Holland 1997), the once bright economic and social future of this city must feel itself like a dream—albeit one which was for a time very real to many people (Ferguson 1999).

Lusaka is certainly not a player in the "major economic processes that fuel economic growth in the new global economy" (Sassen 1994: 198). But copper is still exported, as are agricultural goods and opportunities for investment as state assets are privatized. Despite the lack of foreign currency (and sometimes because of it) all sorts of links and connections to the global economy persist. From the World Bank, to aid agencies, international political organizations and trade in second-hand clothing and other goods and services, Lusaka is still constituted and reproduced through its relations with other parts of the country, other cities, and other parts of the region and globe (see, for example, Hansen 1994, 1997). The city continues to perform its functions of national and regional centrality in relation to political and financial services, and operates as a significant market (and occasionally production site) for goods and services from across the country and the world.

It is one thing, though, to agree that global links are changing and that power relations, inequalities, and poverty shape the quality of those links. It is quite another to suggest that poor cities and countries are irrelevant to the global economy. When looked at from the point of view of these places that are allegedly "off the map," the global economy is of enormous significance in shaping the futures and fortunes of cities around the world. For many poor, "structurally irrelevant" cities, the significance of flows of ideas, practices, and resources beyond and into the city concerned from around the world stand in stark contrast to these claims of irrelevance. As Gavin Shatkin writes about Phnom Penh, "In order to arrive at a proper understanding of the process of urbanization in LDCs [less developed countries], it is necessary to examine the ways in which countries interface with the global economy, as well as the social, cultural and historical legacies that each country carries into the era of globalization" (1998: 381).

The historical legacies of these cities, it is clear from his account, are also products of earlier global encounters. Even the poorest cities have long histories of interactions and contacts with other places and have over time been drawn into the global economy in different roles, for trade, production, extraction or cultural exchange. These connections, perhaps transformed, can remain vital components of contemporary urban dynamics.

[. . .]

THE CASE FOR ORDINARY CITIES

So far . . . I have identified the importance of thinking about cities without privileging the experiences of only certain kinds of cities in our analyses; the value of learning how to think differently about cities by exploring different ways of life in other cities; and the benefits of a cosmopolitan approach to cities, including attending to the wider circulations and flows that shape them in order to appreciate the potential creativity and dynamism of all cities. A number of tactics—dislocating ethnocentric accounts, deploying comparative and cosmopolitan approaches—have been drawn on to move us towards a post-colonial form of urban theorizing. At the same time, they have brought into view the ordinary city. Instead of seeing some cities as more advanced or dynamic than others, or assuming that some cities display the futures of others, or dividing cities into incommensurable groupings through hierarchizing categories, I have proposed the value of seeing all cities as ordinary, part of the same field of analysis. The consequence of this is to bring into view different aspects of cities than those which are highlighted in global and world cities analyses.

First, ordinary cities can be understood as unique assemblages of wider processes—they are all distinctive, in a category of one. Of course there are differences amongst cities, but I have suggested that these are best thought of as distributed promiscuously across cities, rather than neatly allocated according to pre-given categories. And even when there are vast differences, between very wealthy and very poor cities, for example, I have suggested that scholars of these cities have much to learn from one another.

[. . .]

Second, and learning much from global- and world-cities approaches, ordinary cities exist within a world of interactions and flows. However, in place of the global- and world-cities approaches that focus on a small range of economic and political activities within the restrictive frame of the global, ordinary cities bring together a vast array of networks and circulations of varying spatial reach and assemble many different kinds of social, economic and political processes. Ordinary cities are diverse, complex, and internally differentiated.

The consequences of thinking of cities as ordinary are substantial, with implications for the direction of urban policy and for our assessment of the potential futures of all sorts of different cities. Amin and Graham (1997), setting out their account of "The Ordinary City," suggest that thinking about cities as distinctive combinations of overlapping networks of interaction leads very quickly to an account of the capacity of cities to foster creativity. In Western policy circles, they note, there has been a rediscovery of "the powers of agglomeration," and an excitement about cities as creative centers. Agreeing that many accounts of cities highlight only certain elements of the city (finance services, information flows) or certain parts of the city—both leading to a problem of synecdoche—they rather describe (all) cities as "the co-presence of multiple spaces, multiple times and multiple webs of relations, tying local sites, subjects and fragments into globalizing networks of economic, social and cultural change. [...] as a set of spaces where diverse ranges of relational webs coalesce, interconnect and fragment" (Amin and Graham 1997: 417–18).

It is the overlapping networks of interaction within the city—networks that stretch beyond the physical form of the city and place it within a range of connections to other places in the world—which, for Amin and Graham (1997), are a source of potential dynamism and change. The range of potential international or transnational connections is substantial: cultural, political, urban design, urban planning, informal trading, religious influences, financial, institutional, intergovernmental, and so on (Smith 2001). To the extent that it is a form of economic reductionism (and reductionism to only a small segment of economic activity) that sustains the regulating fiction of the global city, this spatialized account of the multiple webs of social relations that produce ordinary cities could help to displace some of the hierarchizing and excluding effects of this approach. For with so many different processes shaping cities and so many potential interactions amongst them, it would be difficult to decide against which criteria to raise a judgment about rank.

Categorizing cities and carving up the realm of urban studies has had substantial effects on how cities around the world are understood and has played a role in limiting the scope of imagination about possible futures for cities. This is as true for cities declared "global" as for those that have fallen off the map of urban studies. The global-cities hypothesis has described cities such as New York and London as "dual cities," with the global functions drawing in not only a highly professional and well-paid skilled labor force, but also relying on an unskilled, very poorly paid and often immigrant work force to service the global companies (Sassen 1991; Allen and Henry 1995). These two extremes by no means capture the range of employment opportunities or social circumstances in these cities (Fainstein et al. 1992; Buck et al. 2002). It is possible that these cities, allegedly at the top of the global hierarchy, could also benefit from being imagined as "ordinary." The multiplicity of economic, social and cultural networks that make up these cities could then be drawn on to imagine possible paths to improving living conditions and enhancing economic growth across the whole-city.

[...]

CONCLUSION

Global- and world-cities analyses have been enormously productive in refocusing urban studies on the wider processes and networks that shape cities; and they have announced a new, more inclusive geography of the role of cities in globalization. But they have left intact earlier assumptions about hierarchical relations amongst cities, with potentially damaging consequences especially, but not only, for poorer cities. They have, in fact, consigned a large number of cities around the world to theoretical irrelevance. Building on global- and world-cities approaches, but mindful of these criticisms, other writers have turned to the ordinary city—diverse, contested, distinctive—as a better starting point for understanding a world of cities. Ordinary cities also emerge from a post-colonial critique of urban studies and signal a new era for urban studies research characterized by a more cosmopolitan approach to understanding cityness and city futures. This can underpin a field of study that encompasses all cities and that distributes the differences amongst cities as diversity rather than as hierarchical categories. It is the ordinary city, then, that comes into view within post-colonialized urban studies.

More than this, the overlapping and multiple networks highlighted in the ordinary city approaches can be drawn on to inspire alternative models of urban development. These would be approaches that see the potential for productive connections supporting the diverse range of economic activities with varying spatial reaches that come together in cities. Approaches that explore the diversity of economic activities present in any (ordinary) city (Jacobs 1961: 180–-81) and that emphasize the general creative potential of all cities could help to counter those that encourage policy-makers to support one (global) sector to the detriment of others.

[...]

REFERENCES

Abu-Lughod, J. (1995) "Comparing Chicago, New York and Los Angeles: testing some world city hypotheses." In P. Knox and P. Taylor (Eds.), *World Cities in a World-System*, Cambridge: Cambridge University Press.

Allen, J. and Henry, N. (1995) "Growth at the margins: Contract labour in a core region." In C. Hadjimichaelis and D. Sadler (Eds.), *Europe at the Margins*, London: Wiley.

Amin, A. and Graham, S. (1997) "The ordinary city," *Transactions of the Institute of British Geographers* 22: 411–29.

Beaverstock, J., Taylor, P. and Smith, R. (1999) "A roster of world cities," *Cities* 16(6): 444–58.

Beaverstock, J., Taylor, P. and Smith, R. (2000) "World-city network: a new metageography?" *Annals of the Association of American Geographers* 90: 123–34.

Buck, N.I. et al. (2002) *Working Capital: Life and Labour in Contemporary London*, London: Routledge.

Clark, J. with Allison, C. (1989) *Zambia Debt and Poverty*, Oxford: Oxfam.

Fainstein, S., Gordon, I. and Harloe, M. (Eds.) (1992) *Divided Cities*, Oxford: Blackwell.

Ferguson, J. (1999) *Expectations of Modernity: Myths and Meanings of Urban Life on the Zambian Copperbelt*, Berkeley: University of California Press.

Friedmann, J. (1995a) "The world city hypothesis." In P. Knox and P. Taylor (Eds.), *World Cities in a World System*, Cambridge: Cambridge University Press.

Friedmann, J. (1995b) "Where we stand now: a decade of world city research." In P. Knox and P. Taylor (Eds.), *World Cities in a World System*, Cambridge: Cambridge University Press.

Hansen, K. (1994) "Dealing with used clothing: Salaula and the construction of identity in Zambia's Third Republic," *Public Culture* 6: 503–23.

Hansen, K. (1997) *Keeping House in Lusaka*, New York: Columbia University Press.

Jacobs, J. (1961) *The Death and Life of Great American Cities*, Harmondsworth: Penguin.

Knox, P. (1995) "World cities in a world-system." In P. Knox and P. Taylor (Eds.), *World Cities in a World-System*, London: Routledge.

LDA—London Development Agency (2000) "London development strategy," London: LDA.

Marcuse, P. and Kempen, R. (2000) *Globalising Cities: A New Spatial Order?* Oxford: Blackwell.

Markusen, A. and Gwiasda, V. (1994) "Multipolarity and the layering of functions in world cities: New York's struggle to stay on top," *International Journal of Urban and Regional Research* 18: 167–93.

Marvin, S. and Graham, S. (2001) *Splintering Urbanism: Networked Infrastructures, Technological Mobilities and the Urban Condition*, London and New York: Routledge.

Moser, C. and Holland, J. (1997) *Household Responses to Poverty and Vulnerability: Vol 4*, Confronting Crisis in Chawama, Lusaka, Zambia, Washington, DC: World Bank for Urban Management Programme.

Robinson, J. (2002a) "City futures: new territories for development studies?" In J. Robinson (Ed.), *Development and Displacement*, Oxford: Oxford University Press.

Robinson, J. (2002b) "Global and world cities: a view from off the map," *International Journal of Urban and Regional Research* 2: 531–54.

Sassen, S. (1991) *The Global City: New York, London, Tokyo*, Princeton, NJ: Princeton University Press..

Sassen, S. (1994) *Cities in a World Economy*, Thousand Oaks, CA: Pine Forge Press.

Sassen, S. (2001) *The Global City: New York, London, Tokyo*, 2nd edition, Princeton, NJ: Princeton University Press.

Scott, A. (Ed.) (2001) *Global City-Regions: Trends, Theory, Policy*, Oxford: Oxford University Press.

Shatkin, G. (1998) "'Fourth world' cities in the global economy: the case of Phnom Penh," *International Journal of Urban and Regional Research* 22: 378–93.

Short, J.R. et al. (1996) "The dirty little secret of world cities research: data problems in comparative analysis," *International Journal of Urban and Regional Research* 20(4): 697–717.

Sites, W. (2003) *Remaking New York: Primitive Globalization and the Politics of Urban Community*, London and Minneapolis: University of Minnesota Press.

Smith, M. (2001) *Transnational Urbanism*, Oxford: Blackwell.

Storper, M. (1997) *The Regional World*, London and New York: Guilford Press.

Taylor, P. (2000) "World cities and territorial states under conditions of contemporary globalization," *Political Geography* 19: 5–32.

Taylor, P. (2001) "West Asian/North African cities in the world city network: a global analysis of dependence, integration and autonomy," *GAWC Research Bulletin 58*. Available online: http://www.lboro.ac.uk/gawc

Taylor, P. (2004) *World City Network: A Global Urban Analysis*, London: Routledge.

Varsanyi, M. (2000) "Global cities from the ground up: a response to Peter Taylor," *Political Geography* 19: 33–8.

Young, R. (1988) *Zambia: Adjusting to Poverty*, Ottawa: The North-South Institute.

PART III

The city lived

Figure 23 Windows, Ponte City, Johannesburg 2008–2010 (detail)
Source: © Mikhael Subotzky and Patrick Waterhouse courtesy of Goodman Gallery and Magnum Photos.

SECTION 3
Migratory Fields

Editors' Introduction

People have always moved. The diverse languages we speak with their common roots, for example the Indo-European languages, attest to this. Sometimes these movements are voluntary (relocation), sometimes by force due to social, economic, political, or environmental calamities (displacement). Sometimes moves are temporary, for example, by rural or international migrant laborers, and other times permanent, what is often referred to as immigration. Many historical events and processes have led to large-scale relocation and displacement. For our purposes, two of the most important historically are slavery and colonialism. The slave trade was perhaps the first large-scale forced movement of labor across the globe. Over centuries of slavery millions of people were captured, uprooted, and displaced to work for free for European masters. Between 1500 and 1800 CE it is estimated that 12 to 15 million slaves were shipped across the Atlantic to be sold and traded in market places for work on plantations and other sites (Davidson 1995:4). Under colonialism, settlers relocated to explore and exploit new territories and opportunities. In these processes, generations of white settlers and displaced labor produced and reproduced their built environments and shaped the foundation of urbanization patterns and cities–as Anthony King explains in his contribution to this *Reader*. In addition, in many colonies, people were uprooted from their traditional lands and forced to work in industrial and agricultural enterprises owned by the settlers, often crossing borders to do so.

Albeit with different incentives and trajectories, these movements of people across long or short physical distances have continued in the new global order. They are often referred to as (im)migration, a term that does not fully capture the distinct stories behind people's movements within and across political, social and cultural borders.

In this section of the *Reader*, we focus on contemporary movements of populations and migratory fields, along with their historical roots and affinities. We introduce the broader social, political, and economic context of these population movements in the contemporary era and discuss the extent and scope of these movements; the range of incentives, rationales, and consequences they imply; and how they produce and are produced by the cities of the global South.

KEY ISSUES

Migratory fields

We use the term migratory fields as a way to capture the more complex paths, trajectories, and stories that shape populations' movements. Increasingly we do not specify rural–urban, rural–rural, or transnational pathways; instead we specify the migratory field, as it may include a series of steps from a rural through a smaller urban to a larger urban center, from rural to urban in another nation and so on. Clearly, changes in travel and communication technology play a large role in shaping migration flows. Globalization–the international flow of capital, goods, and services, in particular–has changed the ways in which we are beginning to understand migration. There can now be multiple places of origin (depending on marriage, children being born, etc.) and often multiple destinations. Most importantly, many migrants seem to be caught in a series of oscillating movements. What was once described as a bipolar world is now more often than not referred to as a migratory field, which is expanding in ways that are not always very clear.

Migratory flows and trends are highly context specific and vary according to the broader social, political, economic, and policy environments within which they occur. In the last half-century the movement of populations in the global South has been marked by a set of policies that advocate free markets and trade. These policies that in the 1970s and 1980s were labeled as the new world order of globalization brought about a greater mobility of capital and labor with distinct implications for people and cities (Sassen 1999a, 1999b). Footloose capital moves wherever it is cheaper to produce goods and services, and that includes new locations within or across national borders. In the early 1970s, for example, the movement of US plants across the border to Mexico gave rise to a range of border settlements. *Maquiladoras*, as these border zone plants are called, enticed people, mostly women, from rural as well as other urban areas across Mexico to towns where *maquila* jobs were located. This marked a process referred to as the feminization of labor (Fernández-Kelly 1983). The workers settled in the location of *maquila* industries and created, almost overnight, cities without adequate infrastructure, housing, waste collection or water and sewage systems to handle the new arrivals. Similarly, since the 1990s the export orientation of the economy turned China into the manufacturing plant of the world, and entailed a massive movement of rural populations to cities where local and foreign investors set up manufacturing plants to produce for export. The rural people displaced to cities for economic reasons and moving in search of jobs are referred to as the "floating population." While these migrants (numbering over 200 million) work in the city, they are not recognized as legal residents and are hence denied an urban *hukou* or household registration permit, which would entitle them to urban services. In the first years of the twenty-first century China's urban population experienced an especially sharp increase, rising from 36 percent of the total population in 2001 to over 50 percent in 2012 (UN Population Division 2011, 2012).

The selection by Li Zhang included in this section sheds further light on living conditions of the floating population in Beijing. The similarities in the conditions of Chinese floating populations with those of Mexican migrant laborers working in the United States, or with the black workforce working in white-owned cities and mines under South African apartheid are striking (see Alexander and Chan 2004).

The hyper-mobility of capital and labor over the last couple of decades has markedly increased the total number of international migrants, from an estimated 155 million in 1990 to 213 million in 2010 (UN 2008). Today, the international labor force represents one third of the global labor force (ibid.). We must also note that almost half of all migrants worldwide are women (Mahler and Pessar 2006). Most women migrants are engaged in temporary labor migration, with the Middle East, East Asia, and South East Asia as their main destinations (International Organization for Migration 2008). Countries of the global North, due to their decreasing working-age population (20 to 64), need and receive labor from developing countries but they are not the main recipient of international migrants (ibid.).

Migration is now more widely distributed across more countries and displays varied regional trends. Broadly speaking, the 2008 report by the International Organization for Migration (IOM) sketches the following: in Africa, migrants predominantly move within the continent, with South Africa, the Maghreb, and West Africa being most affected. The selection from Simone that follows this introduction reminds us of the complexity, creativity, and promise of these intercontinental movements in Africa. In Asia there is strong intra-regional flow of migrant workers, particularly the internal movements in China and India. But Asia is also the largest source of temporary contractual migrant workers, many of whom head to the Middle East, which is by far the most important recipient of temporary contractual workers. In a number of Middle Eastern countries, migrants comprise the majority or nearly the majority of the workforce. These include Qatar (87 percent), United Arab Emirates (70 percent), and Jordan (46 percent) (International Organization for Migration 2008).

Europe, as mentioned earlier, also continues to be a recipient of international migration due to its low birth rate and declining native-born populations. But European Union countries tend to heavily regulate migratory flows with a strong push for temporary migrants as opposed to permanent immigration. Migratory flows in the

Americas, in contrast, are characterized by strong south to north movements–to Canada and the US, where there is a growing demand for workers.

In the cities of the global North, work that does not easily lend itself to relocation has been restructured into lowly paid, precarious employment such as informal or contracts jobs, many of which are held by immigrants (Sassen 1999a). These include jobs in the service sector, be it domestic or commercial services; in the care sector, be it care for children, the elderly, or patients (Ehrenreich and Hochschild 2003; Hontagneu-Sotelo 2001), in agriculture, as well as in more hazardous manufacturing sectors such as the meat packing and processing industry (Stull, Broadway and Griffith 1995; Miraftab 2011; Sandoval 2013). Despite this, migrants continue to move to richer countries in search of livelihood opportunities, often at great cost and risk to themselves and their families.

In addition to the dominant patterns of international migration discussed above, there is a less discussed migration trend involving the movement of people from the global North to South. This not only refers to the historical North to South migrations/occupations that constituted European settlements in the colonies, but also the contemporary relocation of retirees to countries of the global South where they improve their lifestyle by crossing national borders. Examples of such migration are referred to as lifestyle migration, and include communities that Italians create in Argentina, Japanese in Peru, or Americans in Mexico. American retirees have been increasingly moving south of the border and shaping exclusive "colonies" or gated communities of expatriates, to afford better living conditions, better care, and higher social status (Sunil, Rojas and Bradley 2007).

Finally, we cannot consider migratory flows and fields within and across national territories without accounting for the displacements that are created by war and conflict, by large-scale development projects (also referred to as development-induced displacements) and by natural disasters. Worldwide numbers for internally displaced populations grew from 21 million in 2000 to 27 million at the end of 2009 (Norwegian Refugee Council 2012). Global climate change also has undeniable implications for shifting the development processes of communities, cities, and regions and causing large-scale displacements. It poses difficult challenges with expected sea-level rise along coastal areas, water scarcity in some regions accompanied by decreasing food security, as well increasing numbers of natural disasters. The impacts of these shifts will be experienced most severely by people positioned at the lower end of the social hierarchy. Tsunamis, hurricanes, and cyclones have much more devastating effects on the more vulnerable populations in the global North and South, and are creating an ever larger number of internal and cross-national population displacements.

Development–migration nexus

An important debate regarding the migratory movements that shape cities of the global South concerns the flow of resources. During the urban transformation in the 1940s–1960s in Latin America, or currently in China, which is undergoing a rapid and massive urbanization through rural migration, urban–rural connections and networks that shaped movement and flows constituted the main focus of research. However, with the growing magnitude of international migration and the international flow of resources through transnational remittances, these debates have increasingly included a focus on the relationship between international migration and development. In 2010, for example, the 213 million international migrants would have constituted the fifth largest country in the world. The size of remittances this population sent home to the global South over the last decade increased exponentially to an estimated $307 billion in 2009 (World Bank 2011), an amount slightly less than peak years due to the current economic crisis. Nonetheless, remittances that immigrants sent to the global South were more than twice as large as official aid and nearly two-thirds of the foreign direct investment (FDI) flows (ibid.).

The effect of remittances on the cities of the global South remains, however, an unsettled debate (Guarnizo 2003). Some scholars argue that remittances lead to increased social inequities based on connections individuals may have to international migrants. This argument also holds that remittances create dependencies and are counterproductive since recipients spend the money on conspicuous consumption (Binford 2003;

Sana and Massey 2005). Others argue that remittances lead to development because recipients invest the money in employment-creating activities such as construction of new homes or the improvement of existing housing. This perspective further contends that resources often flow to families left behind in the country of origin to support education and health care as well as overall social and economic development (see Cohen, Jones and Conway 2005; Durand, Kandel, Parrado, and Massey 1996).

The implication of these academic debates for policy is important. Most governments facilitate the inflow of remittances. Some, like the Mexican government, not only promote this inflow, but also try to direct the remittances toward public works through projects that benefit the community as a whole as opposed to the recipient family alone. The "three to one" program the Mexican government has instituted is precisely to direct the inflow of remittances (among the country's biggest sources of foreign income) toward social or infrastructure development projects. For every dollar immigrants send to Mexico dedicated to a public works project, the federal, state, and local governments each add one dollar–hence the name of the program *tres por uno*–three to one.

ARTICLE SELECTIONS

The selection of readings included in this section of the *Reader* aims to offer diverse perspectives on the range of migratory fields constituting cities of the global South. The first short text comes from a memoir by Mzwanele Mayekiso, a South African academic who was active in the struggle against apartheid. The author shares with the reader a personal account of growing up during apartheid South Africa in an artificially created "homeland" called the Transkei. He details how migrant laborers, spatially confined to the homelands, created the wealth that made the South African cities. His reflections provide a glimpse into the geography of apartheid which very much relied on migrant labor by blacks to produce spaces of privilege for white South Africans.

The second reading is by AbdouMaliq Simone, an influential African and Africanist urban sociologist whose writings question the taken-for-granted ideas about African urbanism (2004, 2009). In the section below (2011) Simone takes us to the in-between-and-across spaces of borders on the African continent. A whole range of zigzag movements exists across countries of the global South that the dominant international migration literature, focused on immigration to the global North, does not capture. Simone, in his paper on the urbanity of movement, focuses on these often ignored spaces made by the complex movements of people and goods across national borders of African countries and the resources, creativity, and promise these movements engage. He questions the conventional notions of population movement predominantly defined in terms of rural to urban or cross-national migration that assume a certain stability in the relationship of population to place–people move from point A and settle in point B. He rather stresses an increasingly elaborate circulation of people within metropolitan regions, across primary and secondary cities, and along elaborated transnational circuits of movement and exchange. He calls for attention to be paid to the kinds of spaces and cities that are articulated through these fluid movements of people, goods, and services.

The third reading is by Li Zhang, a prominent urban anthropologist whose publications concern the social, political, and cultural repercussions of market reform and socialist transformations in contemporary China (2001a, 2010). In the excerpt included below Li Zhang (2001b) takes us to the rapidly growing and privatizing Beijing of the 1990s and discusses the urban fate of the "floating population," who live in a newly emerged migrant enclave "Zhejiang Village" (zhejiangcun). Li Zhang discusses the informal community of these migrant laborers in the wake of Beijing city government's 1995 cleanup campaign, which aimed to demolish and displace this community for political and economic motives. Paradoxically, because this population does not enjoy legal residency in the city they are also less controlled by the authorities, a situation that allowed for their under-the-radar organizing using traditional networks through the social and economic niches they had created in the cities. In this excerpt Zhang reveals the power struggle through which these migrant laborers, once displaced from rural to urban and yet again under the pressures of urban displacement, resist eviction and shape urban space in post-socialist market-oriented Chinese cities.

REFERENCES

Alexander, P. and Chan, A. (2004) "Does China have an apartheid pass system?" *Journal of Ethnic and Migration Studies* 30(4): 609–629.

Binford, L. (2003) "Migrant remittances and (under)development in Mexico," *Critique of Anthropology* 23(3): 305–336.

Cohen, J., Jones, R., and Conway, D. (2005) "Why remittances shouldn't be blamed for rural underdevelopment in Mexico: A collective response to Leigh Binford," *Critique of Anthropology* 25(1): 87–96.

Davidson, B. (1995) *The African Slave Trade*, Boston: Atlantic-Little Brown.

Durand, J., Kandel, W., Parrado, E. A., and Massey, D. S. (1996) "International migration and development in Mexican communities," *Demography* 33(2): 249–264.

Ehrenreich, B. and Hochschild, A. (Eds.) (2003) *Global Woman: Nannies, Maids, and Sex Workers*, New York: Henry Holt and Company.

Fernández-Kelly, M. P. (1983) *For We Are Sold, I and My People: Women and Industry in Mexico's Frontier*, Albany, NY: State University of New York Press.

Guarnizo, L. E. (2003) "The economics of transnational living," *International Migration Review* 37(3): 666–699.

Hontagneu-Sotelo, P. (2001) *Domestica: Immigrant Workers Cleaning and Caring in the Shadow of Affluence*, Berkeley: University of California Press.

International Organization for Migration (2008) "World migration report 2008: Managing labour mobility in the evolving global economy." Available online at http://www.iom.int/jahia/Jahia/about-migration/facts-and-figures/labour-migration-and-demographics/cache/offonce/ (accessed May 1, 2013).

Mahler, S. and Pessar, P. (2006) "Gender matters: Ethnographers bring gender from the periphery toward the core of migration studies," *International Migration Review* 40: 27–63.

Mayekiso, M. (1996) *Township Politics: Civic Struggles for a New South Africa,* New York: Monthly Review Press.

Miraftab, F. (2011) "Faraway intimate development: Global restructuring of social reproduction," *Journal of Planning Education and Research* 31(4): 392–405.

Norwegian Refugee Council (2012) Internal displacement update. Available online at http://www.internaldisplacement.org/8025708F004CFA06/%28httpPages%29/713EC5B3D03718FE8025709A004F7148?OpenDocument (accessed May 2, 2013).

Sana, M. and Massey, D. S. (2005) "Household composition, family migration, and community context: Migrant remittances in four countries," *Social Science Quarterly* 86(2): 509–528.

Sandoval, G. (2013) "Shadow transnationalism: Cross-border networks and planning challenges of transnational unauthorized immigrant communities," *Journal of Planning Education and Research* 33(2): 176–193.

Sassen, S. (1999a) *Globalization and Its Discontents: Essays on the New Mobility of People and Money*, New York: New Press.

Sassen, S. (1999b) "Whose city is it? Globalization and the formation of new claims," in J. Holston (Ed.), *Cities and Citizenship*, Durham, NC: Duke University Press, pp. 178–194.

Simone, A. (2004) *For the City Yet to Come: Changing African Life in Four Cities*, Durham, NC: Duke University Press.

Simone, A. (2009) *City Life from Jakarta to Dakar: Movements at the Crossroads*, London: Routledge.

Simone, A. (2011) "The urbanity of movement: Dynamic frontiers in contemporary Africa," *Journal of Planning Education and Research* 31(4): 379–391.

Stull, D., Broadway, M., and Griffith, D. (Eds.) (1995) *Any Way You Cut It: Meat Processing and Small-Town America*, Lawrence: University Press of Kansas.

Sunil, T. S., Rojas, V., and Bradley, D. E. (2007) "United States' international retirement migration: The reasons for retiring to the environs of Lake Chapala, Mexico," *Ageing & Society* 27(4): 489–520.

United Nations (2008) International migrant stock: The 2008 revision. Available online at http://esa.un.org/migration/ (accessed May 1, 2013).

UN Population Division (2011) Urban migration. Available online at http://www.driversofchange.com/demographics/2009/04/16/urban-migration/ (accessed May 2, 2013).

UN Population Division (2012) World population prospects: The 2012 revision. Available online at http://esa.un.org/wpp/ (accessed November 26, 2013).

World Bank (2011) *Migration and Remittances Factbook 2011*. Available online at http://go.worldbank.org/QGUCPJTOR0 (accessed May 2, 2013).

Zhang, L. (2001a) *Strangers in the City: Reconfigurations of Space, Power, and Social Networks*, Palo Alto: Stanford University Press.

Zhang, L. (2001b) "Migration and privatization of space and power in late socialist China," *American Ethnologist* 28: 179–205.

Zhang, L. (2010) *In Search of Paradise: Middle-class Living in a Chinese Metropolis*, Ithaca: Cornell University Press.

"Township Politics"

Township Politics: Civic Struggles for a New South Africa (1996)

Mzwanele Mayekiso

I was born and raised in a homeland, the Transkei, in the vicinity of the trading center of Cala. There are about 100,000 people living in the general area. Our family, like so many others, was forced into the apartheid labor system. My parents were not formally educated, and my father was an ordinary laborer for the railways, working in Cape Town, a day's journey from Askeaton. He could only return home for three weeks a year. Unions did not exist so when he retired (I was ten years old) there were no benefits and no retirement fund.

As for my mother, she had no paying job, but her work was the most important unpaid labor, raising a large family. As children, we therefore grew very close to my mother. I have six brothers and two sisters still living. Moss, my eldest brother was born in 1948; the youngest, Mzonke, was born in 1968; and I was born in 1964. Four sisters didn't survive: they died in their infancy.

[. . .]

Just walking distance away from Askeaton was Bumbana. That was my family's traditional home, but official policy was to relocate people from their ancestral areas in order to move them closer together. The point, it seemed, was social control. Bumbana was made to disappear, and today its run-down houses are used as a camp for cattle.

[. . .]

On a day to day basis, life could be harsh. Subsistence living involved getting water from far away. We dug furrows to help irrigate the gardens, although this was impossible in the larger maize (corn) fields. Askeaton had very poor soil, and the produce was not good. On top of that, the people of the area were mainly overworked women, along with elderly people, children and people who were injured. The migrant system churns up men and spits them out again, considering them useless when they can no longer work.

Through this system, the women developed a sort of matriarchy, even though the bigger system was patriarchal. The women controlled the household. At harvest time, my mother would sell maize, as well as some fruits from our small orchard. She did well compared to many people in the area. Sometimes my father would send sixty Rands a month [Ed. note: about 14 US dollars based on the 1996 exchange rate], maybe a hundred Rands in a good month. My father was unable to save for himself, though, and lived hand to mouth.

"The Urbanity of Movement: Dynamic Frontiers in Contemporary Africa"

Journal of Planning Education and Research (2011)

AbdouMaliq Simone

Africans have long traveled widely across the region and the world, moving themselves and goods across many obstacles. Yet it is ironic that history and efficacy is still barely utilized as evidence or precedence for a planning and governance framework that could make productive use of such movement as a formal economic resource.

Over the past decades, I have myself been a migrant across many different African cities, doing many jobs, some of which explicitly involved trafficking of various kinds. I have also developed several research and intervention projects in cities attempting to better understand who is involved in the business of migration, how it operates, and experimenting with different small-scale interventions to see how such movement could be more productively integrated into local urban economies.

Movement of people, goods, and information across various parts of Africa seems to operate within many different realities at the same time. Within households, cities, and nations, it is often difficult to discern clear patterns of efficacy—even in terms of who gets visas; who gets to reach particular destinations; what goods get blocked at particular borders; what quantities and kinds of goods manage to get through. There is an arbitrary feel about these dynamics that make strict accounting and measurement often difficult. What shows up on official registers, waybills, immigration forms, and statistical profiles is a weak approximation of what seems to really take place. For some, there are seemingly well organized networks that efficiently tend to each facet of the transshipment of something; for others, everything seems improvised, and it is often that which seems the most provisional or casual that often attains the greatest efficacy. Given these features, I have relied on my own engagements here as a way to try and tease out some of the dynamics lurking behind such apparent obfuscation, recognizing that much is left out in such an approach and that more systematic and inventive methodological processes still need to be pursued.

[. . .]

THE INSTRUMENTALITY OF MOVEMENT

During the mid-1980s, while living in Khartoum (Sudan), I made an arduous journey with a good friend to southern Sudan to attend his wedding in Yei (Sudan). It was a journey of some ten days on train, truck, and foot across treacherous physical and political landscapes with many deviations, many zigzags to circumvent disputed territories and potential interdiction. This was a time of war, of both clear and unclear sides. Shortly after the wedding, on my foolish insistence, we traveled the 100 kilometers, often by foot, across the Congolese border to Aba (Democratic Republic of Congo), an unremarkable frontier town with the exception of the vast quantities of goods and services available. From sacks of coffee and other agricultural produce as far as the eye could see, to hundreds of Toyota trucks, to scores of mechanics, guides, "bankers," arms dealers, prostitutes, seers, healers, and itinerant teachers, and authorities on all subjects, this was in some respects a thriving city, a magnet for the ambitions and urgencies of a seemingly constantly changing population.

Throughout this region, people moved in various directions, coming and going. Some would be moving toward zones of apparent danger and others moving away; some looking for at least momentary stability in order to restart or take respite from former lives and others to make money from uncertainty and conflict. Even the flow of goods had highly variegated patterns. Certainly, some of the consumer items, electronics, and hardware that Somali truckdrivers brought all the way in from Mombasa (Kenya) could have been forwarded on from the town, as well as large volumes of diverted relief supplies intended for Juba (Sudan). But the ambiguity of the town—the fact that one never knew exactly who or what was headed where in any meaningful aggregate manner—meant that traders came to Aba to speculate on the price of things, to try and get good deals, or even change their plans for the destination of the things that they acquired.

Certainly, the region was full of desperate people: people who had lost livelihoods and homes, people who were frightened and on the run, as well as people who stuck it out in terrible conditions. Certainly people carried substantial marks of identity and belonging; they had little difficulty recognizing each other as friend, stranger, enemy, refugee, or combatant. At the same time, there was also a lack of clarity and the anticipation that no one was really who they seemed to be, and that many were available to become something very different from what they were showing you at the moment (Agier 2008).

Here, movement was not simply a matter of running away or running toward, of vacating or reinhabiting. Rather, it was a stepping into a fold created in the moments where the unstable and the sedentary assume an intense contiguity. The process does not create some hybrid mix, but rather a slippery landscape where allegiances always have to be declared and then forgotten; where commitments have to be followed through on, and then turned into opportunities to do something very different. The conditions of underdevelopment and political manipulation are certainly important. But their specificities, especially in a town like Aba, seem to lose relevance as movements take on a life of their own, catapult people into destinations, work, and personalities that are not easy to pin down for long.

Not long after this journey, I left Sudan to return to West Africa, where a university position in Ghana had me working on the long-established dreams of many to maximize the economic potentials of the regional urban corridor running from Abidjan (Cote D'Ivoire) to Lagos (Nigeria). With its combined population of nearly 30 million, encompassing Cape Coast (Ghana), Accra (Ghana), Aflao (Ghana), Lome (Togo), Cotonou (Benin), Porto Novo (Benin), and Ibadan (Nigeria), this "region," stretching across four national borders, has been viewed as potentially the urban motor of Africa. Under the Economic Community of West Africa (ECOWAS) formalities, large numbers of residents of these cities move back and forth, taking advantage of niche markets, differentials in national regulatory structures, and singular economic histories in order to ply various small and medium-scale trades.

Given the long history of well-worn entrepreneurial circuits, urbanized infrastructure, and market density, the existence of thriving markets acting as if they are in the middle of nowhere came as a surprise. While probably now long gone, the market just beyond the town of Ifangi on the road to Sakete north of Porto Novo (Benin) was a riot of nocturnal transactions. Here, almost everything imaginable and then some was for sale, but only during one night a week. The particular date also would arbitrarily fluctuate, compelling reliance on word of mouth. With no readily discernible exceptions, all the goods available in this clearing in the bush were available elsewhere—in Mushin (Nigeria), Cotonou (Benin), and Lome (Togo). There were no surprises, even in the price. Given that the market was set at some distance from the road and inaccessible to any vehicle except motorbike, the costs of hauling the goods to and from the clearing would seem to unnecessarily add to the price and thus curtail the profits. As Ifangi (Benin) is close to the Nigerian border and connected by road to Otta (Nigeria), and then to Lagos (Nigeria), the prevailing assumption was that the frontier here was less regulated than the main Benin-Nigeria crossing located south and from the old capital, Porto Novo (Benin), and thus permitted an easier flow of untaxed and prohibited goods. But this was not always the case.

Unhinged from clear-cut economic efficacy, the market seemed to function as a coupling of various agendas, affect, and aspirations. Here, like in Aba, it was possible to find "expertise" of all kinds—for example, healers, "pharmacists," seers, sorcerers, and schemers. Some traders would spend the night talking about their exploits in distant lands; neighbors would travel here to talk to each other about events they felt were not possible to discuss at home. Politicians, "big men," and "big women" would meet in makeshift shacks for deliberations they could not be seen having elsewhere. Then, there was the continuous party atmosphere, replete with fireworks, music, and other entertainment. This coupling of the surreptitious, the celebratory, and the entrepreneurial was seen as creating opportunities on which it would be difficult to affix a price.

Although established with the veneer of great secrecy, everyone in the region seemed to know about its existence, to the extent that many households viewed it as a kind of family outing. At the same time, the press never mentioned it, and although, as I just indicated, everyone seemed to know about, its existence was never discussed in any detail. People had no idea about who administered it or how, and once the goods were loaded on various trucks parked under the trees lining the main road, no one was sure where they would eventually go. There was always a sense that anything could happen, and while many who frequented the night market certainly garnered much unanticipated information and many opportunities, it always seemed that much more could have transpired than did; that there was a great deal of flash and show. Probably there were layers on layers of dissimulation that I would never apprehend in the few visits I made here. But, nevertheless, it seemed that much effort was expended to make it seem as if a vast world of undercover economic transactions were taking place, when they probably were not.

Indeed many of the logics exhibited in this market have long been incorporated in the major urban markets anyway. Across the massive urban markets in Africa today—from Kinshasa to Adjame (Cote D'Ivoire) to Oshidi (Nigeria), to name just a few—the majority of people who spend their days in markets are not selling or buying anything specific. Rather, they are taking their chances; for markets are spaces of information, impressions, manipulations, and diffuse opportunities. Discrepant goods and services are bundled and sold together in often unfathomable ways, and calculations of price sometimes take into consideration relationships far more wide ranging than supply and demand (Guyer 2004).

But what I think is significant in the existence of places like Aba (Democratic Republic of Congo) and Ifangi (Benin) is their recognition that movement is not simply a reaction to conditions and forces, nor the instigator of still others, but its own "world," and a world that need not make sense in order to exert value.

[...]

In Africa, the inner city districts of Johannesburg; some of the inner districts of Abidjan; towns at the confluence of the Gambian, Senegalese, and Guinea Bissau border; as well as secondary cities such as Aba (a city in Nigeria, different from that mentioned before in the Democratic Republic of Congo), Lumbumbashi (Democratic Republic of Congo), Katsina (Nigeria), and Aflao (Ghana) reflect some of these dimensions. Whatever their composition, their vitality depends on their availability as a space of operation for shifting synergistic effects generated by the sheer presence of nascent residents coming from different places. Rather than having to fit into an agenda or way of operating dictated by the rules or necessities of the local system, new arrivals have an opportunity to use their working-out arrangements for coexisting with others, both new and old residents, as a platform on which to initiate new entrepreneurial activities and residential practices.

These efforts are seldom without conflict and contestation. Experiments come and go, and incursions are often of the "hit and run" variety, rather than anticipations of long-term commitments. In some respects, this volatility is necessary as the compulsion to innovate and to go beyond initial accommodations. Because people are often coming and going, making only short-term investments in place and relationships, this provisionality can take its toll on the built environment. Buildings, markets, and services can be quickly worn down, diminishing the capacity of the place to absorb all that is demanded of it. In part, this process is attenuated through the eventual need of new residents to stake certain claims and to institutionalize their presence in ways that can capitalize on efforts made often with great risk and expense.

So the vitality of these places to be what they are—that is, places that enable the productive conjunction of heterogeneous backgrounds, aspirations, and practices and, at the same time, anchor them in ways that eventually lessen the constant need for labor-intensive improvisation—is a delicate balance. It is one not easily amenable to policy or programmatic intervention. As "high-wire acts," these places are also vulnerable to distortions brought about by skewed economies—such as drug smuggling, property speculation, and other forms of trafficking. These distortions then often bring in the unwanted attention of various authorities, who are more likely to "join the game" in ways that introduce further distortions. The resultant defensive maneuvers of the majority then steer everyday life into dispositions that increasingly mirror the very conditions that propelled people to leave their former homes in the first place.

SMALL STEPS AT REAL FRONTIERS?

While the most dynamic and complicated frontiers of Africa may be located in the uncertainty of what is happening to its cities, there remain the old, lingering frontiers of national borders—many of which states no longer have the political commitment nor resources to protect—but which remain the actants (the living instruments) of claims of sovereignty.

Across the region, national borders are crossed often not only at will, but practically cease to exist as markers of transition or difference. Yet, various regional accords and coordination of the official movement of people, goods, money, and information remain slowed down by preoccupations with sovereignty. Even in the Economic Community of West African States (ECOWAS), residents have long been able to cross the borders of member states but still are not allowed to take formal work or establish formal enterprises in countries other than their own.

Sovereignty remains the clarion cry of stolen elections—for example, Zimbabwe and Cote D'Ivoire—where strongmen can simply shut out compliance to international norms and agreements as threats to national sovereignty. In actuality, sovereignty is the arbitrary decision outside the law to which the law is subject—in the case of many African states, the capacity of those who control instruments of violence to arbitrarily decide what will be subject to the law or to subject citizens to conditions outside the law (Mbembe 2010). It is often estimated that 35 to 40 percent of Africa's economic product moves through unofficial, often illicit channels (ECA 2008; ECOWAS-SWAC/OECD 2006) and that the proceeds from these transactions are acquired by various tiers of brokers, officials, politicians, and entrepreneurs. Here, the capacity to claim and enforce sovereignty is actualized by its very breach—by keeping goods and money outside of official trade conduits, procedures, and budgetary accounts (Bayart and Warnier 2004). Rather than participate in regional accords and coordination that could officially base developments on the clear mobilization, accounting, and deployment of funds generated by official economic activity, "real" sovereignty is maintained by circumventing any sense of national responsibility—a responsibility, that given Africa's economic conditions, infrastructural needs, residential patterns, and migration practices, can only be exercised through efforts to work toward more functional economic integration as outlined in the by now long-standing protocols of the Africa Union's New Economic Partnership for Africa's Development (NEPAD).

The key issue then is how planning processes can be inserted within this dilemma . . . [since] expanding official articulation among African cities may also be largely contingent on developments elsewhere—in this instance, at the concrete frontiers that separate nations. Here, long-integrated economies and communities have simultaneously lived in different times—the time of individuated national belongings with their specific array of customs, tax, and land use procedures; the time of integrated economic activities, common or interrelated cultural practices, continuous exchange and contact; the time of complementary exchanges, where national differences are opportunistically plied in order to maximize the accumulation possibilities of those who live on both sides of the border; and oscillating times of over and under-regulation, of times when borders are alternately porous and sealed; where they are sites of military or police activity, interdiction, and manipulation. These simultaneous realities may be analogous to the simultaneity explored throughout this discussion—of fluidity and constraint, organization and provisionality, and visible and invisible knowledges and mobilizations as it relates to movement.

During the past fifteen years the Prospectives, Dialogues, Politiques section of the international NGO Environmental Development Action in the Third World (ENDA) has conducted groundbreaking studies on interurban economic articulations in West Africa. These included the study on long-term economic complementarities among three medium-size cities located in three contiguous nations: Sikasso (Mali), Bobo Dioulasso (Burkina Faso), and Korhogo (Cote D'Ivoire)—a situation where their distinct national positionings simultaneously impeded and facilitated various facets of their long-term economic articulation, and where this articulation was the critical element for the development of new capacities in infrastructure and social welfare for each of the national sub-regions in which these cities were located. In many ways, it would make sense to eliminate the barriers that ensconce these cities in distinct national procedures—and certainly there would be much economic justification for this in terms of expanding contributions to the fiscus of each nation. But Cote D'Ivoire's preoccupation—given its highly cosmopolitan makeup—with who is a "real" Ivoirian and the subsequent de facto division of the country has deferred any concrete moves toward integration here. A subsequent study focused on the ways in which the trade between the cities of Kano, Katsina (Nigeria), and Maradi (Niger) was crucial to each city's survival, and how their respective markets are thoroughly integrated. Yet again, intranational competition among political elites in Nigeria attenuates the ability of entrepreneurs and public managers to develop protocol, budgetary procedures, and support mechanisms to further develop the capacities of this integration—in this case involving two of Nigeria's largest and most important urban centers.

Given the complexities entailed in this explicitly urban focus, ENDA Diapol (2007) has, in recent years, been concentrating on the rural villages along the Senegambia border—at the confluence of Senegal, Gambia, and Guinea Bissau—working with residents who have long operated by straddling national borders, and where the borders are largely irrelevant to the cohesion of local governance and agricultural production and trade. What has been key in this work has been to develop concrete models of decentralized border governance that can be used by regional bodies in order to develop new, concrete regimes officially capable of practicing and extending economic and social integration among villages in these areas.

At the 29th meeting of ECOWAS heads of state in Niamey in January 2006, agreement was secured to institute the concept of "border zones" to be subject to cross-border cooperation in all sectors, including agriculture, health, trade, peace and security, arts and cultures, as well as the free movement of people, goods, services, and capital (ENDA Diapol 2009). The agreement calls for a multipronged effort to be concluded in 2014 by which time the legal and institutional structures would be established to permit the conjoint governance of adjacent areas separated by national borders through decentralized governance mechanisms. Presently, the accord calls for a series of pilot projects, initially in the Senegambia region, that explores the issues related to bringing the governance of the major dimensions of local economic, social, and cultural life under the rubric of institutions conjointly managed by residents of both sides of a border. Mechanisms include collaborative controls by local police and customs officials based on consultative process with new associations of farmers and traders to work out mechanisms for inputs and exporting based on a comprehensive understanding of costs and needed profit margins. They include conjoint strategic planning for enhancing access to domestic and international markets, access to technical assistance, and the distribution of public moneys across integrated sectors to ensure equivalent access to health and social services and to areas of potentially strategic investment.

All of this takes place at a scale where the interrelationships between local economy, policing, taxation, social welfare, culture, and local authority systems can be discerned and monitored. The issue is not so much on scale here but rather the capacity-tracing patterns of interrelationship and finding ways to address problems as they arise.

MOBILITY AND INFRASTRUCTURE

[...] It is clear that articulations among cities can only be expanded if there is new infrastructure to do so. The composition of this infrastructure investment will inevitably reflect the exigencies and motivations of powerful economic interests, and in doing so, reiterate some of the insularity and tunneling of the past. If this insularity is not attenuated by a heightened commitment to maximize the density of interconnections among territories, people, and activities, a critical opportunity for development will be lost. In many respects, as we have seen throughout this discussion, Africans are leading the way by "their feet"—by their own efforts to thicken the connections between any given location and a wider world of economic activity. The implantation of infrastructure—highways, optic cables, satellite transmission, solar power, hydroelectric systems—creates important opportunities to define and institutionalize significant articulations among places and economic activities. Given the high sunk costs entailed, maximizing the long-term productivity of this infrastructure, in addition to the potentials for eventual cost recovery, necessitate high degrees of consonance between the articulations infrastructure puts in place and the diverse patterns of connectivity being generated by African entrepreneurs, traders, farmers, and retailers themselves.

Of course not every circuit or trajectory can be addressed; infrastructure will alter routes and ways of life and establish new ones. Many existent circuits of transaction are arduous, diffuse and a waste of time. Enhanced and established connectivity may also lessen the vulnerability of certain areas to protracted struggles over resources and territorial control by extrajudicial forces.

But the most important consideration of regional urban development planning would emphasize how the differentiated resource bases, histories, and geoeconomic positions of cities and towns could be most productively connected in order to create regional domains—crossing distinct national territories or rural–urban divides—with a density of synergistic relationships among diverse economic activities (Mimche and Fomekong 2008; United Nations Human Settlements Program 2008; Foster and Bricēno-Garmendia 2010).

Instead of trying to keep people in place, in newly democratic and decentralized localities, perhaps emphasis should be placed on how to make already existent movement more productive and convenient, and to accede to the possibility that urban residents "come to go, and go to come."

REFERENCES

Agier, M. (2008) *On the Margins Of The World: The Refugee Experience Today*, London: Polity.

Bayart, Jean-Francois and Jean-Pierre Warnier (2004) *Matiére á politique: la pouvoir, les corps, et les choses*, Paris: Karthala.

Economic Commission for Africa and the Africa Union (ECA) (2008) *Economic Report on Africa 2008*, Addis Ababa: Economic Commission for Africa.

ECOWAS-SWAC/OECD (2006) *Atlas on Regional Integration in West Africa*, Paris: OECD.

ENDA, Diapol (2007) *Les dynamiques Transfrontalières en Afrique de l'Ouest*, Dakar: ENDA Tiers Monde.

ENDA, Diapol (2009) *Tam-Tam: The Echo of Cross Border News*, Dakar: Enda Diapol.

Foster, V. and Bricēno-Garmendia, C. (eds.) (2010) *Africa's Infrastructure: A Time for Transformation*, Washington, DC: The International Bank for Reconstruction and Development, World Bank.

Guyer, J. (2004) *Marginal Gains: Monetary Transactions in Atlantic Africa*, Chicago: University of Chicago Press.

Mbembe, Achille. (2010) *Sortir de la grande nuit: Essai sur l'Afrique décolonisée*, Paris: Editions La Decouverte.

Mimche, H. and Fomekong, F. (2008) "Dynamiques urbaines et enjeux socio-demographiques en Afrique noire: comprendre la present pour prevoir l'avenir," *Revue Internationale des Sciences Humaines et Sociales* 02, L'Afrique subsaharienne a l'epreuve des mutations, Paris; Yaounde: L'Harmattan, 241–64.

United Nations Human Settlements Program (2008) *State of African Cities Report 2008/9*, Nairobi: United Nations Human Settlements Program.

"Migration and Privatization of Space and Power in Late Socialist China"

American Ethnologist (2001)

Li Zhang

[. . .]

THE FLOATING POPULATION

In the first three decades of socialist China, internal rural-to-urban migration was virtually eliminated by the state through the household registration system (*hukou*).[1] Under this system, the Chinese population was officially divided into two kinds of subjects—urban hukou holders and rural hukou holders. Rural subjects were prohibited from moving into the cities. Their rural residential status denied them access to state-subsidized foodstuffs, housing, employment, and other essential services reserved for urbanites only. This situation, however, has dramatically changed since the beginning of economic reforms in 1978, leading to the rise of a mass labor migration unprecedented in modern China. For example, according to an official survey in 1994, in Beijing alone, the already painfully crowded capital city had 3.29 million migrants (also called *waidiren*, meaning people coming from outer provinces), who came to compete with the 12 million official permanent Beijing residents for limited urban resources and services (Beijing Municipal Planning Committee Research Team 1995). The floating population consists of people with diverse socioeconomic and regional backgrounds, but their primary goals are the same: to get rich by migrating to the cities.[2] Some of these rural transients are able to bring a modest amount of capital with them to start small businesses and exploit the huge potential of the urban consumer market. Indeed, some of them have accumulated a considerable amount of wealth. But the majority of peasant workers who come to the cities have nothing but their labor to sell. The lucky ones manage to find temporary menial work in construction, restaurants, factories, domestic service, street cleaning, and other jobs that many urbanites are not willing to take. Still, there are many others who cannot find anything to do and thus drift hopelessly from place to place. Although migrants in China are not entitled to the same legal rights and social benefits as permanent urban residents and are subject to discrimination and periodical expulsion, they are now allowed to work in the cities on a temporary basis and are able to sustain themselves by acquiring daily essentials through the newly emerged *free* market. In more recent years, numerous migrant enclaves (based on the migrants' common place of origin) have thrived in various Chinese cities, developing into unofficial communities outside government city planning.

Despite the fact that the cheap labor and services provided by rural migrants are in high demand in the cities, the floating population is regarded by city officials and many urbanites as a drain on already scarce urban public resources and frequently blamed for increased crime and social instability. Appearing to be "out of place" and "out of control," this extraordinarily large and mobile population challenges the existing modes of state control that are largely based on the assumption of a relatively stable population fixed in space. Migrants are too far away from their places of origin to be reached by rural authorities but at the same time are not integrated into the urban control system. For many years, local urban officials were unwilling to extend their jurisdiction to rural migrants because they were considered "outsiders" in the cities and thus were not seen as subject to urban regulation. It was this lack of official control that created opportunities for migrants to develop their own social and economic niches in the cities. Later, migrant leaders gained local control through patronage and clientelist networks within these newly emerged migrant enclaves. "Zhejiang Village" (*Zhejiangcun*), the community on which my study is based, is one of the many unofficial communities that appeared in the reform period.

ZHEJIANGCUN: AN EMERGING MIGRANT ENCLAVE

Zhejiangcun, about five kilometers from Tiananmen Square, is the largest and most well-established migrant settlement in Beijing. This enclave was named by Beijing residents after the provincial origin (Zhejiang) of the majority of the migrants living there. Although it is called a "village," this settlement covers a large area across several suburban neighborhoods in the Fengtai district on the southern edge of the city proper. In 1995, it had about 100,000 migrants, and the majority of them (about 60 to 70 percent) were self-employed petty entrepreneurs and merchants from the Wenzhou region (mainly two areas under the administrative control of Yueqing city and Yongjia county) of Zhejiang province. As a relatively wealthy sector of the floating population, Wenzhou migrants mostly specialize in family based clothing businesses in manufacturing and wholesale or retail sales. Most of the remaining migrants in the area come from other provinces such as Anhui, Hubei, Henan, Shandong, and Sichuan and work as wage laborers who perform sewing, sales, or security work for the Wenzhou migrant employers.

Migration into Zhejiangcun began in the early 1980s and soon evolved into large-scale, chain migrations based on kinship ties and native-place networks. Wenzhou migrants' successful clothing businesses have transformed the preexisting suburban community from a once poor and bleak farming area into a dynamic private commercial center for garment manufacturing and trade. It now supplies garments not only to the markets of Beijing but also to other parts of Northern China. International traders from Russia and Eastern Europe come to buy Wenzhou migrants' fashionable yet affordable clothing goods.

The development of Zhejiangcun is highly politicized and closely monitored by city officials for a number of reasons. First, the sweatshop-like, labor-intensive, family based private production most Wenzhou migrants are engaged in is not deemed compatible with what the reform-era Beijing government wishes to promote in the capital: namely, modern hi-tech development, large corporate commerce, managed foreign capital investment, and lucrative international and domestic tourism. Second, based on kinship ties and native-place networks, Wenzhou migrants have begun to constitute themselves as an identifiable community with its own leadership invested in informal yet pervasive patron networks (called *bang*). Third, the perceived political danger of a self-organized migrant group to rival state power is heightened by the fact that this community is located in the capital city, only five kilometers away from China's political center—Tiananmen Square. And fourth, due to the increasing demand for housing, and the use of large sums of private capital accumulated through their family businesses, Wenzhou migrants constructed numerous large walled housing compounds in the early 1990s that consequently altered the spatial organization and power dynamics in this area.

[. . .]

THE DEMOLITION

In November and December 1995, the central Peoples Republic of China (PRC) and Beijing city governments mobilized a campaign to clean up Zhejiangcun. Although the explicit aim was to remove illegal building construction in the area, the ultimate goal was to eliminate what was perceived as a spatialized form of social power outside official control. As one official in the campaign team alluded to me, "A community with its own territorial ground (*dipan*) has the potential to become a separate regime of power." Such official concerns about political control and stability were heightened by the fact that Beijing is the emblematic heart of Chinese political power, a symbol of modernity, and a popular national and international tourist destination. Therefore, government control and scrutiny of non-state-directed communities and social forces are much tighter there than in other parts of the country.

The official apparatus of the campaign was called the "workteam" (*gongzuo dui*), composed of 2,000 diversely positioned government officials. The workteam was directly led by top Fengtai district officials under the supervision of the city government. Campaign activities were reported daily to the mayor and other central party leaders via daily campaign newsletters. The majority of the team members who did the actual day-to-day campaign work were lower level district, township, and sub-district officials, policemen, and village cadres. The campaign was carried out in a "closed" manner because, as some workteam members told me, upper level officials feared that if popular resistance rose during the housing demolition and violent suppression had to be used, media exposure would damage the image of the Chinese party-state. It would also invite severe criticism from the international community concerning human rights.

As specified in the government propaganda materials released to the public, the immediate goal of this campaign was to destroy all the Wenzhou migrants' housing compounds as well as some other illegal buildings put up by local farmers to accommodate migrants. Throughout the campaign, upper and lower level officials had very different views about how to handle the relationship between political control and economic gain. For the former group who did not receive direct economic benefits from Wenzhou migrants, maintaining political stability and state control was more important than anything else. They preferred to sacrifice local level economic interest to ensure stability. In one official's words, "Economic loss cannot be compared with political loss in the larger scheme of things." By contrast, local officials (especially village cadres) regarded economic growth and prosperity as a true source of stability. Thus, there was considerable instability and conflict within the "state" itself.

During this event, unorganized and semi organized popular resistance was widespread among not only migrants but also local farmers and workers who had constructed additional rooms in their houses for rental income. Many of the local residents refused to tear down their housing additions as ordered by the government. Popular rage was also expressed in the burning and destroying of government campaign notices. After encountering such resistance, the Beijing government decided to take harsh measures to move the campaign out of impasse. Pressure was put on local officials and cadres to show concrete signs of progress; they were warned that they could face serious political charges for not fulfilling their duties. Migrants and local farmers were notified that if they refused to flatten their own illegal housing additions, forced demolition would begin soon and they would receive a heavy fine and other punishments. Local cadres and party members were urged by higher authorities to demolish their own illegal structures to provide a good example to the community. Constantly pressed and threatened by workteam members, some local residents began to tear down their housing additions. Once the destruction began in certain areas, some residents began to dismantle their own housing to avoid government persecution because, as they told me, they felt no one could stop the campaign. By early December, 1995, the majority of illegal housing additions built by the locals had been demolished. Migrants who lived in those additions were forced out and many of them moved in temporarily with relatives whose housing had not yet been destroyed.

[. . .]

A once lively community with a flourishing private economy suddenly resembled the bombed out remnants of a war zone. Nearly 40,000 migrants lost their homes and were forced out of the settlement. The majority of them dispersed to rural counties outside the city proper, and some went to Tianjin and other parts of Hebei province (such as Baoding, Langfang, and Sanhe). Only a small portion of them managed to hide in run-down suburban factories, out-of-business motels, and remote villages.

But this was not the end of the story. Three months after this political hurricane had dwindled, the majority of displaced migrants began to return to what had been Zhejiangcun to rebuild their community and businesses. Since most available space within local villages was already rented out, returning migrants had to create new ways of locating housing. Some sought rentals in neighborhoods adjacent to Zhejiangcun or moved into nearby high-rise residential buildings where they paid higher rents than before. Others lived in housing compounds disguised in one of two ways. One was to borrow the title of a Beijing government agency and start a seemingly legitimate business, such as a clinic, and then build apartments on the acquired space. The other was to turn the existing space in local state-owned enterprises into migrant residences. This was made possible by recent state enterprise reform. Since the mid-1990s, the crisis of failing state enterprises had reached the point where many large and mid-size firms were forced to close down or dissolve. In Zhejiangcun, several state-run factories were facing shut downs. This situation presented an unusual opportunity for migrants to form new economic alliances with the factories struggling on the edge of a late socialist society. In 1996, some Wenzhou return migrants began to rent the abandoned factory buildings; the factory managers used the rent money to support their laid-off workers. (See Zhang 1999.) In this way, four factories in the area were turned into covert migrant housing compounds. Disguising housing compounds was done mainly to avoid the scrutiny of upper level officials and could only occur because local officials who knew what was going on turned a blind eye to this practice. Or, as the Chinese saying goes: "Opening one eye and closing the other" (*zheng yizhi yan, bi yizhi yan*).

[. . .]

CONCLUSION

Managing the floating population and migrant enclaves has long been regarded by certain government officials as the most difficult task facing the post-Mao political regime. Though the socialist state had successfully penetrated local Chinese society through a grassroots cadre system in the past, given a relatively stable population, it has not been able to create a new mediating social stratum to regulate the much more mobile, fluid, and culturally diverse migrant population. My analysis suggests that the tension between the floating population and the state is partly derived from the state's refusal to grant a legitimate status to the mediating social stratum between migrant masses and the state.

[. . .]

NOTES

1 The *hukou* system was enacted by the Chinese central state in 1958. Its primary goal was to block rural to urban migration in order to avoid what government officials perceived as a pathological growth of oversized cities and to ensure the agricultural production of grain to supply those working in industry. Restricting people's spatial mobility was also an important strategy used by bureaucrats to maintain socialist stability at that time. For detailed information regarding the origin, function, and socioeconomic consequences of this complex system, see Cheng and Selden 1994; Dutton 1999; Solinger 1999; and Zhang 1988.

2 It is worth stressing that the floating population is not a homogenous group. Rather, Chinese internal migrants' urban experiences vary greatly depending on a number of factors such as gender, place of origin, previous capital accumulation, and social networks. As a result, their senses of social belonging are also very different.

REFERENCES

Beijing Municipal Planning Committee Research Team (1995) "Beijingshi liudong renkou guanli duice yanjiu" (A Study of Regulation Strategies Regarding Beijing's Floating Population), *Shoudu Jingji* (The Capital's Economy) 5: 14–17.

Cheng, T., and Selden, M. (1994) "The origins and social consequences of China's *hukou* system," *China Quarterly* 139: 644–668.

Dutton, M. (1999) *Street Life China*, Cambridge: Cambridge University Press.

Solinger, D. (1999) *Contesting Citizenship in Urban China: Peasant Migrants, the State, and the Logic of the Market*, Berkeley: University of California Press.

Zhang, L. (1999) "Reconfigurations of wealth and urban citizenship in late socialist China," Paper presented at the Annual Meeting of the American Anthropological Association, Chicago, November 20.

Zhang, Qingwu (1988) "Basic facts on the household registration system," *Chinese Economic Studies* 22(1): 22–85.

SECTION 4
Urban Economy

Editors' Introduction

The city as marketplace and as part of a regionally and globally networked economic system provides a functional definition for the basis of the urban economy. Within this marketplace, a threshold figure for the percentage of employment outside agriculture, for example, remains the basis for deciding which places are "urban" in many parts of the world. The urban economy, however, is broader than just markets and employment; it includes the institutional milieu, regulatory structures, territorial endowments, interrelated economic sectors from manufacturing to services, and the labor force. Moreover, in our global world, production complexes are not contained by geographic or political boundaries. Networks of financial flows, mobile pools of experts and managers, low-wage service workers and repeated economic crises are becoming the norm.

In the introduction to this section of the *Reader* we explore the topic of the urban economy within the cities of the global South, given a free-flowing yet sticky capitalist global economic system. We start by examining the definition and characteristics of the informal economy, which is central to Southern economies and particularly to the life of the poor. Next, we examine the impacts of globalization on urban economies in the South, looking not just at global city theory but also at a range of other issues such as the role of special economic zones, the financialization of urban development, and remittance economies. We conclude with two article selections, each of which focuses on a particular aspect of the urban economy.

KEY ISSUES

Informal economy

The term informal sector was popularized by Keith Hart, an English anthropologist, who studied informal income opportunities and urban employment among the immigrant Fafras in the Nima district of Accra, Ghana, in the late 1960s. In making the case that the "'reserve army of underemployed and unemployed' ... possess some autonomous capacity for generating growth [instead of constituting a] passive, exploited majority" (Hart 1973: 61), he was among the first to emphasize the importance of small-scale indigenous entrepreneurs. Hart made a definitional distinction between *formal income opportunities, that is, wage-earning employment* in *public and private entities and informal income opportunities, both legitimate and illegitimate, mostly focused on self-employment* [emphasis ours] (1973: 69–70). This paralleled the distinction being drawn at that time between the unorganized and the organized sections of the urban labor force.

Subsequently, the International Labor Organization (ILO) popularized the concept of informal work in a well-publicized report on Kenya. The emphasis on informality as *self-employment* in small enterprises outside the purview of the state remained dominant. Theoretical models of the time expected the informal sector to shrink with increasing industrialization and economic growth. Contrary to assumptions, the informal sector continued to grow. In 1989, the sociologists Alejandro Portes, Manuel Castells, and Lauren Benton edited a volume, *The Informal Economy: Studies in Advanced and Less Developed Countries*, which pulled together two decades of work in a range of disciplines to emphasize the idea that the informal economy was not just about self-employment, and that it was not restricted to poor or third world countries. Wage employment and micro-entrepreneurial work (a term that came into vogue in the 2000s) were understood as important components of work in the informal economy. In their introduction to the book, Castells and Portes further emphasized that the informal economy was not a fixed object separate from the formal economy; instead it

> [S]imultaneously encompasses flexibility and exploitation, productivity and abuse, aggressive entrepreneurs and defenseless workers, libertarianism and greediness. And above all, there is disenfranchisement of the institutionalized power conquered by labor, with much suffering in a two-century-old struggle. (1989:11)

Castells and Portes's reference to labor's struggles to unionize and institutionalize rights is particularly important in the context of expanding informality and globalization of production in the current era.

A great deal of work exists on the informal economy, and its impacts on particular groups of workers such as women who make up a disproportionate share of informal economy workers (for a comprehensive bibliography, see WIEGO 2013). Women are also heavily represented in particular sectors such as domestic work (Ally 2010), construction (Agarwala 2007) or street vending (Brown 2006; Cross and Morales 2007; Tinker 1997). The broader spatial, social, economic, political, and policy consequences of informalization have also been explored extensively (Rakowski 1994; Tripp 1997; Roy and AlSayyad 2004; Kudva and Beneria 2005).

In the 1990s, as neoliberal development policy became entrenched, Fordism declined and globalization picked up speed (see Goldman, Part II, this volume). This coincided with a rapid expansion of informalization in most parts of the world. The economist Jacques Charmes (2000) carefully demonstrated that the share of employment and economic output generated from within the informal economy was increasing. Even in advanced industrialized economies like the United States, the informal economy in places like California produced about a quarter of all economic output (Aspen Institute 2003) in sectors as varied as software contract work, domestic services, agri-business, construction, and childcare. Current estimates for the size of the informal economy vary greatly depending on how the informal economy is operationalized. The ILO (n.d.) estimates that worldwide 35–90 percent of employment in each country is in the informal economy.

There is general agreement in the vast literature on the informal economy that three main distinguishing characteristics need to be emphasized.

- *The informal economy is not something separate from the regular economy* and, neither can survive without the other. Processes of flexible production, outsourcing, or "autoconstruction" in the context of housing emphasize interlinkages, allowing the formal to mimic what informal economy activities excel at: flexibility and cost-effectiveness leading to higher profits. As important are the ways in which the working lives of workers in the informal and formal economies intersect; for example, street food vendors feed formal sector wage-earners in the business districts of cities across the global South.
- *The informal economy is not just the preserve of the poor and the destitute.* Whether a worker is seen as being within the informal economy or not depends on her relationship to forms of production; whether she is poor or not is linked to the process of distribution of gains. A moonlighting software consultant, a childcare worker, or a machinist on a temporary contract could be classified as an informal worker. While this opens up conversations around elite forms of informality, there is an important caveat: the work of the majority of poor people, especially in the global South, falls in the space of the informal economy.
- *The informal sector is not regulated by the state.* Labor laws, taxation, workplace safety, mandated benefits, and insurance policies are rarely, if ever, enforced. However, while the informal economy is not regulated, it is often policed and is a lucrative source of income (by way of bribes and protection money) for government officials, the political system, and the underground economy. With the spread of video and the rise of the internet, a range of other issues have begun to emerge around intellectual property, piracy, and the media, all of which highlight the extent to which the formal and informal economies intersect and coexist (Sundaram 2009). Another important crossing-over is occurring in the ways in which corporate actors who once looked to invest in industrial sectors are moving into areas like real estate and land development, where informal

relationships dominate (see Mitra, this volume). Outsourcing and subcontracting arrangements in production have also resulted in the state and corporations being actively complicit in hiring workers from the informal economy in a range of occupations including municipal waste-collectors, garment manufacturers, or the makers of components for sophisticated high-tech manufacturing (Miraftab 1996, 2004).

The absence of regulation, however, does shape firms' practices and the status and rights of labor in particular ways. Workers are much more vulnerable to employer whims and dictates: they are often paid lesser wages and lack benefits such as health insurance, sick leave, or compensation for job-related accidents. Worker vulnerability is not random, as several researchers have pointed out; it is tied to certain social characteristics and those with little status in society–minorities, women, children, immigrants–are more vulnerable than others (see Hunter and Skinner 2003). However, workers with low education and few extra resources rely on the minimal barriers to entry in the informal economy, even if the conditions of labor are often worse and hazardous conditions affecting hygiene and safety more common. Also important are organizational practices and forms of management of firms in the informal economy where cash transactions and tax evasion is the norm. In most cases, employers in the informal economy have their own ways of regulating and controlling their workers. Moreover, sophisticated systems and trust-based norms often prevail. For example, Harvard Business School case writers have detailed the work of Mumbai's "dabbawalas" who deliver home-prepared lunches for thousands of office workers in downtown Mumbai with less than a 1 percent failure rate every day (Thomke and Sinha 2013).

Researchers have also documented the efforts of labor organizers and activists attempting to expand the rights of workers and change government policies toward informal enterprises (Agarwala 2013; ILO, n.d.). A prime example is the trade union, Self-Employed Women's Association in India (SEWA), which has over a million members. SEWA began work in the 1970s under the leadership of Ela Bhatt and has helped inform the creation of multiple networks across the world like StreetNet in South Africa (Chen, Jhabvala, Kanbur, and Richards 2007). Both SEWA and StreetNet are part of worldwide networks like WIEGO (Women in Informal Employment: Globalizing and Organizing) that are working "to increase the voice, visibility, and validity of the working poor, especially women, in the informal economy" (WIEGO 2013).

Informality is also present in housing markets, in the production of infrastructure and in other sectors as various sections in this *Reader* demonstrate. It shapes social and political life in the city. The informal spaces of housing, for example, tend to overlap with spaces that have a concentration of informal economy production (Kudva 2009). This overlap reproduces not just the segregated spatial structure of the colonial city but also creates new forms of spatial segregation as informally produced housing and services are located in left-over spaces or in areas close to new developments for the increasing numbers of middle classes and increasingly in the fast growing peripheries of cities in Asia and Africa (Simone 2009). The co-production of the formal and the informal is an area currently being explored (Rumbach 2011) while others, like Ananya Roy (2011), are making the argument that the widespread prevalence of informality and the dominance of the slum in Southern cities demands new categories and theorization.

But even as the urban economy is shaped by informality from the ground up it is equally a product of global forces, a topic we consider next.

Globalization and the urban economy

In the boom years of the early 2000s, when globalization and explosive economic growth across China and to a lesser extent across the other BRICS was the talk of the world, *The New York Times* columnist Thomas Friedman (2005) announced that the world had become flat, and that geographical and political boundaries did not matter. He argued that everyone armed with a personal computer, access to the Net and work-flow software (not just multinational companies as in earlier phases of globalization) had equal opportunity to compete and to enlarge their prospects. He was not the only one to have his fingers on the pulse of a major economic shift that had begun many decades earlier, rooted in networks of trade and empire established over many centuries of history (Arrighi, Hamashita, and Selden 2003). There was, however, much less recognition of the discontents of globalization and the fact that even as globalization was leading to explosive growth in some parts of the world, it was increasing poverty and inequality everywhere as states retreated and markets gained ground. Globalization combined with neoliberal economic policies was not the leveler that its fervent proponents such as Friedman claimed, and as the bust of 2009 later made even clearer.

The sociologist Saskia Sassen (1991[2001]), who studied international migration and economic restructuring, was one of the first to take John Friedmann and Goetz Wolff's (1982) characterization of the world city and describe the global city as the command and control center for the new global economy. Through the cases of New York, London, and Tokyo, she emphasized the importance of cities, and therefore of place and territory, in the placeless world of global flows and delocalized production chains that were being discussed at that time. Sassen's work, summarized most recently in *Cities in a World Economy* (2012), is part of a longer tradition of understanding the restructured role of cities as economies of agglomeration and innovation in larger systems–regions, nations and now, the world.

Sassen's global city thesis set off a storm of research and the global-ness of cities began to be categorized based on their degree of centrality to various networks of finance, services, commodity production, and more. A better known effort along these lines was the GaWC (Global and World Cities Research Network), founded by geographer Peter Taylor and his colleagues. Some researchers sought to identify the points at which a city's economy had a global insertion (helping make the city competitive in economic terms within a global urban hierarchy). Others looked to see the spatial and social impacts of such insertions, which created enclaves characterized by First World consumption patterns and produced a mobile social, cultural class of city residents delinked from the cities and nations around them. Scholars continue to unpack the complexity of the processes through which the world's cities strain to follow particular policy and planning strategies to become more attractive to global capital (see Robinson, this volume; Mitra, this volume; Shatkin 2013). A considerable body of this scholarship focuses on the great costs to the majority of citizens such an approach entails. The desires and place-making strategies for becoming global include building malls and gated enclaves, "starchitect" designed mega-projects and high-tech office campuses; high-tech infrastructure systems for a paying minority; and special economic zones and enclaves that suspend labor laws and regulations giving capital subsidized access to land and labor (Roy and Ong 2011). The article selection by Mitra in this section focuses attention on three of these place-making strategies. What remains poorly theorized, and has only begun to be discussed, is the economy in smaller cities and towns that house the majority of the world's urban population but are not the focus of global investments.

An important aspect of globalization has been the rise of the remittance economy. Recent calculations of remittance flows estimate them as more than twice that of official development assistance (Ratha and Shaw 2007; World Bank 2011). While countries like India receive the largest proportion of remittances from diaspora labor, small countries like Cape Verde or Bangladesh are much more dependent on remittances for development. The physical and social transformation of both rural and urban places through remittance-led growth that spans the formal-informal divide has become the subject of considerable study (see Editors' Introduction to Migratory Fields, this volume, for more details).

These segregated yet deeply interlinked economic processes in the formal and informal economies, and at local, regional, and global scales, are the background for economic life in the city. Focusing on employment, however, is one way of grasping the complexities of these intersecting processes and their outcomes.

ARTICLE SELECTIONS

Employment, the relations between worker and employer and the relation between a worker and her place of employment, cuts across the divisions we laid out earlier between the formal and informal economy, between globalized capital and localized production and politics, all of which are inextricably linked to each other. Employment allows us to focus instead on the lives of workers and to understand the relationship of the economy and work to the spaces of the city across theoretical and analytical divides. Hence, we selected an article by geographer and planner Ray Bromley. He provides a framework for characterizing employment relationships while taking into account appropriation of worker's labor, autonomy, and flexibility of work regimes as well as the stability and security of work opportunities. Though he wrote the article in the mid-1990s using field work from the late 1970s, the street occupations and conditions Bromley describes continue to be highly visible in the everyday life of the city and remain crucial to the livelihoods of contemporary city-dwellers, particularly the urban poor.

The second article, commissioned for the *Reader*, shifts scales to analyze the urban effects of transnational financial flows and changing geopolitical and institutional alignments in the contemporary moment. Sudeshna Mitra, who has worked internationally as a development consultant on land-based economic development projects, focuses here on the ways cities are accommodating global sectors such as finance, services, and IT. Here she considers three types of hypermodern urban spaces–technology parks, privatized cities, and new business districts–to understand the anchoring of transnational capital through public–private negotiations, and the shift from economic investments to land speculation and real estate development (see also Hsing 2010). As Mitra points out, such spaces of exception in large Asian and Gulf cities are providing new benchmarks for the global city of the twenty-first century. They involve place-making strategies that pass over entire cities and populations and further entrench social and spatial dichotomies.

REFERENCES

Agarwala, R. (2007) "Resistance and compliance in the age of globalization: Indian women and labor organizations." *Annals AAPSS*, 610: 143–159.

Agarwala, R. (2013) *Informal Labor, Formal Politics and Dignified Discontent in India*. Cambridge and New York: Cambridge University Press.

Ally, S. (2010) *From Servants to Workers: South African Domestic Workers and the Democratic State*. Ithaca, NY: Cornell University Press.

Arrighi, G., Hamashita, T., and Selden, M. (Eds.) (2003) *The Resurgence of East Asia: 500, 150 and 50 Year Perspectives*. London and New York: Routledge.

Aspen Institute. (2003) "The informal economy and micro enterprise in the United States," *FIELDForum*, 14, March. Washington, DC: The Aspen Institute. Available online at http://fieldus.org/Publications/Field_Forum14.pdf (accessed May 7, 2013).

Bromley, R. (1997) "Working in the streets of Cali, Colombia: Survival strategy, necessity, or unavoidable evil?" In J. Gugler (Ed.), *Cities in the Developing World: Issues, Theory, and Policy*. Oxford and New York: Oxford University Press, pp. 124–138.

Brown, A. (Ed.) (2006) *Contested Space: Street Trading, Public Space, and Livelihoods in Developing Cities*. Rugby: ITDG Publishing.

Castells, M. (1989) *The Informational City: Information Technology, Economic Restructuring, and the Urban-Regional Process*. London: Blackwell Oxford.

Castells, M. and Hall, P. (1994) *Technopoles of the World: The Making of the 21st Century Industrial Complexes*. New York: Routledge.

Castells, M. and Portes, A. (1989) "World underneath: The origins, dynamics and effects of the informal economy." In A. Portes, M. Castells, and L. A. Benton (Eds.), *The Informal Economy*. Baltimore: Johns Hopkins Press, pp. 11–37.

Charmes, J. (2000) "Informal sector, poverty and gender: A review of empirical evidence." Background paper for the 2001 World Development Report. Washington, DC: World Bank.

Chen, M., Jhabvala, R., Kanbur, R., and Richards, C. (Eds.) (2007) *Membership-Based Organizations of the Poor*. London: Routledge.

Cross, J. and Morales, A. (Eds.) (2007) *Street Entrepreneurs: People, Place and Politics in Local and Global Perspective*. London and New York: Routledge.

Friedman, T. (2005) *The World is Flat: A Brief History of the Twentieth Century*. New York: Farrar, Strauss, and Giroux.

Friedmann, J. and Wolff, G. (1982) "World city formation: An agenda for research and action." *International Journal of Urban and Regional Research*, 6(3): 309–344.

Hart, K. (1973) "Informal income opportunities and urban employment in Ghana." *The Journal of Modern African Studies*, 11(1): 61–89.

Hsing, Y. (2010) *The Great Urban Transformation: Politics of Land and Property in China*. New York: Oxford University Press.

Hunter, N. and Skinner, C. (2003) "Foreign street traders working in inner city Durban: Local government policy challenges." *Urban Forum*, 14(4): 301–319.

ILO (n.d.) "Informal economy." Available online at http://www.ilo.org/emppolicy/areas/informal-economy/lang--en/index.htm (accessed July 2012).

Kudva, N. (2009) "The everyday and the episodic, understanding the spatial and political impacts of informality in two Indian cities." *Environment and Planning A*, 41(7): 1614–1628.

Kudva, N. and Beneria, L. (Eds.) (2005) *Rethinking Informalization: Poverty, Precarious Jobs and Social Protection.* Ithaca, NY: Internet First University Press. Cornell OAR. Available online at http://hdl.handle.net/1813/3716 (accessed May 25, 2014).

Miraftab, F. (1996) "Space, gender and work: Home-based workers in Guadalajara, Mexico." In E. Boris and E. Prugl (Eds.), *Homeworkers in Global Perspective: Invisible No More*. New York: Routledge, pp. 63–80.

Miraftab, F. (2004) "Neoliberalism and casualization of public sector services: The case of waste collection services in Cape Town, South Africa." *International Journal of Urban and Regional Research*, 28(4): 874–892.

Portes, A., Castells, M., and Benton, L. (Eds.) (1989) *The Informal Economy*. Baltimore: Johns Hopkins University Press.

Rakowski, C. (Ed.) (1994) *Contrapunto: The Informal Sector Debate in Latin America*. Albany, NY: State University of New York Press.

Ratha, D. and Shaw, W. (2007) "South–South migration and remittances." World Bank Working Paper No. 102, Washington, DC: World Bank.

Roy, A. (2011) "Slumdog cities: Rethinking subaltern urbanism." *International Journal of Urban and Regional Research*, 35(2): 223–238.

Roy, A. and AlSayyad, N. (Eds.) (2004) *Urban Informality: Transnational Perspectives from the Middle East, Latin America, and South Asia*. Lanham, MD: Lexington Books.

Roy, A. and Ong, A. (Eds.) (2011) *Worlding Cities: Asian Experiments and the Art of Being Global*. Chichester: Wiley Blackwell.

Rumbach, A. (2011) "The city vulnerable: New town planning, informality and the geography of disaster in Kolkata, India." Unpublished doctoral Dissertation: Cornell University. Ithaca, NY.

Sassen, S. (1991[2001]) *The Global City: New York, London, Tokyo*. Princeton, NJ: Princeton University Press.

Sassen, S. (2012) *Cities in a World Economy* (4th edn). Thousand Oaks, CA: Sage Publications.

Shatkin, G. (Ed.) (2013) *Contesting the Indian City: Global Visions and the Politics of the Local*. Chichester: Wiley Blackwell.

Simone, A. (2009) *City Life from Jakarta to Dakar: Movements at the Crossroads*. London: Routledge.

Sundaram, R. (2009) *Pirate Modernity: Media Urbanism in Delhi*. London and Delhi: Routledge.

Thomke, S. H. and Sinha, M. (2013) "The Dabba-wala system: On-time delivery, every time." Harvard Business School Case 610-059. (Revised from original February 2010 version.)

Tinker, I. (1997) *Street Foods: Urban Food and Employment in Developing Countries*. New York: Oxford University Press.

Tripp, A. M. (1997) *Changing the Rules: The Politics of Liberalization and the Urban Informal Economy*. Berkeley, Los Angeles, and London: University of California Press.

WIEGO (Women in Informal Employment: Globalizing and Organizing) (2013) Publications and Resources. Available online at http://wiego.org/publications-resources (accessed May 7, 2013).

World Bank (2011) *World Bank's Migration and Remittances Factbook 2011*. Available online at http://go.worldbank.org/QGUCPJTOR0 (accessed May 7, 2013).

"Working in the Streets of Cali, Colombia: Survival Strategy, Necessity, or Unavoidable Evil?"

Cities in the Developing World: Issues, Theory, and Policy (1997)

Ray Bromley

This essay reviews some of the theoretical, moral, and policy issues associated with the low-income service occupations found in the streets, plazas, and other public places of most African, Asian, and Latin American cities. These "street occupations" range from barrow-pushing to begging, from street-trading to night-watching, and from typing documents to theft. They are often grouped together in occupational classifications and they are generally held in low esteem. The street occupations are frequently described by academics and civil servants as "parasitic," "disguised unemployment," and "unproductive," and they are conventionally included within such categories as "the traditional sector," "the bazaar economy," "the unorganized sector," "the informal sector," "the underemployed," and "subproletarian occupations." It seems as if everyone has an image and a classificatory term for the occupations in question, and yet their low status and apparent lack of developmental significance prevent them from attracting much research or government support.

This discussion of "street occupations" is based mainly on 1976–1978 research in Cali, then Colombia's third largest city with about 1.1 million inhabitants. The occupations studied in the streets and other public places of Cali are remarkably diverse, but they can be crudely described under nine major headings:

Retailing: the street-trading of foodstuffs and manufactured goods, including newspaper distribution.

Small-scale transport: moving cargo and a few passengers for payment, using *motocarros* (three-wheel motorcycles), horse-drawn carts, bicycles, tricycles, handcarts, or direct human effort as porters.

Personal services: shoe-shining, shoe repair, watch repair, the typing of documents, etc.

Security services: night-watchmen, car-parking attendants, etc.

Gambling services: the sale of tickets for lotteries and *chance*, a betting game based on guessing the last three digits of the number winning an official lottery.

Recuperation: door-to-door collection of old newspapers, bottles, etc.; "scavenging" for similar products in dustbins, rubbish heaps, and the municipal tip; and the bulking of recuperated products.

Prostitution: or, to be more precise, soliciting for clients.

Begging.

Property crimes: the illegal appropriation of movable objects with the intention of realizing at least part of their value through sale, barter, or direct use. This appropriation can be by the use of stealth (theft), by the threat or use of violence (robbery), or by deception ("conning").

Of these nine categories, "retailing" was the largest in 1977, accounting for about 33 percent of the workforce in the street occupations, followed by small-scale transport and gambling services, each accounting for about 16 percent of the workforce. The six remaining categories each accounted for 2–10 percent of the workforce in street occupations.

With the exception of small-scale public transport, all these occupations can be conducted in private locations as well as in public places, *and* private locations are generally considered more prestigious. Private premises give a business an aura of stability and security which is not available to most businesses conducted in streets plazas, and other areas of public land. Those who work in public places may try to obtain a degree of stability by claiming a fixed pitch, and by building a structure there to give them some protection, but their tenure is almost always precarious, and their investment in "premises" is likely to be very limited. Thus, the street occupations are classically viewed as "marginal occupations," as examples of how the poor "make out," or as the "coping responses" of the urban poor to the shortage of alternative work opportunities and the lack of capital necessary to buy or rent suitable premises and to set up business on private property.

Even though they are an integral part of the street environment and interact strongly with the street occupations, four major groups of economic activities are not considered as street occupations *per se*: the off-street private shops, supermarkets, market stalls, etc., which open on to the street; government and company employees who are responsible for building, maintaining, and cleaning the streets; the police and soldiers who are responsible for law and order on the streets; and the operators of larger scale public transport vehicles such as buses, trucks, and taxis. These occupations are considered in relation to the street occupations, but not as part of the street occupations. All of them have a strong off-street base, most having working regimes and relationships rather different from those prevailing in the street occupations, and many have much higher levels of capital investment in premises, equipment or merchandise.

Because they are neither practiced in conventional (off-street) establishments, nor in the homes of the workers, the street occupations are usually severely under-represented or excluded altogether in statistics based on sample surveys of establishments or households. In spite of their under-representation in most official statistics, however, their highly public location ensures them a prominent place in the urban environment and popular consciousness. Even if they have no direct dealings with those who work in the street occupations, the general public can hardly fail to be aware of their existence.

The streets of the city serve a wide variety of interrelated purposes: as axes for the movement of people, goods, and vehicles; as public areas separating enclosed private spaces and providing the essential spatial frame of reference for the city as a whole; as areas for recreation, social interaction, the diffusion of information, waiting, resting, and, occasionally, for "down-and-outs and street urchins," sleeping; and as locations for economic activities, particularly the "street occupations" (Anderson, 1978: 1–11 and 267–307). Within the functional complexity of the street environment, the street occupations are both strongly influenced by changes in other environmental factors, and also contributors to general environmental conditions. Thus, for example, street-traders and small-scale transporters depend upon the direction, density, velocity, and flexibility of potential customers' movements, and are immediately affected by changes in traffic flows and consumer behavior. At the same time, they influence patterns of movement and overall levels of congestion.

[. . .]

As well as low status and low average incomes, the street occupations are characterized by relatively low inputs of capital in relation to labor, and by low "formal" educational requirements. In 1976–1978 most street enterprises operated with a total working capital equivalent to less than US$100 in terms of equipment and/or merchandise.[1] Many porters, watchmen, scavengers, beggars, and thieves incurred no monetary expenditures in order to be able to work, beyond the costs of their normal clothing, food, and transport to a workplace. [. . .] Basic literacy and numeracy are generally useful to participants in the street occupations, but even these relatively low educational standards are rarely required. Instead, the occupations are characterized by skills learned outside the government educational system, such as hard bargaining, quick-wittedness, manual dexterity, good memory, an engaging personality, and physical endurance. On-the-job experience and effective utilization of social networks are particularly important in the street occupations, together with such difficult-to-alter variables as "an honest face," physical toughness, or beauty.

[. . .]

The most basic need of the urban poor is an income in goods and/or money to provide for food, drink, housing, clothing, and other necessities. An income may come from government or private charitable institutions, from investments, moneylending and renting, from windfall gains, or from work. For the poor, work is the normal way to obtain an income [. . .]. Ideally, work should be both enjoyable and rewarding, yielding an income and a sense of personal achievement and satisfaction. Instead, to most people, including those working in the street, work is boring and exhausting, and even dangerous or degrading. Furthermore, work opportunities are usually scarce and insecure, and work is often inadequately remunerated, leading to poverty and deprivation. All of the occupations under consideration here have some of these negative characteristics of work, and together the street occupations form a complex of low-status, poorly remunerated, insecure forms of work. Although many who work on the streets comment that their occupations are less exhausting than heavy manual labor like cutting sugarcane and carrying bricks on construction sites, the street occupations usually require long hours and uncomfortable conditions. [. . .]

WORK, ILLEGALITY, AND INFORMALITY

Work is defined here as "any activity where time and effort are expended in the pursuit of monetary gain, or of material gain derived from other persons in exchange for the worker's labor or the products of such labor." In other words, work is the labor involved in producing goods and services for exchange, and it is "income-generating." The category of "work," thus defined, excludes the equally important category of "expenditure-reducing" activities which can be described collectively as "subsistence labor,"—for example, growing food for household consumption, self-help house construction and repair, unremunerated housework and child-minding, voluntary unpaid help given to friends and neighbors, and walking or cycling to places of work or recreation so as to avoid paying transport fares. Any form of work which is regularly performed by a given person may be described as an occupation.

Under these definitions, such classic lumpenproletarian occupations as begging, prostitution, and theft can all be viewed as work, and hence can be analysed together with the remaining street occupations. The presence of these illegal, disreputable, or public nuisance activities in the category of "street occupations" serves to emphasize the high degree of differentiation which exists within this category. Of course, illegality and disrepute extend much further than begging, prostitution, and theft, as a few traders deal in contraband, stolen, or falsified merchandise. Various forms of street gambling are illegal, and substantial numbers of persons working in transport and gambling services and the overwhelming majority of street-traders do not possess the licences and documents required by official regulations.

Illegality is widespread in the street occupations, and serious problems and suffering may be associated with theft, robbery, conning, and prostitution. Most cases of illegality in the street occupations, however, are either trivial or common to many other groups across the social spectrum. [. . .]

Many street enterprises provide useful services, but do not have some or all of the permits they are required to have, or fail to meet official specifications on receipting, taxation, equipment, uniforms, hygiene, etc. Just like big business corporations, wealthy taxpayers, and real-estate dealers, workers in the street occupations often cut corners and break a few rules here and there in order to make a living or a little extra profit. Complying with each and every official regulation can be costly and time-consuming. For minors and for adults who have lost or never obtained key identity papers, compliance may be impossible. In addition, many of the regulations are rarely enforced, and some officials will accept a bribe to stop enforcement.

Hernando de Soto and his collaborators (de Soto, 1989) have developed a simple, clear concept of "informality" to explain the socio-legal status of most street occupations, numerous other small enterprises, and self-help housing through squatting and illegal subdivision. [. . .]

Using a simple means/ends criterion, the de Soto approach divides economic activities into three basic groups: formal, where both means and ends are legal; informal, where ends are legal but means are nominally illegal; and, illegitimate, where both ends and means are illegal and/or anti-social. Formal enterprises obey the spirit and the letter of the law. Informal enterprises obey the spirit of natural law but not the letter of formal law; they perform useful functions and provide necessary services, but do not obey every official regulation applying to their activities. Illegitimate activities contravene the principles of natural law—they are anti-social and/or criminal, whether or not they are officially proscribed by formal law. The means/ends criterion is reinforced by a second criterion, social utility, that the people involved and the society as a whole are better off if the law on these activities is broken than if it is obeyed. Hence, "an activity is informal when it neither produces a deterioration in the social situation nor an antisocial result when the law and the regulations applicable to it are disobeyed" (Ghersi, 1991: 46). This second criterion excludes such morally questionable economic activities as prostitution, child labor, and begging from the category of informality.

De Soto portrays informal activities as manifestations of the vitality and entrepreneurial dynamism of the poor. They break a few rules here and there, but only to support themselves and their dependents, to make a living, and to avoid crime or destitution. [. . .] His writings tend to focus on "informality" by the poor, but the logic of his analysis suggests that the rich and big business may break "formal" laws just as much as the poor do.

The socio-legal problems that de Soto focuses on are societal problems of enormous importance: the role of the state, the rights and responsibilities of the citizen, and the nature of the legal system. Whether or not we agree with all his views, they do provide some key pointers for the analysis of street occupations and the formulation of policies affecting those occupations. The street occupations are highly visible and often scapegoated, but many of the problems they pose and face are posed and faced by less visible and notorious off-street activities. They are societal problems which should be addressed systematically rather than by scapegoating one occupational group.

EMPLOYMENT RELATIONSHIPS

In this analysis, the term "work" has a different meaning from the term "employment." "Employment" is used to denote a relationship between two parties, an "employer" and an "employee," the former paying the latter to work on the former's behalf for a significant period of time (at least a working day), or for lesser periods on a regular basis.

When there is a direct two-tier employer–employee relationship based on some form of contract (an oral or written agreement), there are two main forms of working relationship: "on-premises working," when the employee works at a site owned, rented, or operated by the employer; and, "outworking," when the employee works away from the employer, usually in his own home, in the streets, or in some door-to-door operation. An employee may be paid wages per unit of time worked, per unit of "output," or by some combination of the two. When work is remunerated wholly or partly by the unit of time worked, whether as "on-premises working" or "outworking," it is generally recognized as a form of employment. When it is remunerated solely per unit of "output," however, it is usually only viewed as employment if it is conducted "on premises." When conducted off-premises, piece-work is conventionally viewed as a form of self-employment, and this conception is embodied in Colombian labor legislation and in the perceptions of most middle- and upper-class Colombians. In contrast to such views, and in keeping with our definition of employment, "outwork on a piece-work basis" is here as a form of employment remunerated by a piece-wage. Because it is not officially or widely recognized as such, it may be viewed as "disguised wage-working," as distinct from the more widely recognized forms of wage-working which take place "on-premises" or which involve off-premises work remunerated per unit of time worked.

When a worker is not employed by someone else either overtly or disguised through off-premises piece-working, two alternative working relationships are commonly found: "dependent working" and "self-employment" (Bromley and Gerry, 1979: 5–11). Dependent workers are not employees and have no fixed margins and commissions, but they have obligations which take a contractual form and which substantially reduce their freedom of action. These obligations are associated with the need to rent premises, to rent equipment, or to obtain credit, in order to be able to work. Although the appropriation of part of the product of the worker's labor is not as clear and direct in the dependent-working case as in the disguised wage-working case, there is normally an appropriation process through the payment of rent, the repayment of credit, or purchases and sales at prices which are disadvantageous to the dependent party in the relationship. In contrast, true self-employment has no such relationships; the workers work on behalf of themselves and other persons that they choose to support. Self-employed workers must, of course, rely on inputs provided by others, on the receipt of outputs by others, and on a system of payment. However, the bases of their self-employment are that they have a considerable and relatively free choice of suppliers and outlets, and also that they are the owners of their means of production. They are dependent upon general socio-economic conditions and on the supply and demand conditions for their products but they are not dependent upon specific firms for the means to obtain their livelihoods.

The employment relationships described above form a continuum ranging from wage-work to self-employment. Two major variables, "relative stability and security of work opportunities," and "relative autonomy and flexibility of working regimes and conditions," can be used to divide up the continuum into six major categories whose relationships and characteristics are set out in Figure 24. The two extremes of the continuum, described as "career wage-work" and "career self-employment," have relatively high levels of stability and security, and they are not encountered among the street occupations of Cali. The four intermediate relationships have relatively low levels of stability and security, can be described collectively as "casual work," and are found in varying proportions among the street occupations. The six categories of the continuum are not commonly distinguished, and ambiguous cases will arise as in all classifications, but they do provide a much richer and more appropriate focus for studies of contractual relationships, economic linkages, and changes in employment structure than the more conventional "dualist" distinction between wage-work and self-employment with no intermediate categories.

In the streets of Cali, self-employment is a much less common phenomenon than might at first glance seem apparent, and there are signs that it is diminishing in significance in the face of the expansion of disguised wage-working and dependent working (Birkbeck, 1978; Bromley, 1978; Gerry, 1985; Gerry

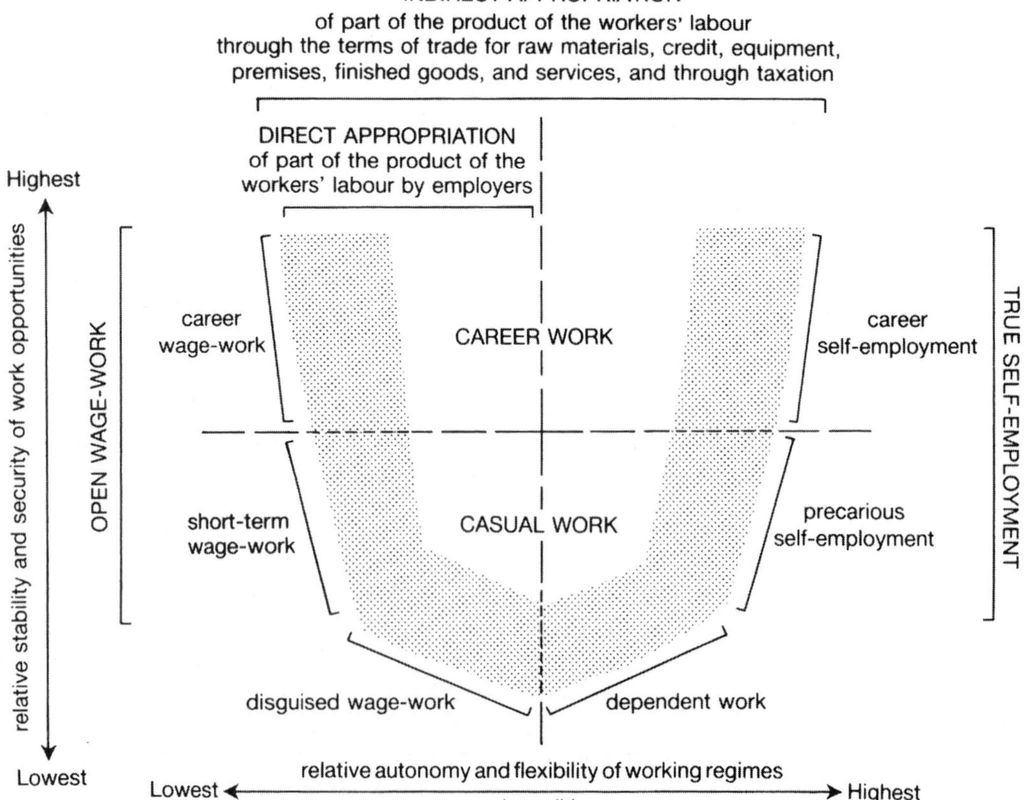

Figure 24 The continuum of employment relations.

and Birkbeck, 1981). According to our estimates, only 40–45 percent of those working in the street occupations were in precarious self-employment, 39–43 percent were disguised wage-workers, 12–15 percent were dependent workers, and only about 3 percent were short-term wage-workers.

Disguised wage-workers and dependent workers have a variety of obligations to employers, contractors, suppliers, property-owners, and usurers, yet do not have the employees' rights specified in government labor legislation. They are "disenfranchised workers." Thus, for example, *motocarro* drivers notionally own their vehicles, but a substantial minority are buying them on hire-purchase, or have borrowed money for the full purchase, so that they are really dependent workers. Similarly, most nightwatchmen who work on the streets watch over a single property, or a group of neighboring properties, guarding the property of the same owner(s) night after night. Though they are officially viewed as self-employed, they are paid a fixed sum by the owner(s) every day, week, or month, and their working relationship is effectively one of disguised wage-working. As a further example, those street-traders who have kiosks or fairly sophisticated mobile stalls usually rent the units or buy them on some form of credit (dependent working), or, as in the case of soft drinks company kiosks and ice-cream company carts, they have been lent these units by companies on the condition that they sell company products on a commission basis (disguised wage-working). Even in the street-trading operations requiring least capital and yielding the lowest incomes, such as door-to-door newspaper selling by small boys and small-scale fruit and vegetable selling in the street markets, dependent working relationships are formed when, because of their own property, the sellers have to obtain the merchandise on credit from other sellers or wholesalers. The condition of this credit is that they must sell that merchandise and pay for it before they can have another lot of merchandise on credit.

The low status and economic dependence of many of the people working in street occupations, combined with the intensely competitive, individualist mentality associated with most of these occupations, contributes to a lack of political and economic organization. The street occupations are moreover fractionalized by the high socioeconomic differentiation within and between occupations, low general levels of education, insecurity, instability, official ignorance and persecution, and conflicting loyalties to a variety of different "masters." Although trade unions and cooperatives exist, these associations are usually small, unstable, and relatively ineffective. Worse still, they are frequently corrupt and/or personalistic, there are often several different rival organizations within a single occupation, and only a small proportion of the total work-force belongs to any organization. Even among lottery-ticket sellers, the most unionized group in Cali's street occupations, only half the total workers were paid-up members of trade unions or occupational associations in 1977. They were divided between five different organizations. Only about one-eighth of street-traders were in unions and associations, and they were split between six different organizations (Bromley, 1978: 1167). Among most of the other street occupations, the proportion of workers in such organizations was even lower, and several occupations had either never had a "formal" organization, for example theft, begging, and knife-grinding, or had only defunct organization which no longer held meetings or collected subscriptions, for example shoe-shining and garbage-scavenging. Viewing the street occupations as a whole, worker solidarity is ephemeral or non-existent, and group interests tend to be subordinated to individual concerns and ambitions.

[. . .]

REGULATION AND PERSECUTION OF THE STREET OCCUPATIONS

It is not surprising, and in total accordance with the legal system, that clearly illegal activities such as property crimes, trading in contraband goods, and the sale of prohibited gambling opportunities are persecuted occupations in Cali. Indeed, many complain that these occupations are not persecuted enough. It is also hardly surprising that such occupations as prostitution and begging, viewed as "disreputable" by most of the population, are officially regulated and suffer from periodic police harassment. (See Bromley, 1981.) In the case of prostitution, however, official attitudes are decidedly ambiguous, and there are many complaints that organizers of prostitution and upper-class prostitutes are free from harassment, while lower-class prostitutes are frequently persecuted.

The intervention of the authorities in occupations which are not clearly criminal, immoral, or anti-social according to conventional, elite-defined standards, is much more complex and diverse. In general, the concern is to regulate activities by introducing checks and controls on prices, standards, and locations, and by limiting entry to the occupations. Government personnel are appointed to enforce these regulations, and penalties are prescribed for offending workers. Thus, for many street occupations, e.g. the operation of a street stall; the commercial use of a *motocarro*, horse drawn cart, or a handcart; and shoe-shining, registration procedures have been introduced, and regulations have been made as to when, where, and how these occupations should be practiced. [. . .] In reality, however, these regulations are excessively complex, little known, and ineffectively administered, resulting in widespread evasion, confusion, and corruption.

[. . .]

In general, those who administer law and order on the streets complain that there are so many people working in the street occupations, and that there is such widespread ignorance and disrespect for the official regulations, that controls must be very selective. The main objective is usually "containment": to hold down the numbers of people working in the streets of such priority areas as the central business district, the upperclass shopping-centers and residential zones, and the main tourist zones.

Although there are occasional cases of assistance by the authorities to street occupations, as when help was given in improving street stalls and providing uniforms for street-vendors and shoe-shiners at the time of the Pan-American Games in Cali in 1971, official intervention in the street occupations is essentially negative and restrictive, rather than supportive. The basis of government policy is that "off-street occupations" should be supported in the hope that they will absorb labor from the street occupations. However, this objective has not been achieved because insufficient investment funds have been mobilized, and because investment has concentrated in areas which generate relatively few work opportunities. In the meantime, the street occupations have often been persecuted, and opportunities to improve working conditions in these occupations have generally been neglected.

Street occupations conflict strongly with the prevailing approaches to urban planning. Although Cali has a warm, dry, and congenial climate for economic activities and social interaction in the open air, city planners have been concerned with reserving the streets for motorized transport and short-distance pedestrian movement, and concentrating economic activities into buildings. In general, no special provision has been made for street occupations, and restrictions on *motocarros*, non-motorized transport, and the sale of goods and services in the streets have been imposed to reduce traffic congestion, road accidents, and the incidence of these occupations. Cali is officially twinned with Miami and has strong links with other North American cities. Urbanistically, Cali is being planned along North American lines as a sprawling automobile-dependent metropolis, and the street occupations are, from the planners' point of view, an unfortunate embarrassment.

[. . .]

In summary, therefore, working in the streets encompasses a very wide range of service occupations, almost all of which can be described as survival strategies for those who work in them, many of which can be described as necessities for those they serve, and for the efficient functioning of the contemporary national economy, and a few of which can be described as unavoidable evils. In policy terms, the most urgent requirement is to generate more work opportunities both outside the street occupations, and in the more necessary street occupations, so as to divert manpower from anti-social and/or parasitic occupations, and to improve the general range of income opportunities available to the urban poor. It is also necessary to adopt a more positive series of policies towards those street occupations which are not clearly anti-social or parasitic, simplifying rules and regulations and administering them more equitably, providing work-places and sources of credit and training for workers, and encouraging the formation of workers' organizations without co-opting them into the web of government. Unfortunately, of course, there can be no assurance, either in Colombia or in most other Third World countries, that the sort of government which would adopt such policies will assume power in the near future. In such circumstance increased worker consciousness and the mobilization of those working in the street occupations are crucial to press for more favorable government policies and to develop alternatives to the present dependent links with employers, contractors, suppliers, usurers, and the owners of sites and equipment.

NOTE

1 Editors' Note: While working capital required for street enterprises varies widely by place and nature of work, and costs would have gone up since this article was written, low costs of entry remain very important.

BIBLIOGRAPHY

Anderson, S. (1978) *On Streets*, Cambridge, MA: MIT Press.

Birkbeck, C. (1978) "Self-employed proletarians in an informal factor: the case of Cali's garbage dump," *World Development* 6: 1173–85.

Bromley, R. (1978) "Organization, regulation and exploitation in the so-called 'urban informal sector': the street traders of Cali, Colombia," *World Development* 6: 1161–71.

Bromley, R. (1981) "Begging in Cali: Image, reality and policy," *International Social Work* 24, 2:22–40. Reprinted (1982) in *New Scholar* 8:349–70.

Bromley, R. and Gerry, C. (1979) "Who are the casual poor?" in R. Bromley and C. Gerry (eds.), *Casual Work and Poverty in Third World Cities*, Chichester: John Wiley, pp. 3–23.

De Soto, H. (1989) *The Other Path: The Invisible Revolution in the Third World*, New York: Harper and Row.

Gerry, C. (1985) "Wagers and wage-working: Selling gambling opportunities in Cali, Colombia," in Ray Bromley (ed.), *Planning for Small Enterprises in Third World Cities*, Oxford: Pergamon, pp. 155–69.

Gerry, C. and Birkbeck, C. (1981) "The petty commodity producer in Third World cities: Petit bourgeois or disguised proletarian?" in B. Elliott and F. Bechhofer (eds.), *The Petit Bourgeoisie: Comparative Studies of the Uneasy Stratum*, London: Macmillan, pp. 121–54.

Ghersi, E. (1991) "El otro sender o la revolucion de los informales," in G. Marquez and C. Portela (eds.), *Economia Informal*, Caracas: Ediciones IESA, pp. 43–64.

ILO (International Labor Office) (1972) *Employment, Incomes, and Equality: A Strategy for Increasing Productive Employment in Kenya*, Geneva: ILO.

Sethuraman, S. V. (1976) "The urban informal sector: Concepts, measurement and policy," *International Labor Review* 114, 1: 69–81.

"Anchoring Transnational Flows: Hypermodern Spaces in the Global South"

Essay written for *Cities of the Global South Reader* (2015)

Sudeshna Mitra

INTRODUCTION

Increasingly cities in the South are disrupting contemporary imaginations of modernity rooted in the global North by creating futuristic, hypermodern spaces and built form. Dubai has become (in)famous for materializing real estate fantasies and creating impossible landscapes of land on water and snowscapes in a desert. Rio, Beijing and Cape Town have won bids to host the World Cup and the Olympics and followed through with a rewrite of their urban landscapes to match and exceed expectations of spectacle associated with these events. These cities are competitively, even chauvinistically, jumping forward on the development trajectory. The commentary from the North remains skeptical and eager to expose deficiencies in the South's new facade (particularly evident during the Beijing Olympics), while claims and projects in the South are ready to be post-history and write their legacy in terms of the future, rather than the past. These claims and counter-claims highlight (1) the spatial/social dichotomies inscribed within competitive practices of hypermodernity and (2) the reemergence of place and place-making in contemporary globalization research. This chapter analyzes place-making efforts underlying production of hypermodern spaces in the global South and highlights that these spaces reveal public and private negotiations to anchor transnational flows of capital, and institutional preference given to elite desires at the cost of social displacements, urban fragmentation and increasing inequality.

In the late 1960s and 1970s, countries in the global North faced a crisis of "over-accumulation" (Harvey 1981, 2001) characterized by production overcapacities, mass redundancies and domestic investments stagnating and yielding lower profits. This encouraged industrial relocation to the global South, where labor was cheaper and new untapped markets offered the potential for increased profits. The "classic" international division of labor, in which products for the international market were manufactured in the industrialized nations of the world, gave way to the New International Division of Labor (NIDL) where profitable production for the world market became possible in developing countries (Fröbel et al. 1982). Efficient management of production networks dispersed across the world were facilitated by advances in information and communication technologies (ICTs) and made available in the global South via "technology transfers" (Ramachandraiah et al. 2008). Present claims to hypermodernity in the global South may be traced back to these diffusions of industrial and technological capacities. This chapter begins by describing how ICTs affected urbanization in the global South, then discusses three types of hypermodern space that are now becoming important: (1) technology parks, (2) privatized cities and (3) aspirational new urban skylines and business districts. All three types of space represent state attempts to spatially anchor transnational flows of money through spatial and fiscal guarantees, including land acquisition. Large-scale displacements and compensation disputes have made many such projects controversial. These spaces highlight increasing financialization in the global economy, a shift from economic investments (predominant in technology parks) towards real estate investments (predominant in privatized cities and new futuristic downtowns) and growing aspirations in the global South to create destinations that represent the advance guard of globalization, rather than replicating the spaces and trajectories of the global North. The economic reasoning for creating certain types of exclusionary spaces such as technology parks and futuristic downtowns is often explicit; nonetheless, there are also strong real estate motivations behind the production of these spaces. The line between economic objectives and real estate motivations is often strategically fuzzy in both policy discourse and project marketing. These spaces highlight the emergence of a new class of elites in the global South, who value transnational mobility and are keen to participate in a "global" economy and enjoy a "global" lifestyle, including low density housing and work environments, landscaped areas, higher quality of infrastructure and consumer-friendly governance (public and quasi-public). International consultancies, private developers and corporate lobbies have reinforced such global inter-referencing. Governments have accommodated elites via new enclaves and new exclusions imposed on urban public spaces.

ICTS AND URBANIZATION IN THE GLOBAL SOUTH

Advances in ICTs made management of post-Fordist, global commodity chains (Gereffi 2001) efficient. Many imagined the end of place-based agglomeration and a future of virtually connected "telecommuters" (Castells 1989). In reality, some cities in the global North assumed nodal higher order service functions (Sassen 2001) and cities in the South joined the network with time. Singapore, Hong Kong, Dubai as entrepôts are early examples. However ICTs also made space "slippery" (Markusen 1996), especially for lower-order urban centers, acting as manufacturing hubs. Substituting one manufacturing location with another became easier with standardized production processes and buyer driven commodity chains, where retailer–manufacturer relationships were mediated by contracts and third party buyers. Enclaved industrial zones, designed to increase competitiveness of locations in the global South, through infrastructure, competitive rentals, business-friendly regulations and tax concessions, encouraged investment mobility by reducing place-based "frictions."

However, ICTs are more than just management tools. They comprise an independent set of goods and services, with significant potential for innovation and profits, as the IT-led economic boom of the 1990s highlighted. Standardized processes to manage global production using ICTs encouraged mobility in traditional manufacturing, but set the stage for IT "offshoring," following Vernon's (1966) international product cycle model for diffusion of entrepreneurial activities from the global North to the global South. This created new opportunities for the South, in sectors typically considered the domain of "developed" economies. It also led to a new social division of labor, and the emergence of a new generation of elites in the global South.

The Asian Tiger economies, Taiwan, South Korea, Hong Kong and Singapore, joined global high-technology goods and services production in the 1960s and 1970s. China opened up to foreign investments in the 1980s. However, most high-technology manufacturing and/or knowledge economy jobs moved to the global South in the last two decades of the twentieth century. Imaginations regarding the global South's development possibilities irretrievably changed when resource-poor countries, such as India, emerged as globally significant IT offshoring destinations. Theories of global convergence (Friedman 2006) and agrarian economies "leapfrogging" directly to service-led economies (Amsden 1989), bypassing traditional manufacturing, proliferated and became part of mainstream globalization discourse.

Now with over four decades of public and private investments many "developing" countries possess competitive advantages in emerging high-tech industries, such as green technologies. The National Science Board (2012) highlights that the US is rapidly losing ground to Asia in new research, patents and jobs. Between 2000 and 2010 high-tech manufacturing in the US declined by 28 percent. Many of these jobs moved to the "Asia 8," that is, China, Taiwan, India, Indonesia, Philippines, South Korea, Thailand and Malaysia. This highlights a change in the world economy, which is historically significant, and neither spontaneous nor inevitable. Governments in the global South pursuing competitive policies have played a central role in this global industrial shift, and contemporary claims to hypermodernity in the global South reveal a sophistication of the same project. However, governments are now targeting not just production capital, but also transnational finance capital. Technology parks, privatized cites and new downtowns are now designed to offer higher order financial services and facilitate greater transnational capital and personal mobility.

Place-making has been important to the project of wooing transnational investors, and governments have used various forms of hypermodern space to incentivize global capital to invest in particular places. Spectacular urban form and exceptional legislative frameworks reveal public–private collaborations underlying the production of these spaces. Increasing financialization in the world economy has meant that governments have had to move beyond industrial incentives in their place-making efforts and offer advanced financial services, easy entry/exit to capital and different forms of debt and equity financing. It has meant a greater role for real estate, with higher potential for profits and risk hedging. In the following sections, three forms of hypermodern space, technology parks, privatized cities and new business districts are analyzed, highlighting shifts in governments accommodating global capital. They also reveal how institutions of inequality are being inscribed into the most globally competitive projects in the global South, creating increasing social and spatial polarity in cities of the South.

HIGH-TECHNOLOGY ENCLAVES

Technology parks or planned "technopoles" (Castells and Hall 1994) to replicate "natural" clusters, such as Silicon Valley, have been central to regional economic strategies to attract technology firms around the world by (1) offering conducive physical and fiscal environment (specialized infrastructure, affordable land, tax concessions) and (2) promoting innovation through clustering. Innovation is a primary motive in the global North and high-technology enclaves are often linked to research universities. Science Parks in Taiwan, Singapore, Hong Kong and now China share these innovation objectives and incorporate research universities. However, in many countries of the South, when surrounding urban infrastructure and civic amenities are poor, infrastructure rather than innovation remains the key advantage of technology parks. High-tech enclaves allow focused public investments, over long periods of time, to raise infrastructure levels to globally competitive levels. For example, the 1,000 acre Hsinchu science park in Taiwan represents an investment of US$483 million over a 15-year period (PricewaterhouseCoopers India 2008). Innovation is also a low priority in technology parks servicing the offshoring market. Offshoring typically comprises routine and standardized software services, Business Process Outsourcing (BPO) and IT-enabled services, such as call centers, and accounts for a very small percentage (5% of expenditures) of the global software and IT services market. However, it is significant for countries such as India, Israel and Ireland and more recently Eastern Europe, Vietnam and China (Carmel and Tjia 2005).

The US-style technoburb is an important spatial blueprint for technology parks in the global South. Characteristics such as low densities, campus-style developments and gated cities targeting a specific economic sector that emerged from a particular US history of suburbanization and technological innovation have become "best practice" norms with the global diffusion of IT companies based in the United States, a "reverse brain drain" (Saxenian 2002); mobility of IT workers; and inter-referencing. A report on "best practices" for IT Parks (2008) commissioned by the Global ICT Department of the World Bank Group, in partnership with Pricewaterhouse Coopers, makes a case for inter-referencing. The report contends: Given the potential complexity of these projects and scope of required investments, the growing interest of governments in developing and transition economies in designing and promoting such projects, and their needs for policy advice and financial support from the donor community, there is a pressing need for a synthesis of best practices and lessons learned (both from success and failure).

International development agencies encourage the use of enclave economies to promote "non-traditional" sectors in developing countries (UNDP 2004).

However, materializing US-style technoburbs in the peripheries of large and very dense cities of the global South has been both difficult and controversial, requiring the use of Eminent Domain and necessitating large-scale displacements, intensive investments and new authoritarian regulations to enforce physical separation from surrounding areas via enclaving. Numerous cases of public protests and violence have been recorded (Bunnell 2002; McDonald 2008;

Shatkin 2011). Technology enclaves are also designed for economic separation from the local economy. Economic exchanges between enclaves and surrounding regions are regulated and monitored, preventing local backward and forward linkages. Meanwhile global connectivity is encouraged via "smart" technologies and upgraded physical connections to international airports. Legislation accompanying technology parks has been designed to permit greater transnational mobility and flexibility to accommodate the "global" technology worker (Bunnell 2002; Van Grunsven 2008). The MSC Malaysia and the "One-North" in Singapore (Van Grunsven 2008) are both examples of technology enclaves with special provisions that facilitate hiring of knowledge economy workers from around the world. Massey et al. (1992) argue that ICTs have reinforced a more unequal and polarized social division of labor, with high-technology workers enjoying a privileged status, because of their specialization and relative transnational mobility, which separates them from manufacturing workers and other service sector workers. Massey's argument regarding technology workers may be extended to elites in the high-end FIRE sectors, that is, finance, insurance and real estate.

Massey et al. (1992) also argue that science parks are primarily real estate ventures. This argument may be extended to other forms of high-technology enclaves as well. In the global South many high-tech enclaves are created through public–private partnerships and private developers enjoy incentives such as easier access to land, tax concessions, more ready access to loans and simplified processes for project permissions. Thus developers are in the unique position of being able to target a high-profile corporate rental market, while enjoying government support for marketing to the right profile of clients, as well as government incentives. Sometimes high-tech enclaves are self-contained projects, with high-end housing, retail, etc., along with office space, which further enhances the potential of these projects acting as exclusive real estate developments, enjoying government guarantees and incentives. Many Indian IT corporates have ventured into real estate development to diversify their investment portfolio, leveraging on their preferential relationship with government agencies. Prominent national/regional developers are involved in the development of technology parks and/or "smart" cities across Asia, Africa and Latin America.

PRIVATIZED CITIES

Privatized cities are another form of hypermodern space becoming increasingly common in the global South, receiving both investor attention and government support. The underlying design objective of privatized cities is to create self-sufficient urban enclaves, separated from the chaos and congestion of the main city for a target market that can afford to pay for a higher quality urban experience. In Latin America privatized townships are advertised as "crime free" zones. They range in size from a few hundred acres to several thousand acres and are often located on urban peripheries. "Self-sufficiency" ranges from captive access to water and power sources, to elaborate mixed-use developments, combining residences, work areas, shopping, hospitals, schools, etc. (Shatkin 2011). New tech-cities across Asia and Africa, combine work-leisure-home functions, along with hotels, conference centers and golf courses. Cities targeting demand for second homes, high-end transnational tourism, expatriates and retirees interpret "self-sufficiency" even more elaborately. The exclusive Aamby Valley township (10,000 acres) in western India has its own airport, which was personally inaugurated by India's minister of civil aviation (Singh 2010).

Facilitating global connections and maintaining local exclusivity are important features of privatized cities. They reveal desires to secede from the local region (Graham and Marvin 2001). Governments facilitate such secession by creating separate development areas, separate planning and development authorities, and allow their "borders" to be protected by zoning, public policing and private security. Global connectivity is reinforced. In India, new privatized cities on the peripheries of major cities are connected to international airports via high speed and exclusive roads. These roads circumvent the main city's traffic, urban forms and mix of formal and informal economies. The sophisticated illusion of convergence that such design normalizes is evident in Friedman's (2006) evocative thesis regarding the "World is Flat," based on his experiences of Bangalore's "globalized" peripheries. The concept of "aerotropolis" (Kasarda and Lindsay 2011), taking root across various locations such as Dubai, Durban, Hyderabad and Hong Kong, imagines a future of self-contained "smart" cities around airports, maximizing connectivity for a mobile group of elites and facilitating greater segregation with the main city and those not directly linked to the global economy.

Government support for privatized cities is explicit when they are vehicles for investments. State guarantees may include tax incentives, off-site infrastructure and help with land acquisition. City-scale Special Economic Zones, such as Shenzhen in China, and regional initiatives, such as Malaysia's multimedia super corridor (Bunnell 2002), were early examples of governments willing to create large-scale zones of exclusion for investors, with simplified systems of governance. "Charter Cities," a concept floated by Paul Romer (2010), highlights the current transition towards "laissez-faire" cities, created through government dispensation, where normal rules of law are suspended for a select few. In 2012, the Honduras Supreme Court ruled against a Memorandum of Understanding (MOU) between the investment group MGK and the Honduran government to build three privatized cities based on the charter city concept, all with their own police, laws, government and tax systems. The cities were part of a new policy to create Special Development Regions (REDs). Shatkin (2011) argues that with private developers taking a central role in "planning, development and regulation," privatized cities represent a privatization of planning.

Privatized cities highlight both North–South flows and new South–South flows of global capital. Goldman (2011) highlights that post-recession, private equity including US hedge funds have become interested in opportunities in the global South. In Russia, new townships are being proposed by Chinese investors. Renaissance Partners, a Russian property developer, has announced plans to build a private city outside Nairobi (Ombok 2012). South Africa is attracting Chinese, Indian and Russian investors (South African News Agency 2013). Land has become a tool to secure financing, either as collateral or through securitization of future real estate returns, creating new transnational patterns of financial growth and crises. For example, the Emaar group, one of Dubai's largest developers, has delayed repayments on debts in excess of $1.2 billion in India, in the wake of the real estate market crash in Dubai (Karmali 2009; Seth 2013).

Privatized townships are speculative. They use government guarantees to target the highest end of the market. However, investors, particularly corporates typically not involved in construction, are also attracted to these projects as they offer opportunities for diversifying investment interests and hedging of risk in investment portfolios by including land, which has a relatively stable long-term intrinsic value. Such long- and medium-term strategies mean that land may be consolidated and built-up space constructed without users actually buying and/or occupying the space. Also investors may be influenced by each other's actions and projections, increasing the likelihood of speculative bubbles, driven by market sentiment rather than tangible demand. Fears of a real estate bubble in China have escalated since a change in government policy allowed people to buy homes. Dozens of new cities, such as Zhengdong New District in central China, are being built with government support, using foreign design consultants and architects. These cities remain mostly unoccupied as middle- and upper-income families in China buy multiple apartments for investment purposes, lacking other viable investment options (Xue et al. 2013). Infrastructure maintenance is both a question and a challenge in these cities. Further, Shatkin (2011) highlights that many private cities in Asia, such as the Dankuni township, outside Kolkata, remain undeveloped as actual demand does not materialize and/or projects encounter protests from those being displaced.

These cities reveal new facets of urbanization in the South. They highlight a new generation of spatial demands from urban elites, with lifestyle aspirations and fear of crime and theft, which overlap with those of urban elites around the world. Spatial imaginations of a "good" living environment, materialized in these cities, reveal transnational inter-referencing. The term "privatized" highlights a contradiction—that even though many of these new ventures are private, their implementation is not restricted to the private realm. State support for these cities reveals a new chapter in neoliberal urban development. As "laissez-faire cities" they are zones of exception, with exceptional privileges for those that can afford it. Meanwhile in the main cities, informal sectors and sectors not on the cutting edge of global competition are marginalized and pushed towards illegality, as their access to legitimate spaces that they can use and occupy is controlled and cut off. Privatized cities reveal new flows of global finance and private equity and new compromises that governments are willing to negotiate to attract and spatially anchor external investors and create destinations attractive to a new class of urban resident, with global aspirations regarding life and work.

NEW URBAN SKYLINES AND BUSINESS DISTRICTS

New iconic skyscrapers and revamped/new business districts are some of the most visually stunning, explicit and public statements of globality and hypermodernity in the global South. Competitive construction of buildings claiming to be the "tallest" in the world provide a window into the global South's aggressive project to claim hypermodernity. Of the ten tallest buildings in the world, only Willis Tower in Chicago is located in the global North. In May 2013 the One World Trade Center, New York City, joined the list at third place. As of 2013 the Bhurj Khalifa in Dubai, at more than 2,700 feet, was the tallest building in the world. The next generation of the world's tallest buildings will be in Shenzhen, Beijing, Wuhan, Tianjin, Kuala Lumpur, Seoul, Jakarta and Jeddah (Galante 2012). Downtown rewrites are evocative testaments of elite desires to inscribe cultures of consumption and spectacle into the urban fabric of cities and claim a place on the global stage. They reveal government intentions to create global destinations, private intentions to claim global presence and strategies, both public and private, to fulfill particular global economy functions. Dubai, Singapore and Hong Kong have emerged as elite shopping destinations. Cape Town, Rio, Mumbai and Hong Kong are regional gateways. Bangalore was the first IT offshoring hub in India and Abu Dhabi is attempting to become the cultural capital of the Emirates. The extent and success of downtown rewrites also stand testimony to the relative degree to which public and private intentions become successful. Through the example of the Petronas Towers (once the tallest building in the world) in Kuala Lumpur, Ong (2011) highlights Prime Minister Mohamed Mahathir's attempts to "hyperbuild" a skyscraper to proclaim Malaysia's global relevance and bring in investors. With the communication technology that defined the towers becoming obsolete and Malaysia's cyber-aspirations remaining mostly unfulfilled, the towers have become an ironic testimony to a particular conjuncture of Malaysia's history.

These rewrites also represent exclusionary decision-making processes and often become lightning rods for conflicts between competing urban desires regarding possible urban futures. Ong (2011) analyzes the new CCTV tower in Beijing and highlights the fractures between the "global" design sensibilities of Koolhaas, materialized through the patronage of the Chinese government, and popular cultural interpretations. In a country where public symbolism is very important, Ong highlights that the CCTV tower's imposing and iconic design has attracted widespread critique against state power and the role of "starchitects" furthering an autocratic regime's propaganda. Often downtown rewrites are made possible by violently suppressing alternate claims to produce urban space and physically erasing spaces produced by social, political and economic factors with longer local histories. This is visible in the remaking of Cape Town for the soccer World Cup and militarization of the police against Rio's favelas, which are standing both visually and philosophically in the way of Rio claiming its place as a global destination by hosting the World Cup and the Olympics. Remade downtowns offer a sharp contrast to the rest of the city and highlight dualities between old and new economies. The practice of inter-referencing is relevant here too. However, envisioning new hypermodern downtowns often engages regional benchmarks. This is particularly true in Asia, where Shanghai (Lee 1999; Roy 2009), Singapore and Dubai have become preferred benchmarks. Regional emulation highlights the eagerness of countries to break from trajectories set in the global North. It also highlights the growing importance of South–South capital flows. Singapore has become an important regional adviser for urban projects (Huat 2011) and Dubai-based developers are important investors in South Asia.

Revitalized downtowns are often at the heart of political projects to insert cities into new financialized flows of capital and capture gains. Investments in "emerging" economies are, however, often seen as high-risk, high-return ventures, especially where real estate is involved. The risk is allayed in part by governments in the global South being more willing to accommodate new forms of investing as they pursue capital. Projects implicated in the East Asian financial crisis offer an example of high-risk ventures in which governments were implicated. The crisis has been analyzed both as a failure of due-diligence on the part of national governments to sift out speculative real estate ventures, as well as an outcome of political "cronyism," which allowed risk to be harbored within urban markets via state patronage. Both analyses focus on the role of the government in underwriting high-risk investment, even as the political nature of harboring investments to gain global relevance and concentrate capital in the hands of a small elite remains at the forefront. Newly made downtowns often accommodate the most ambitious and riskiest public–private ventures.

CONCLUSION

Diffusion of ICTs, IT offshoring and the mobility of finance capital have all contributed to the production of hypermodern spaces in the global South. However, as discussed, these spaces are not "natural" market responses to increased flows of technology and finance. They are material elements of competitive economic and political agendas of local, regional and national governments across the South, strategically spatializing transnational flows to gain global relevance, in sectors perceived to have the highest growth and value increase within the global economy. Public guarantees, including infrastructure, user-friendly regulatory processes, zoning and policing, tax concessions and new debt-related project financing support these spaces. These spaces reveal competitiveness, but also aspirations to break from linear development trajectories and create new global destinations. However, as presented above, these state strategies have not always been coherent and often have comprised multiple spatial and policy initiatives at cross-purposes, resulting in partial success and/or failures. The production of hypermodern spaces has been closely linked to segregation, speculation and protest. Investment-rich enclaves juxtaposed against severely under-invested urban areas are not just the incidental landscapes of the global South, but rather strategically instituted dichotomies, implemented through infrastructure and investment priorities, public–private partnerships and new development regimes. These instituted patterns of development contradict claims that we are collectively moving towards a future where the "world is flat" (Friedman 2006). Graham and Marvin (2001) argue that dualistic infrastructure policies created to accommodate the needs of investors, while ignoring basic needs of the poor, have led to "splintered urbanism." Goldman (2011) uses the term "speculative urbanism" to highlight the willingness of governments to partner with investors on risky, but ambitious ventures in "order to overcome the 'stasis' of urban life with a globalized and virtuous entrepreneurialism" (2011: 230). The instituted dualities mean new systems of surveillance. Yet, ultimately projects of hypermodernity remain only partially successful, under constant challenge from alternate claims to produce urban space and stymied by internal risks and over-leveraged ambitions.

REFERENCES

Amsden, A. (1989) *Asia's Next Giant: Korea and Late Industrialization,* New York and Oxford: Oxford University Press.

Bunnell, T. (2002) "Multimedia Utopia? A Geographical Critique of High-Tech Development in Malaysia's Multimedia Super Corridor," *Antipode,* 34(2): 265–295.

Carmel, E. and Tjia, P. (2005) *Offshoring Information Technology: Sourcing and Outsourcing to a Global Workforce,* Cambridge: Cambridge University Press.

Castells, M. (1989) *The Informational City: Information Technology, Economic Restructuring, and the Urban-Regional Process,* London: Blackwell.

Castells, M. and Hall, P. (1994) *Technopoles of the World: The Making of the 21st Century Industrial Complexes,* New York: Routledge.

Friedman, T. L. (2006) *The World is Flat: A Brief History of the Twenty-First Century,* New York: Farrar Straus & Giroux.

Fröbel, F., Heinrichs, J. and Kreye, O. (1982) *New International Division of Labour: Structural Unemployment in Industrialised Countries and Industrialisation in Developing Countries,* Cambridge: Cambridge University Press.

Galante, M. (2012) "Move Over, Burj: The Next Generation of Skyscrapers is Coming," Business Insider. Available online at http://www.businessinsider.com/tallest-proposed-buildings-in-the-world-2012-5?op=1 (accessed 12 May 2013).

Gereffi, G. (2001) "Shifting Governance Structures in Global Commodity Chains, with Special Reference to the Internet," *American Behavioral Scientist,* 44(10): 1616–1637.

Goldman, M. (2011) "Speculating on the Next World City," in A. Roy and A. Ong (eds.), *Worlding Cities: Asian Experiments and the Art of being Global,* Hoboken, NJ: Wiley-Blackwell.

Graham, S. and Marvin, S. (2001) *Splintering Urbanism, Networked Infrastructures, Technological Mobilities and the Urban Condition*, London: Taylor & Francis.

Harvey, D. (1981) "The Spatial Fix—Hegel, Von Thunen and Marx," *Antipode*, 13(3): 1–12.

Harvey, D. (2001) "Globalization and the 'Spatial Fix'," *Geographische Revue*, 2(2): 23–30.

Huat, C. B. (2011) "Singapore as Model: Planning Innovations, Knowledge Experts," in A. Roy and A. Ong (eds.), *Worlding Cities: Asian Experiments and the Art of Being Global*, Hoboken, NJ: Wiley-Blackwell.

Karmali, N. (2009) "Indian Trouble for Emaar," Forbes.com. Available online at http://www.forbes.com/2009/12/03/dubai-emaar-india-development-investigation.html (accessed 12 May 2013).

Kasarda, J. and Lindsay, G. (2011) *Aerotropolis: The Way We'll Live Next*, New York: Farrar, Straus and Giroux.

Lee, L. O. (1999) *Shanghai Modern: The Flowering of a New Urban Culture in China, 1930–1945*, Cambridge, MA: Harvard University Press.

McDonald, D. A. (2008) *World City Syndrome: Neoliberalism and Inequality in Cape Town*, London: Routledge.

Markusen, A. (1996) "Sticky Places in Slippery Space: A Typology of Industrial Districts," *Economic Geography*, 72(3): 293–313.

Massey, D., Quintas, P. and Wield, D. (1992) *High-Tech Fantasies: Science Parks in Society, Science and Space*, London: Routledge.

National Science Board (2012) *Science and Engineering Indicators 2012*. Available online at http://www.nsf.gov/statistics/seind12/c0/c0s3.htm (accessed 12 May 2013).

Ombok, E. (2012) "Renaissance Ready to Start Building Kenya's $5 Billion Tatu City," *Businessweek*. Available online at http://www.businessweek.com/news/2012–02-09/renaissance-ready-to-start-building-kenya-s-5-billion-tatu-city.html (accessed 12 May 2013).

Ong, A. (2011) "Hyberbuilding: Spectacle, Speculation, and the Hyperspace of Sovereignty," in A. Roy and A. Ong (eds.), *Worlding Cities: Asian Experiments and the Art of being Global*, London: Wiley-Blackwell.

PricewaterhouseCoopers India (2008) "International Good Practice for Establishment of Sustainable IT Parks: Review of Experiences in Select Countries (Vietnam, Russia and Jordan)." Available online at http://www.infodev.org/en/Publication.557.html (accessed 12 May 2013).

Ramachandraiah, C., Van Westen, A. C. M. and Prasad, S. (2008) "Introduction: High-Tech Urban Spaces in Perspective," in C. Ramachandraiah, A. C. M. Van Westen and S. Prasad (eds.), *High-Tech Urban Spaces: Asian and European Perspectives*, New Delhi: Manohar.

Romer, P. (2010) *Technologies, Rules, and Progress: The Case for Charter Cities*. Available online at http://dspace.cigilibrary.org/jspui/handle/123456789/27610 (accessed 12 May 2013).

Roy, A. (2009) "The 21st-Century Metropolis: New Geographies of Theory," *Regional Studies*, 43(6): 819–830.

Sassen, S. (2001) *The Global City: New York, London, Tokyo*. Princeton, NJ: Princeton University Press.

Saxenian, A. (2002) "Transnational Communities and the Evolution of Global Production Networks: The Cases of Taiwan, China and India," *Industry and Innovation*, 9(3): 183–202.

Seth, D. (2013) "Emaar-MGF Targets Delivery of Old Projects in FY13," *Business Standard Online*. Available online at http://www.business-standard.com/article/companies/emaar-mgf-targets-delivery-of-old-projects-in-fy13-112051800064_1.html (accessed 12 May 2013).

Shatkin, G. (2011) "Planning Privatopolis: Representation and Contestation in the Development of Urban Integrated Mega Projects," in A. Roy and A. Ong (eds.), *Worlding Cities: Asian Experiments and the Art of being Global*, London: Wiley-Blackwell, pp. 77–97.

Singh, B. (2010) "Aamby Valley Airport Inaugurates its First Flight." Available online at http://www.mid-day.com/news/2010/feb/040210-aamby-valley-first-flight.htm (accessed 12 May 2013).

South African News Agency (2013) "SA Attracts Chinese, Indian, Russian Investors." Available online at http://www.sanews.gov.za/south-africa/sa-attracts-chinese-indian-russian-investors (accessed 12 May 2013).

UNDP (2004) *World Urbanization Prospects: The 2003 Revision*, New York: United Nations Publication. Available online at http://www.un.org/esa/population/publications/wup2003/WUP2003Report.pdf (accessed 12 May 2013).

Van Grunsven, L. (2008) "Remaking the Economy and High-Tech Spaces in Singapore: A Consideration of One-North," in C. Ramachandraiah, A. C. M. Van Westen and S. Prasad (eds.), *High-Tech Urban Spaces: Asian and European Perspectives*, New Delhi: Manohar.

Vernon, R. (1966) "International Investment and International Trade in the Product Cycle," *The Quarterly Journal of Economics*, 80(2): 190–207.

Xue, C. Q. L., Wang, Y. and Tsai, L. (2013) "Building New Towns in China—A Case Study of Zhengdong New District," *Cities*, 30: 223–232.

SECTION 5
Housing

Editors' Introduction

Ask any world traveler about an important marker of urban landscapes in the global South and they will likely mention the vast areas of the city that are built by people with makeshift material in areas that seem not to be planned by city officials, professional planners, and urban planning agencies. These areas sometimes occupy up to two thirds of a city. Such areas are referred to by different names in different cities and shaped through distinct histories and strategies in different contexts. In Brazil for example, they are called *favelas*; in Mexico *colonias*; in Peru the common term is *ciudades perdidas* (lost cities); in Turkey it is *gecekondu* (built overnight); while in Iran people refer to these areas as *halabi-abad* (built by tin) just as they are referred to as *katchi-abadis* (temporary settlements) in Pakistan. Sometimes these houses and communities are the result of spontaneous acts of individuals; other times they might be the outcome of pre-organized collective action by hundreds or even thousands of households; or by a well-organized "developer" operating outside state regulatory structures. In many instances politically or economically affluent intermediaries play an important role in the formation of these "informal settlements."

The existence of these informal settlements is largely the result of a human need for shelter on the part of people who cannot afford entry into the formal housing market. They shelter themselves wherever and however they can. In many instances, they set up their own shelter on land that is not theirs and is not designated for housing at all. Perhaps the most difficult question is: why doesn't someone do something to improve this situation? Or to put it another way, why do these spaces persist in urban landscapes of the global South? What relationships do these informal settlements have with housing for the middle and upper classes? Where do the people who have benefited disproportionately from the massive shifts in global economic development and urbanization processes live?

This section of the *Reader* attempts to answer these questions and introduce the complex social, historical, institutional and market relations and processes involved in one of the most important aspects of urban spatial formation in the global South–housing.

KEY ISSUES

Why do informal settlements persist?

Urban analysts diverge in explaining the persistence of informal housing in global South cities. An economic explanation stresses that informal settlements make housing available at low cost. Since residents often pay lower than market prices for housing, they can survive on less money. This means employers can pay workers a lower wage and garner more profits for themselves. Thus, according to this explanation, the state tolerates informal settlements as a form of housing subsidy that attracts employers, and domestic and international investment (Burgess 1978). They also result in the spatial co-production of informal and formal settlements–where the poor who offer domestic services and labor that support their middle- and upper-class employers live alongside each other within many cities. More politically oriented explanations see the illegality of informal settlements as a valuable resource for often corrupt government officials and electoral candidates (Harriss 2007; Gilbert and Ward 1985; Eckstein 1977). Politicians use these settlements as their "vote bank" (Solomon 2008). They exchange provision of basic services or enhancing the legal status of the settlements for political and electoral favors. Hence, in this explanation, the state political machinery has a symbiotic relationship with the vast and unmet needs of urban dwellers in these areas and no incentive to get rid of informal settlements. A third explanation for the persistence of informal housing emphasizes how the social relationships that shape and maintain these settlements rely on the free labor of women, and gendered ideologies and practices (Barnes 1999; Miraftab 1996; Moser and Peake 1987). For example, in patriarchal societies women are expected to be the caregivers regardless of whether they work or not and even when men are unemployed. Women and girls have the duty of walking long distances carrying heavy buckets of water to their homes and suffer the major consequences of a lack of basic services–from toilets to electrical connections. Because the absence of services most directly affects women's daily lives, they often take charge of organizing the neighborhood and bringing the community together to respond to these problems (Miraftab 2010). Poor women internalize the cost of non-existent urban services and raise their families by investing their labor at no cost for social reproduction at household and neighborhood levels–what Bakker and Gill (2003) have called the privatization of social reproduction within the domestic sphere by women.

Another key question that has been the subject of considerable debate is whether the persistence of informal settlements is a positive or a negative development. Does the entrepreneurial energy invested in creating housing from below express the power of a positive outcome? The 1970s mark an important shift in housing debates, when academics, policy makers, and practitioners from both ends of the political spectrum started to ask why should these settlements not persist? Informal settlements offer an efficient and affordable shelter to the poor (De Soto 1989); they allow people to build at their own pace and according to their own cultural needs and interest (Turner 1977). Instead of wishing them away, our formal policies and institutions should learn from the informal practices of the poor. Shifting the lens, what was seen as a problem became the very solution. Strategies the poor use in creation of informal settlements were then labeled in policy circles as self-help or incremental housing. These strategies, as Harris explains in the article that follows, also found institutional support at national and international levels through Sites and Services projects that provided the site and the points of connection for basic services, but let people build whenever and however they wished.

Critics, however, were quick to argue that the support for self-help housing offered an alibi for self-exploitation by the poor, where governments wash their hands of their responsibilities and capital gets the cheap labor it needs for accumulation (Burgess 1978). The Sites and Services projects also caught the attention of feminist critics. In an influential anthology published in 1987, Moser and Peake compiled the first systematic gender analysis of aided housing projects in various cities of the global South. The volume demonstrated that male dominated and "gender-blind" projects had indeed limited women's access to housing through their design and loan requirements. This was particularly problematic considering research in Latin America and Africa that indicated a growing "feminization of poverty" (Pearce 1978) and a concentration of female-headed households that Chant (1985) brought into focus as *Las Olividadas* (the forgotten ones).

Contestations from below

With the onset of neo-liberal development policies and practices in the 1980s, the institutional landscape for housing interventions changed. The shift to market-based strategies is discussed in detail in Richard Harris's article that follows this introduction. A range of new actors and forces emerged. They included non-governmental organizations as well as a large number of grassroots groups, many of them led by women, who have become the protagonists of urban housing for the poor. The relationship between these "new" and "old" actors in the realm of housing for the poor varies and makes for a complex site of power struggle. Some like the Shack Dwellers International (SDI) brought together hundreds of shack dwellers organizations from around the world to inform and influence formal institutions concerned with housing such as UN Habitat or the World Bank (see Harris, this volume; Appadurai, this volume)–a close collaboration that brought them financial support from formal institutions and some argue subordinated their autonomy (see Pithouse 2013). Other grassroots groups formed strong social movements with international links to resist evictions and displacements that were spurred by elitist policies that local and national governments advocated and implemented in their drive to modernize and beautify cities (see Leaf, this volume; Baviskar, this volume; Zhang, this volume).

Today, considering the crisis of housing worldwide, the interaction and solidarity across grassroots movements is stronger than ever before. As people in the global North have experienced their own crisis of foreclosures, homelessness, and eviction, they have increasingly turned for inspiration to the global South with their long histories of organizing for urban land and shelter. The Anti-Eviction Campaign in Chicago and its relationship with the Western Cape Anti-Eviction Campaign in South Africa is a case in point. A similar instance is the International Alliance of Inhabitants (IAI) which was first an umbrella for the global South organizations, but during the 2010 Social Forum held in Detroit IAI extended to include grassroots groups' struggle for housing in the US and Canada (see Right to the City discussion in Smith and Guarnizo, this volume). A global network of grassroots social movements involved in housing is active in a range of activities: from occupying empty houses in Chicago, Philadelphia, and New York City to land takeovers in Brazil and resisting unjust evictions in Cape Town. These organizations, with women often at the forefront, are the protagonists of housing solutions for the poor. They not only bring a pragmatic capacity to address problems of shelter and basic infrastructure for the poor, but also shape new forms of democracy and citizenship and build political capacity.

Housing for the middle and upper classes: forms of enclave urbanism and their production

Even as slums and squatter settlements are common, gated communities and enclave urbanism are also becoming increasingly noticeable across Southern cities as household incomes and inequalities are rising, and institutional landscapes that facilitate housing markets are changing (Harris, this volume). The new forms of enclave urbanism include enclosed city neighborhoods to which public access is controlled, especially at night, by gates and booms erected across existing roads; small housing developments with walls, entrances and security guards; and large housing estates that offer not just a manicured controlled environmentally sound enclave but also security and an appropriate lifestyle package (Salcedo and Torres 2004; Wu 2010; Douglass, Wissink and van Kempen 2012; Hirt 2012; Pow 2012). The early forms of gated community building were linked closely to the desire of elites to segregate and protect themselves from the unruly poor, an experience noted in the North as well (Caldeira 2000; see Blakely and Snyder (1997) and Low (2001) for US examples). More recently it is linked as much to a desire for privately controlled infrastructure provision or "club goods" (Webster 2001; Glasze, Webster, and Frantz 2006; Huang and Low 2008) and to the particular ways in which capital is moving into real estate development (Wu 2000; Hsing 2010a; Goldman, this volume; Mitra, this volume).

Along with enclaves in existing city centers and peripheries, there are examples of entire cities being built anew. The rise of speculative cities as capital moves into real estate development is being noted across the world. Martin Murray (forthcoming) documents projects across African countries where the new city is completely independent of preexisting urban structures. These "City Doubles" are managed and governed as private entities, with their own private services, including security. While banned as unconstitutional in some places–Honduras being the most famous example in 2013–in many countries, especially across Africa, these cities are aggressively pursued by transnational developers to capture the disposable cash and investments not only among international and speculative investors, but also among local elite who will then be able to congregate in the city doubles, oblivious to the other world they leave behind–the world of ordinary cities.

Improvisational urbanism

Despite the bifurcation in the literature between informal housing for the poor versus enclaves for the elite, most parts of the city encompass many types of housing. Moreover, as any keen observer of Southern cities would agree, and as Michael Leaf notes in the article excerpt included in this section, gated communities exist alongside a range of other housing types, often amidst mixed uses. Just as with housing for the poor, they can be located in the city center or in the expanding peripheries and fringes, in what could arguably be called a uniquely Southern suburban type (Harris 2010; Rao 2013). The strict separation that research on housing suggests is less evident at the neighborhood and city level as emergent scholarship on housing for the large numbers of middle and lower middle classes indicates. Examples of such housing include localities such as Ciudad Nezahualcoyotl with a population of over a million in Mexico City (Montejano Castillo 2008), peripheral settlements in Maputo, Mozambique (Nielsen 2011), Hezbollah-dominated areas of Beirut, Lebanon (Fawaz 2009) and auto-constructed (self-built in Portuguese) ownership-settlements in Sao Paolo that Holston (1991, 2010) and Caldeira (2009) describe. As with the informal settlements noted earlier in this introduction, these are all settlements built over time though intricate negotiations. The presence of informal small-scale developers (or a political party as in the case of Beirut) who selectively mimic urban planning principles for plotting and zoning, and install infrastructure and basic services is notable. Auto-construction in lower middle and middle class areas is thus a mode of creating the urban environment that produces an improvisational urbanism whose politics and aesthetics rely on middle class aspirations and on the resolution of everyday problems through negotiation and over time (Caldeira 2009).

Will these forms of improvisational urbanism produce different politics in the city? The question remains wide open. Despite "precarious material and legal circumstances" under which such urbanism emerges in Brazil, Holston (2010) argues that perceived ownership is crucial, since it gives residents a right to the city that allows a different type of insurgent politics to emerge. In China, residents of informal but not necessarily poor neighborhoods, many of which are constructed in city peripheries, are coming together to mount limited challenges against an authoritarian state (Hsing 2010b; Read 2008). How the rising middle classes will or will not change the Southern city–as opposed to the very elite who can afford to wall themselves off–remains a matter of debate (for arguments playing out in one county, India, see Chatterjee 2004; Harriss 2007; Nijman 2006; Kudva 2013; Weinstein 2008).

ARTICLE SELECTIONS

Housing is a multifaceted and complex issue, which we have complicated further by including housing for the majority poor and others into one section. Three readings are included in this section: a commissioned piece by Richard Harris that summarizes the prolific scholarship around low-income housing since World War II; an excerpt from a classic article by Caroline Moser (1992) that focuses on issues of gender; and an early article on suburbanization in Jakarta, Indonesia, by Michal Leaf (1994) that captures many of the debates that continue to structure current work on enclave urbanism.

In the reading that follows, Richard Harris, a prominent Canada-based urban sociologist, synthesizes the debates and challenges that have surrounded the issue of low-income housing in the global South over the last six decades. Harris focuses on the meaning and role of housing, the importance of land markets that shape different housing scenarios in different contexts, and the institutional relationships that shape housing policy outcomes and implementation. He reviews the range of housing possibilities for the poor and the strengths and limitations of what governments have done in the housing field. He also discusses the recommendations that international agencies have made concerning housing, and how these have changed since 1945. The discussion of urban land markets looms large in his review, because as he argues, whether by design or by accident, what happens in one part of the market, for example for middle and upper classes, usually affects all others. In his words, "What happens on the housing scene depends on policies and developments that affect the availability, cost, and regulation of urban land" (Harris, this volume).

While the commissioned selection by Harris offers an updated synthesis and overview of the shifting institutional landscape of housing and land market for low-income city dwellers from World War II to the present, the reprinted article by Moser, now a classic, discusses the foundational issues that make for a gender-sensitive understanding of the housing problem and formal intervention for poor populations. She offers a conceptual framework for understanding women's needs in and contributions to self-help housing projects. Here she examines the assumptions that underlie government housing policies and international agency housing schemes, to show why it is necessary to consider gender. Moser discusses the different needs of women and men in low-income communities, recognizing the varied contributions women make in this process. She builds on Molyneux's (1985) notion of strategic and tactical gender needs to discuss how housing projects addressing women's immediate needs for shelter and services also allow the introduction of broader strategic notions of gender equality. She calls on future policy makers to recognize that housing schemes that deny gender differences (also referred to as "gender-blind") or rely on stereotypes about women's role as "homemakers," ultimately hurt women by limiting the access of low-income women to housing schemes.

The third article is by Michael Leaf, another Canada-based architect/planner. Leaf sought to understand Jakarta's suburbanization as a result of government policy, elite political class collusion, and modernist and developmental ideology that maintained Jakarta and the elite at the center of power relations. Leaf was interested in the gated communities that were being developed in Jakarta's peripheries for the middle and upper classes, and the types of segregated and environmentally unsustainable urban patterns they were producing. The article makes the argument that gating and suburbanization, from the first phase of expansion in the North in the 1920s and 1950s to its spread in the global South starting in the 1980s, should be interpreted "as a product of colonialist and neo-colonialist processes in the world economy." Many of the issues that scholars of enclave urbanism engage with today are present in Leaf's discussion, including voluntary segregation. Leaf is interested in the processes by which the idea of the suburb becomes global even as urbanization patterns everywhere had started to become suburban, a word that is currently eschewed by many researchers in favor of the idea of the periphery (Simone 2010).

REFERENCES

Bakker, I. and Gill, S. (2003) "Global political economy & social reproduction." In I. Bakker and S. Gill (Eds.), *Power, Production and Social Reproduction*, London and New York: Macmillan-Palgrave.

Barnes, T. (1999) *We Women Worked So Hard: Gender, Urbanization and Social Reproduction in Colonial Harare, Zimbabwe, 1930–1956*, Portsmouth: Heinemann.

Blakely, E. J. and Snyder, M. G. (1997) *Fortress America: Gated Communities in the United States*, Washington, DC: Brookings Institution Press.

Burgess, R. (1978) "Petty commodity housing or dweller control? A critique of John Turner's views on housing policy," *World Development*, 6(9–10): 1105–1133.

Caldeira, T. P. R. (2000) *City of Walls: Crime, Segregation, and Citizenship in São Paulo*, Berkeley: University of California Press.

Caldeira, T. P. R. (2009) "Opening remarks, 'At the Peripheries: Decentering Urban Theory,'" University of California at Berkeley, February 5–7.

Chant, S. (1985) "Single-parent families: Choice or constraint? The formation of female-headed households in Mexican shanty towns," *Development and Change*, 16(4): 635–656.

Chatterjee, P. (2004) "Are Indian cities becoming bourgeois at last?" In *Politics of the Governed: Reflection on Popular Politics in Most of the World*, New York: Columbia University Press, pp. 131–148.

De Soto, H. (1989) *The Other Path: The Economic Answer to Terrorism*, New York: Basic Books.

Douglass, M., Wissink, B. and van Kempen, R. (2012) "Enclave urbanism in China: Consequences and interpretations," *Urban Geography*, 33(2):167–182.

Eckstein, S. (1977) *The Poverty of Revolution: The State and Urban Poor in Mexico*, Princeton, NJ: Princeton University Press.

Fawaz, M. (2009) "Hezbollah as urban planner? Questions to and from planning theory," *Planning Theory*, 8(4): 323–334.

Gilbert, A. and Ward, P. (1985) *Housing, the State and the Poor: Policy and Practice in Latin American Cities*, Cambridge: Cambridge University Press.

Glasze, G., Webster, C. and Frantz, K. (Eds.) (2006) *Private Cities, Global and Local Perspectives*, Abingdon, UK, and New York: Routledge.

Harris, R. (2010) "Meaningful types in a world of suburbs." In M. Clapson and R. Hutchison (Eds.), *Suburbanization in Global Society,* Bingley, UK: Emerald.

Harriss, J. (2007) "Antinomies of empowerment: Observations on civil society, politics and urban governance in India," *Economic and Political Weekly*, June 30: 2716–2724.

Hirt, S. (2012) *Iron Curtains: Gates, Suburbs, and the Privatization of Space in the Post-Socialist City,* Oxford: Wiley-Blackwell.

Holston, J. (1991) "Autoconstruction in working class Brazil," *Cultural Anthropology*, 6(4): 447–466.

Holston, J. (2010) *Insurgent Citizenship: Disjunctions of Democracy and Democracy in Brazil,* Princeton, NJ: Princeton University Press.

Hsing, Y. (2010a) *The Great Urban Transformation: Politics of Land and Property in China,* Oxford: Oxford University Press.

Hsing, Y. (2010b) "Urban housing mobilizations." In Y. Hsing and C. K. Lee (Eds.), *Reclaiming Chinese Society*, London and New York: Routledge.

Huang, Y. and Low, S. (2008) "Is gating always exclusionary? A comparative analysis of gated communities in American and Chinese Cities." In J. R. Logan (Ed.), *Urban China in Transition*, Oxford, UK, and Malden, MA: Blackwell.

Kudva, N. (2013) "Planning Mangalore: Garbage collection in a small Indian City." In G. Shatkin (Ed.), *Contesting the Indian City: Global Visions and the Politics of the Local*, Hoboken, NJ: John Wiley.

Leaf, M. (1994) "The suburbanisation of Jakarta: A concurrence of economics and ideology," *Third World Planning Review*, 16(4): 341–356.

Low, S. M. (2001) "The edge and the center: Gated communities and the discourse of urban fear," *American Anthropologist*, 103: 45.

Miraftab, F. (1996) "Revisiting informal sector homeownership: The relevance of household compositions for housing options of the poor," *International Journal of Urban and Regional Research*, 21(2): 303–322.

Miraftab, F. (2010) "Contradictions in the gender-poverty nexus: Reflections on the privatisation of social reproduction and urban informality in South African townships." In S. Chant (Ed.), *The International Handbook on Gender and Poverty: Concepts, Research and Policy*, Northampton, MA: Edward Elgar Publishers, pp. 644–648.

Molyneux, M. (1985) "Mobilization without emancipation? Women's interests, the state and revolution in Nicaragua," *Feminist Studies*, 11(2): 227–254.

Montejano Castillo, M. (2008) "Processes of consolidation and differentiation of informal settlements, case study of Ciudad Netzahualcoyotl, Mexico City," Unpublished Doctoral Dissertation, Faculty of Architecture and City Planning, University of Stuttgart.

Moser, C. (1992) "Women and self-help housing projects." In C. Moser and L. Peake (Eds.), *Beyond Self-help Housing*, New York: Mansell, pp. 53–73.

Moser, C. and Peake, L. (Eds.) (1987) *Women, Human Settlements, and Housing*, London and New York: Tavistock Publications.

Murray, M. (forthcoming) "'City-doubles': Re-urbanism in Africa." In F. Miraftab, D. Wilson and K. Salo (Eds.), *Cities and Inequalities in a Global and Neoliberal World*, New York: Routledge.

Nielsen, M. (2011) "Inverse governmentality: The paradoxical production of peri-urban planning in Maputo, Mozambique," *Critique of Anthropology*, 31(4): 329–358.

Nijman, J. (2006) "Mumbai's mysterious middle class," *International Journal of Urban and Regional Research*, 30(4): 758–775.

Pearce, D. (1978) "The feminization of poverty: Women, work, and welfare," *Urban and Social Change Review*, 11(1–2): 28–36, 78.

Pithouse, R. (2013) "NGOs and urban movements: Notes from South Africa," *City*, 17(2): 253–257.

Pow, C. P. (2012) *Gated Communities in China: Class, Privilege and the Moral Politics of the Good Life*, London: Routledge.

Rao, N. (2013) *House but No Garden: Apartment Living in Bombay's Suburbs, 1868–1964*, Minneapolis: University of Minnesota Press.

Read, B. L. (2008) "Property rights and homeowner activism in new neighborhoods." In A. Ong and L. Zhang (Eds.), *Privatizing China: Socialism from Afar*, Ithaca, NY: Cornell University Press, pp. 41–56.

Salcedo, R. and Torres, A. (2004) "Gated communities in Santiago: Wall or frontier?" *International Journal of Urban and Regional Research*, 28(1): 27–44.

Simone, A. M. (2010) *City Life from Jakarta to Dakar: Movements at the Crossroads*, London: Routledge.

Solomon, B. (2008) "Occupancy urbanism: Radicalizing politics and economy beyond policy and programs," *International Journal of Urban and Regional Research*, 32(3): 719–729.

Turner, J. F. C. (1977) *Housing by People: Towards Autonomy in Building Environments*, London: Marion Boyars.

Webster, C. (2001) "Gated cities of tomorrow," *Town Planning Review*, 72(2): 149–169.

Weinstein, L. (2008) "Mumbai's development mafias: Globalization, organized crime and land development," *International Journal of Urban and Regional Research*, 32(1): 22–39.

Wu, C. T. (2000) "Diaspora capital and Asia Pacific urban development." In G. Bridge and S. Watson (Eds.), *A Companion to the City*, Oxford, UK, and Malden, MA: Blackwell, pp. 207–223.

Wu, C. T. (2010) "Gated and packaged suburbia: Packaging and branding Chinese suburban residential development," *Cities*, 27(5): 385–396.

"International Policy for Urban Housing Markets in the Global South since 1945"

Essay written for *Cities of the Global South Reader* (2015)

Richard Harris

Our media regularly remind us that millions of city dwellers survive without a decent home, or indeed a home at all. As I was writing this essay, newspapers and magazines were carrying reviews of *Behind the Beautiful Forevers* (Boo 2012), a graphic report from inside a Mumbai slum. Such accounts imply that attempts to improve housing in the developing world have failed, perhaps inevitably. Some writers have explicitly reached this conclusion. For example, Mike Davis (2006) has argued that we live in a "Planet of Slums" and that, if anything, the policies of agencies, such as the International Monetary Fund and the World Bank, have made things worse. He holds out little hope that things will change.

The number of people who are ill-housed is almost incomprehensibly large, and growing. The global estimates produced by the United Nations Human Settlements Programme (UN-HABITAT) indicate that, in 2010, 830 million people in the developing world, or about 33 percent of the urban total, lived in slums (UN-HABITAT 2010a: 32).[1] The number was up from 657 million in 1990 and 767 million a decade ago. But the proportion went down dramatically: from 46 percent in 1990 and 39 percent in 2000. This apparent paradox is explained by the fact that the number of people living in urban areas has itself ballooned even as rapid economic change is taking place in some parts of the world. On the average, conditions are improving. In China, for example, over roughly this period, the average amount of dwelling space available to urban residents has increased fourfold, from 6.7 square meters to 28.3 square meters (Man 2011: xi). Such averages must be interpreted with caution: they can be skewed by the experience of a small minority of households that have rapidly become very rich. But certainly most households in urban China now have much more personal space than a generation ago. The sheer size and rate of urbanization makes China exceptional but still hugely significant. And so there are positive narratives, offered by investigative reporters Robert Neuwirth (2005) and Doug Saunders (2010), which suggest that resourceful and ambitious slum-dwellers have commonly been successful in improving their lot, sometimes with the assistance of sound policy.

Ultimately the truth lies somewhere between the pessimistic and optimistic accounts. In their enthusiasm to make their case, Davis, Neuwirth, and Saunders overplay their hands. A more dispassionate view would likely reach more moderate and less striking conclusions. But whether measured in terms of the sorts of statistics quoted above, or of the emerging dynamics of communities, governments, organizations and agencies involved in housing processes, the optimistic account comes closer to the truth.

To appreciate the grounds for optimism we must bear in mind three contexts. First, we need to know something about the actual scene on the ground. What matters more than physical conditions are the processes that produced, and still shape, them. With few exceptions the key processes are those of markets. Accordingly, the first section of this essay sketches the distinctive character of housing markets in the developing world. Second, to understand the role of international agencies we need to consider their varied character, and their evolution in response to geopolitical considerations (section 2). Third, to assess what governments have done in the housing field we need to know the range of possibilities (section 3). This survey includes programs designed to help the middle classes as well as the poor because, whether by design or by accident, what happens in one part of the market usually affects all others. Finally, the fourth section focuses on the international agencies' recommendations on the subject of housing, and how these have changed since 1945. Implicitly' and sometimes explicitly, all sections refer to a fourth context: what happens with housing depends on the availability, cost, and regulation of urban land. The land question and still-wider influences are assessed in a concluding discussion.

The present sketch is neither comprehensive nor fully nuanced. The subject is huge and local experiences vary by nation, region, city size, social class, ethnicity, and gender. I refer to local and national experiences, but my main intention is to highlight broad themes, provoke questions, and indicate where interested readers may find additional, authoritative sources.

URBAN HOUSING MARKETS IN THE DEVELOPING WORLD

In most cities, decisions about who lives where, and under what conditions, are mediated by markets: property rights that affect construction and occupancy are defined and traded, contracts are enforced, commonly by the state, and most strive to minimize transaction costs (Angel 2000: 67–80; Malpezzi 1999; Mooya 2011; Tibaijuka 2009: 20–2). Where a commodity is scarce, rights tend to be carefully defined. This is a characteristic of housing in urban as opposed to rural areas. But there is no single "housing market." Markets are local and varied. Housing and land are immobile, and so the balance of demand and supply varies, with major effects on conditions and price. Locally, housing submarkets—notably, those for the rich and the poor—can operate almost independently. Even more fundamentally, the written and unwritten rules that govern markets vary: each community has norms; each municipality its own regulations; and each nation a legal system. Norms and regulations frame markets, making it possible for them to exist. The state cannot leave the housing market to its own devices: the question is not whether it should be involved, but how.

These observations apply equally in Montreal, Canada, as in Mombasa, Kenya. But Kenyan cities, like others in the developing world, differ systematically from those in North America, Japan, or Europe. Obviously, conditions are poorer in terms of both the dwellings and services such as clean water and waste disposal that make homes safe and livable (and that UN-HABITAT uses to help define slums). Many Montrealers face housing problems but slums barely exist there. For example, in urban Canada in 2001, 99.8 percent of dwellings contained an inside flush toilet (UN-HABITAT 2003: 257). If conditions in the developing world are on average worse, they are also more varied. What is most shocking about the urban scene in cities like Rio de Janeiro, New Delhi, or Jakarta is the juxtaposition of wealth and poverty, penthouses and shacks. To be sure, in much of the developed world, income inequality is considerable and has been growing. Great wealth enables some families to own several large homes, each with many bedrooms, each with its own en suite. But this differentiates them by degree only. Even low-income households—the great majority, anyway—have a dwelling, a bedroom, and a toilet. In the global South, however, prosperity separates the fortunate from the destitute: it produces differences in kind.

Basic contrasts in living conditions reflect equally fundamental differences in market processes. The most important concern house building, finance and, above all, land (Angel 2000: 192–203). Legal ownership of land that is recognized by the state through systems of court-enforced registration offers greater security of tenure, though this does not necessarily extend to men and women equally (Moser 1987: 17–18). In theory such systems provide peace of mind and also enable buyers to use mortgage debt and owners to use property as collateral (De Soto 2000). In practice this may not work since lenders are wary. Because housing is such a large expense, mortgages are the only way that many households can acquire property. Once a dwelling is owned, remortgaging enables owners to cover exceptional expenses that may arise, in the event of job loss or a health crisis, or to acquire capital for a small business investment. Such advantages are undercut where proof of ownership is hard to come by, or disputed. These are common problems in the developing world.

More problematic is the fact that extensive areas—including many but not all slums—are occupied illegally. There are two main types of illegal occupancy: squatting, where residents have no prior right of usage, and semi-legal ("irregular" or "unauthorized") settlement, where they have purchased rights, of ownership or tenancy, in settlements that violate planning or building regulations (UN-HABITAT 2003: 168–72). The latter is most common at the suburban fringe. These forms of illegal settlement have sometimes occurred in the developed world, and occasionally still do (Neuwirth 2005: 177–249; Harris 2012; Mukhija and Monkkonen 2007). But they are uncommon, whereas throughout the developing world they are widespread.

Many Western experts once supposed that irregular and, especially, illegal settlements are inherently anarchic and dysfunctional. Some are. But research has shown that informal methods of regulation and enforcement can sustain effective markets even without state recognition and enforcement—although they can also create opportunities for slumlords (Perlman 1987; Malpezzi 1999: 1795, 1822–4). Recognition may be effectively implied by the government's provision of services or communities may devise their own ways of recognizing ownership and enforcing contracts. For example, in the Pasagarda *favela* of Rio de Janeiro, property transfers and disputes were handled effectively by a residents' association which supplemented state law by using oral testimony from local residents (Santos 1977). Markets require regulation, but this does not have to be provided by the state. Indeed, where informal systems are more responsive to local needs they may be more democratic. The politics of informal settlements, however, is varied (e.g. Benjamin 2004). Where residents have the vote, they acquire clout with civic or national politicians, giving them greater security which may be formalized through "regularization" (Tibaijuka 2009: 172). Of course, this also points to a way in which politicians can buy votes.

If anything, security is even more important for tenants than for owners. High rates of urbanization in the developing world since 1945—first in Latin America, then in Asia and now in sub-Saharan Africa—have depended on rural–urban migration (Saunders 2010). Migrants bring ambition but few savings, and at first most rent or squat. Some may prefer such arrangements. If they have insecure jobs, it can make sense to rent on a short-term basis, rather than buying a property that may take time or be difficult to sell. (Recently this has been a consideration in the United States, too.) Others, especially women in female-headed households, find it convenient to live in centrally located districts where jobs and social services are more accessible, but where property ownership is more difficult (Miraftab 1998). Everywhere tenants commonly make informal, temporary arrangements with landlords, either by choice or through compulsion. One important, if neglected, way in which housing markets in the developing world are distinctive is the extent to which rental contracts are managed informally rather than being subject to state law (Gilbert 2011).

Where so many live in poverty and where land rights are unclear, the scope for commercial building is limited (Angel 2000: 204–19; Keivani and Werna 2001; Tibaijuka 2009: 91–5). Owner- (or self-) builders play the dominant role in all squatter settlements, and in most irregular settlements. For example, in Brazil, 1964–86, self-help accounted for about three quarters of all new dwellings, and it is still responsible for 30 percent of new construction in Latin American cities (McBride and French 2011: 29). Owner-builders scrounge or make building supplies however they can, sometimes refashioning and repurposing materials such as plastic sheets and cardboard packaging. With low and uncertain incomes, and lacking access to credit, they build in stages. A one-room shack acquires a separate bedroom or kitchen. With land at a premium, expansion may become vertical, with an upper room accessed by ladder, a pattern more common in Asia and Latin America than in Africa. Additions are anticipated. Where concrete construction is possible, roofs sprout rebar, awaiting further investments of money and time. Eventually, as in some of the *gecekondus* of Istanbul, multistory dwellings turn owner-built areas into dense, urban neighborhoods.

The results are infinitely varied. In the developed world, we often see entire subdivisions, or at least whole blocks, appear in a single building season. The results look raw at first, but uniform and finished. Where owner-builders reign, no two dwellings are alike: walls may not meet at 90 degrees; buildings are placed variously on each lot; in denser settlements the very existence of distinct building lots is unclear; narrow, meandering streets turn into paths and blind alleys. Such settlements vary one from another. Builders use the cheapest materials available, which may be wood, cement, earth, sun-baked bricks, or the ubiquitous cardboard and plastic sheet. Such settlements are commonly squeezed onto undesirable terrain: steep slopes that are subject to landslides; low-lying swamp susceptible to flooding, or rising sea levels; polluted sites, adjacent to (or actually on) garbage dumps, railroads, and the like. And so the imposing, new international airport of Delhi, India's booming capital, is ringed by shack settlements that crawl up beyond the barbed wire on the 12-foot perimeter wall, declaring their individuality in varied colors, textures, and silhouette profile.

The great theme running through discussions of urban housing in the developing world is the importance of informal processes. This contested and slippery concept reflects a complex reality, but may be defined as activities that occur beyond the purview of the state (UN-HABITAT 2003: 100–4; McBride and French 2011: 74). Sometimes, agents hide transactions within the "black market." In India, for example, because the government levies a non-trivial stamp tax on real estate transactions, buyers and sellers collude and declare only a fraction of the actual sale price. Contractors and building tradesmen encourage customers to pay cash so that they may avoid declaring income. Builders often evade building or zoning regulations, notably on illegally occupied land. This is easiest in the suburbs, where regulations are fewer, and weakly enforced (Harris 2010). Overall, the informal sector is probably responsible for the majority of new dwellings erected globally every year. The existence of informal processes reveals the limitations of state power; by definition, it may seem, they lie beyond the margins of what policy can affect. But it may suit government to turn a blind eye, for example by allowing parallel systems of registration and regulation. This saves both administrative and political trouble. Either way, the informal sector is shaped by the state. The question is not whether governments have policies with respect to this sector, but whether they acknowledge the fact. In a general way, then, these are the circumstances in the housing field that international agencies have sought to shape and improve.

THE NATURE, ROLE, AND REACH OF INTERNATIONAL AGENCIES

We should not take for granted the idea that a foreign agency may offer advice, incentives, or sanctions. Policy is the job of nation states; the legitimacy of international agencies is always open to question. This is especially so because the first such agencies in the global South were colonial powers. Colonial rule, mostly by European powers, entailed the settlement of British, French, German, Belgian, or (briefly) Italian administrators, and so there were two housing policies, one for representatives of the colonizing nation and another for the indigenous population. Administrators tried to keep these two groups segregated, typically on the grounds of promoting public health, but with mixed success (see King, this volume). The most significant of these powers was Britain, notably in India from the late 1800s. By the 1930s, when colonial rule was being questioned, the British defined a new role as promoters of "colonial development," in which housing played a part (Harris and Giles 2003: 169–71). In 1940, this policy was formalized, and in 1947 a development corporation established. The discourse of development—the language and the paternalistic practices that went with it—persisted after most colonies gained independence. It has guided the way academics, politicians, and peoples think about the "developed" and the "developing" world, and about how the former might help the latter to "develop." In modified fashion, this assumption still informs the thinking of international agencies.

Colonial powers could coerce and sanction. In East Africa, for example, from the 1920s to the early 1950s the British discouraged Africans from moving to towns, a policy that influenced conditions in urban and rural areas. Nations and international agencies in the postcolonial era have rarely coerced, but some have offered powerful incentives. For many decades, through trade, investment, and saber-rattling the United States has exerted considerable influence, above all in Latin America; more recently China has made its presence felt in Africa. Influence has also been exerted through international agencies. Some are bilateral: based in, and funded by, a single Western nation, they advise and assist other, specific nations. The most significant was once the United States Agency for International Development (USAID) formed in 1960 as part of President Kennedy's anti-communist Alliance for Progress. This was the height of the cold war, when the Cuban Missile Crisis underlined the immediacy and proximity of the Soviet threat. Other bilateral organizations include the British Department for International Development (DfID), the Canadian International Development Agency (CIDA), the Deutsche Gesellschaft für Internationale Zusammenarbeit (GIZ), the Agence Française de

Développement (AFD) and the Swedish International Development Cooperation Agency (SIDA). These organizations influence national policies by offering loans, advice, and in-kind aid. All have shown interest in housing and related issues. But the most important agencies are multilateral. The first regional agency was the Inter-American Development Bank (IDB). Founded in 1959 in response to regional pressures, the IDB has 48 members and is authorized to make loans to 26 states in Latin America and the Caribbean. As with USAID, United States support was driven by geopolitical considerations. The IDB sought to limit the influence of the Cuban social model in a rapidly urbanizing Latin America, where new urban squatters showed radical tendencies. It advocated a basic needs policy approach, which raised the profile of housing. By the mid-1960s, it was providing two thirds of the international capital for housing across the developing world. Since the 1970s, however, it has been dwarfed by the World Bank, which is global in scope. The World Bank was established in 1944 as a Bretton Woods partner to the International Monetary Fund (IMF). The agenda was to reconstruct postwar Europe. By the late 1950s, it refocused on the developing world. Its mandate has always been to fund capital investment; reflecting the wisdom of the era, at first it ignored housing. After the World Bank's views changed in the early 1970s, and with lobbying from prospective clients, it soon became dominant in the international housing field. Although multilateral, the World Bank speaks with an American accent. It is based in Washington, DC, and has always been headed by a person from the United States. As the communist threat waned, and then effectively disappeared, in the late 1980s and 1990s it promoted a neoliberal agenda, advocating privatization, trade liberalization, and a market orientation. Aligned with the IMF, also Washington-based, and the US Treasury, its pronouncements became part of the so-called Washington Consensus (Kumar 2008: 261; Goldman, this volume). In the housing field, its thinking has sometimes been influenced by US housing programs. For example, in the 1990s, some economists were impressed by the perceived successes of the US housing allowance program (Renaud 1999). Although not a US agency then, its policies have meshed broadly with the perspectives of the United States government.

Other organizations have fewer resources but, being seen as more objective, have wielded moral influence. For example, many non-governmental organizations (NGOs) have taken an interest in housing (Tibaijuka 2009: 159–63). Most have been based in the global North. An early example was the Quaker American Friends Service Committee, active from the 1930s. More recently, Habitat for Humanity, founded by ex-US President Carter, has become active. Increasingly, however, such organizations have emerged in the developing world. Most are local, or at most national, in scope. One of the earliest was the National Slum Dwellers' Federation (NSDF), established in Bombay in 1974. It now claims a membership of about 750,000, two percent of India's slum population (Kumar 2008: 273; Tibaijuka 2009: 156). But a few groups have made global connections. In South Africa in 1996, several national organizations, including the NSDF, formed the Shack/Slum Dwellers' International (SDI) (Kumar 2008: 277; UN-HABITAT 2003: 161). This now links groups in twelve countries. Like other organizations, a major concern has been microfinance, coupled with slum upgrading. Such organizations as these, and the Asian Coalition for Housing Rights (ACHR), based in Bangkok, are loose-jointed, and sometimes work with bilateral agencies such as SIDA (McBride and French 2011: 36). Although NGOs appear altruistic, they have attracted criticism. Davis (2006: 75–6), for example, claims that they articulate an outsider's point of view. Locals learn to manipulate the system, grow dependent on Western agenda and funds, and become agents of soft imperialism. Needless to say, the agencies involved dispute this interpretation.

In a different category is the United Nations, a dispassionate and, along with the World Bank, an especially well-informed agency on the housing scene. The UN comprises a Secretariat and several specialized agencies that deal with health, housing, and so forth. From 1948, soon after its inception, the organization took an interest in urban housing (Harris and Giles 2003: 171). Its commitment has grown, in step with the postwar trend of urbanization. In 1976, the first global conference on human settlements (Habitat I) was held in Vancouver, Canada. This led to the establishment of the UN's Centre for Human Settlements (UNCHS) in Nairobi in 1978. In 2001, this was reorganized as UN-HABITAT, with full program status, a sign that housing's profile had risen. Compared with the World Bank, UN-HABITAT and its consultants place greater emphasis on social criteria as opposed to economic efficiency, but the outlook of the two agencies has often run closely parallel. Inevitably, it too has had its critics (Davis 2006).

UN-HABITAT cannot compel governments to follow its advice. The same is true for other agencies. The World Bank, however, does have "tremendous leverage" (Davis 2006: 70). The funds it controls are a drop in the bucket in terms of global investments, but they can make a significant marginal difference, especially when coupled with IMF guidelines. There are instances where World Bank recommendations have had a significant impact on national housing policy, as happened with the establishment of Calcutta's Metropolitan Development Agency in 1971, and then of sites and services in Madras (now Chennai) (Pugh 1997: 1572; Pugh 2001). More generally, from the 1970s, they have helped shift the terms of international debate in favor of a market-enabling approach, as discussed below (e.g. Datta and Jones 1998: 3). The World Bank's indirect impact may be greater, China being a case in point. The Chinese began opening their economy in 1978, but arguably the week-long Bashan boat conference, organized by the World Bank in 1985, gave this policy a boost and a new focus. It was associated with a report that favored home ownership, soon a government priority. Recently, with the Development Research Center of the State Council, an official Chinese think-tank, the World Bank has prepared another report for China. Typical of the agency's current thinking, this does not highlight housing, but it does include sweeping recommendations about suburban land development and deplores the excessive powers of local governments over rural–urban land conversion as well as the dependence of those governments on the use of land as collateral for borrowing (World Bank 2012: 31). The impact of these recommendations is uncertain. But clearly the World Bank has the ear of even the most powerful governments in the global South.

Despite the power and influence of the World Bank, governments much smaller and weaker than China's have chosen to ignore agency recommendations: in the 1980s, Buenos Aires rejected synchronized advice from the World Bank, the IDB, USAID, and the United Nations (Cohen 2001: 48). In general, governments follow broad shifts in expert advice, but never slavishly (Hardoy and Satterthwaite 1981). While international organizations influence national policies, housing policy is a national responsibility, and is routinely modified or contradicted locally (UN-HABITAT 2003: 178–80). Municipalities enact and enforce many regulations that frame housing markets. They tax property and/or transactions to provide vital housing-related services, just as they enforce building and land use regulations. The successes and failures of national policy depend on such actions, whose successes and failures may in turn shape national policy. In short, politics at the local level matters. Apart from intergovernmental transfers, municipalities rely on property-related revenues, and act accordingly. Urban development can yield large profits and, depending on whether these are reaped by municipalities or by developers, housing markets can be profoundly affected. For all its international profile, then, UN-HABITAT cannot even determine what happens next door to its head office in Nairobi. At most, as one employee suggested, it provide facts and moral ammunition for civil society organizations to "put pressure on their governments" (Neuwirth 2005: 242).

THE POSSIBLE PURPOSES AND FORMS OF HOUSING POLICY

Governments can influence housing markets in many ways. Their choice of policy and program depends on their purpose. In the developing world, it may seem that their main purpose should be to improve conditions for the poor, and that has been a major consideration. But governments must pay attention to the needs of the more prosperous, influential, and sometimes more numerous middle class, including its own civil service. Since at least the Great Depression, governments have used the housing sector to promote or revive the economy. The house building industry is large and labor-intensive, with multiplier effects on furnishings, supplies, and services; it is a lever for growth (Tibaijuka 2009). Most housing policies can serve more than one of these purposes, but not all equally. The following discussion assumes that they are directed primarily at the poor, but it also notes wider effects.

An obvious type of program is to promote new construction, most directly by erecting subsidized public housing (Kumar 2008: 256–7; Malpezzi 1999: 1830; Tibaijuka 2009: 39–40). Its advantage is that the physical conditions of those housed can be immediately improved, emancipating the poor from the unfeeling logic of the market; a secondary consideration, sometimes primary, is that such action is highly visible. The disadvantages are the high per unit cost, prohibitive in housing the very poor. Except in unusual circumstances—as in Hong Kong and Singapore—governments cannot afford to help more than small numbers in this way. The result is horizontal inequity: people in identical situations are treated differently. Public tenants receive a large subsidy while other poor households get nothing. Associated problems arise with tenant selection: politicians may reward their own supporters, kin, or ethnic communities, regardless of income, and use construction contracts as patronage. Comparable programs involve the partnership of state agencies with NGOs or other private-sector groups which can bring additional funding to a program.

Another option for the government is to help amateurs build their own homes. Such programs are known as "aided self-help," which entails the provision of advice, materials, and credit, or "sites-and-services," a related program which provides a serviced building site. By harnessing the owner's labor, per unit costs are reduced; the "buy-in" from occupants is tangible. Neighborly cooperation may also foster community. But there are problems of scaling up from individual projects, and again of horizontal inequity (Cohen 2001: 44; Kumar 2008: 258; Tibaijuka 2009: 41). Another problem arises because projects of owner-occupied homes with secure title are attractive. The resale of properties to affluent households—a type of gentrification, filtering up, or middle-class "raiding"—means that projects may provide only short-term benefits even to those who are directly involved.

Instead of creating a specific project, a different approach is to alter the business environment. The "enabling" approach aims at systemic change (Tibaijuka 2009: 172–7; Pugh 2001: 406–11; World Bank 1993). In some instances this may target the building industry, encouraging the use of new tools and materials, or the adoption of new organizational forms. In other cases the emphasis may be fostering new lending methods or institutions, promoting land registration, or changing how land use and construction are regulated.

The enabling approach may enhance security of tenure, improve industry efficiency, cheapen credit, reduce housing costs, and help the middle classes and the poor. In reality, governments struggle to acquire the information on which sound policy must be based, and to secure the necessary political support. They may be tempted to act precipitously, on the basis of simplified, ideological mantras: de-regulate, privatize, let "the market" decide. The losers will be the poor, because enabling implies the allocation of dwellings on the basis of income, not need. This ruthless logic can be mitigated, however, by demand-side subsidies, where the poor receive transfers that help them rent or purchase better housing, and this type of subsidy is endorsed by moderate advocates of the enabling approach (Angel 2000: 11–22).

Governments' increasingly preferable option is to upgrade existing housing. Even in a rapidly growing city, new construction rarely increases the housing stock by more than five percent a year. Redevelopment, like public housing, is expensive and may force residents to move to distant fringe locations, disrupting communities and provoking political opposition. Depending on circumstances, upgrading is cheaper on a per unit basis, while preserving social networks and access to existing jobs. A limitation is that it does little to reduce high residential densities. Moreover, where successful, improved neighborhoods invite "raiding" by the middle class while they do little to increase the stock of housing.

Most governments rely on several strategies. Such pragmatism may reflect a well-judged attempt to extract maximum benefit from each program. More commonly, it accumulates piecemeal, as governments respond to shifting domestic pressures and advice from international agencies.

THE EVOLUTION OF INTERNATIONAL HOUSING POLICY SINCE 1945

A number of writers have outlined the evolution of international housing policy since 1945 (e.g. Cohen 2001; Harris and Giles 2003; Kumar 2008: 256–72; Pugh 1997, 2001; Tibaijuka 2009: 35–55; UN-HABITAT 2005; McBride and French 2011: 5–10; Majale *et al.* 2011: 4–6). The present sketch modifies the conventional chronology, and suggests a broadly optimistic interpretation of the long-term trend.

From 1945 until the late 1960s, the most visible housing program in the developing world was public housing. It was prominent in Latin America, the region urbanizing most rapidly. Partly in response to public projects, and to land invasions that implied their inadequacy, John Turner developed his ideas (Turner 1968). A British architect active in Peru,

Turner became, and perhaps remains, the most influential postwar commentator on housing in the global South. He praised squatters for housing themselves, and celebrated their control over living space. Adapted, and arguably distorted, by the early 1970s, his arguments were used by development agencies to justify a new emphasis on sites and services.

Turner's arguments were not novel. While colonial regulations had discouraged building in stages or the use of cheap, indigenous materials, they also helped build publicly owned housing, more for government workers than the indigenous poor. But after 1945, insofar as the British Colonial Office, the United States government, or the United Nations had coherent housing policies, their experts criticized public housing and favored aided self-help (Harris and Giles 2003: 174–8; Tibaijuka 2009: 84–5). A few aided self-help projects were carried out, notably in Puerto Rico during the 1940s. But most governments in the developing world ignored agency advice. This was true of colonial governments, such as Kenya's in the 1950s; postcolonial governments, such as in India after 1947; nations such as Brazil, whose colonial experience was long past; and above all communist governments, which made mass housing a symbol of socialist achievement. These countries copied what Western nations were doing (Tibaijuka 2009: 169–71, 193–8). Until rising costs and political opposition halted it in the late 1960s, public housing was the main program that European and North American governments directed at the poor. Only in the 1970s did international agencies begin to seriously influence housing policy in the developing world.

The impact of these agencies increased in the 1970s for several reasons. Drawing on Turner's arguments, they emphasized aided self-help, repackaged as "sites and services" (Kumar 2008: 257–9; Tibaijuka 2009: 171). Sites and services were cheaper and politically more acceptable. And with the entry of the World Bank into the housing field in 1972, sites and services became the most visible single housing program.

By the 1980s, however, the limitations of such projects were becoming obvious. In China, strict migration controls were restraining urban growth, but elsewhere in Asia and in Africa urbanization dwarfed the impact of site-and-service schemes. International agencies had long noted the need for systemic changes in how housing markets worked, but these calls never amounted to much. In the early 1980s, however, they became louder. This was signaled by development of the UN's Global Strategy for Shelter in the late 1980s (Pugh 2001: 407; Kumar 2008: 261; UNCHS 1989), and sealed by an influential statement from the World Bank (1993). By then, all agencies were advocating a "whole sector" approach, one which would reshape the whole market, with the United Nations emphasizing the social aspects of housing, and the World Bank focusing on the economic (Pugh 2001: 406–11; Tibaijuka 2009: 172–9). Housing finance was put on the agenda (Datta and Jones 1998). So, too, was the idea of promoting security of tenure in informal settlements by providing owners with legal title. This idea was promoted through a Global Campaign for Secure Tenure, launched by UNCHS in 1999, and was further boosted by the arguments of Hernando de Soto (2000). Together, these policies promoted homeownership over tenancy, encouraging owners to think of homes as market assets as well as basic needs.

Published, programmatic statements imply that the enabling approach and land titling were pushed single-mindedly, part of a neoliberal approach that attracted much criticism. In practice, however, the application of such ideas made allowance for local circumstances. By the late 1990s, local failures, especially in Africa, led World Bank economists to acknowledge the importance of flexibility (Kumar 2008: 256, 261–3; Tibaijuka 2009: 43). Introducing a policy review, a World Bank vice-president spoke ruefully about "less certitude" (Buckley and Kalarickal 2006: viii). A subtler, messier, and more balanced approach came to the fore (Payne, Durand-Lasserve, and Rakodi 2009; Buckley and Kalarickal 2006: viii). This recognized that tenure security mattered more than titling per se and gave more prominence to upgrading and paid more attention to local circumstances.

International agencies have also learned how to work with local agencies (Cohen 2001). As early as 1986, the UN and World Bank established an Urban Management Program to improve municipal government. In the late 1990s, they helped to form "a donor action group," the Cities Alliance, which facilitates cooperation of agencies and cities (Buckley and Kalarickal 2006: xiii; Kumar 2008: 267–8; UN-HABITAT 2005: 27). Agencies now routinely work with NGOs (e.g., Majale *et al.* 2011: 50).

Advocacy of public–private partnerships, pioneered in India, has become standard (Kumar 2008: 265–6). Apart from topical and regional surveys, UN-HABITAT now prepares manuals for housing and municipal administrators. The Asian manual written by ACHR in 2008, for example, argues against public housing and includes a one-page feature on "why is on-site upgrading often the best option of all?" (e.g., UN-HABITAT 2008: 14, 33). Cooperation extends to the organization of local slum surveys (UN-HABITAT: 2010b).

Informed observers still criticize gaps and imbalances. For half a century, agencies have promoted homeownership as the best tenure (Gilbert 2011; Kumar 2008). They still neglect tenants, the poorest of the poor, together with the motives and circumstances of the small landlords who provide much of the rental stock. They recognize that rent controls, one of the few tools widely deployed in the rental sector, are sometimes disastrous (Malpezzi 1999: 1836). Agencies, and governments, still tend to overlook the home as a place of employment, and its connections with labor markets (Tipple 1993). This is a legacy of the enabling approach, whose advocacy of institutional change led to a "dematerialization" of housing policy that overlooked the physical character of urban environments (Cohen 2001: 56). The building industry, including retailers and producers, has rarely received the attention it deserves, except in Mexico and to a lesser extent Brazil (see Harris 2012). The key issue of urban land remains vexed. International agencies know land matters, but its availability and price, together with ownership patterns and forms, vary locally, making generalized advice unhelpful (McBride and French 2011: 48). It is largely because of rising land costs that housing is often so unaffordable, which is why the World Bank flagged land as a key issue for China (Buckley and Kalarickal 2006: xiii; World Bank 2012). In the early 2000s, in the developed world, house prices were about four times average household incomes; in Latin America, six times; in Asia, seven to ten times; and in Africa higher still (UN-HABITAT: 2005). Recent estimates for urban China indicate 8.8 times, and climbing (Man 2011: 11). Bound up with land, but extending into construction and finance, is the question as to how the formal and informal sectors are intermixed. Disentangling this ball of twine is arguably the single greatest, unmet challenge for researchers and policymakers alike.

CONCLUDING DISCUSSION

It is easy to be pessimistic or, worse, cynical about the urban housing prospects in the developing world, especially for the poor. Sensational media accounts feed this tendency, and some have argued that use of pejorative terms like "slum" makes things worse (Gilbert 2007). No one who has observed, and smelled, garbage pickers at work in a Kolkata *bustee* can dismiss the problems of world urbanization. More must be done. Governments do not do all they might, and why should we expect otherwise? Everywhere, they respond to political pressure, not just human needs. Some pressure comes from below: when the poor organize, or when the middle class perceives that its interests lie in improving conditions for everyone. Pressure also comes from outside, from larger geopolitics as well as the domestic and international institutions that shape housing programs. This is a condition that leaves ample scope for pessimism.

Furthermore, urban conditions are shaped by policies not framed with housing in mind. Here the story is more encouraging, migration being a case in point. All countries now regulate immigration: the mass movements that populated North America, and relieved European cities, in the late nineteenth century are now impossible. But international migration is still a force, at source and destination. After they leave, immigrants send back remittances to family members that can make a big difference. In recent years, for example, they have accounted for 13 percent of the GNP of the Philippines, and almost nine percent for Bangladesh (Majale *et al.* 2011; cf. Tibaijuka 2009: 156–7). After they arrive, migrants may increase pressure on the housing stock in the short run, but they routinely contribute to economic growth. Migrants commonly prosper, and this helps everyone (Saunders 2010).

Of course migrations, housing, and much else are shaped by forces beyond the influence of any policy. Here, again, the trend is generally encouraging. The rise of the Asian Tigers and the BRICS has helped rebalance the world economy. In the global North, this has contributed to stagnation in real incomes for the middle and working classes, growing inequality, and fiscal challenges, as Western multinationals outsource physical and clerical work. In the global South, too, this has brought growing inequalities, but also rising incomes for an emerging middle class. Inequalities within each country remain a major concern, but rising national prosperity in many parts of the developing world is transforming the housing scene.

But what about the specific contribution of international policy? Here, it is important to be skeptical. The statements of international agencies are replete with impressive declarations, but their ability to change the world is circumscribed. Many observers were dubious when, in 2000, the UN spoke about a future of "cities without slums" and announced, as the first of its new Millennial Development Goals, the eradication of "extreme poverty and hunger." The specific goal was to halve between 1990 and 2015 the proportion of people living on less than $1/day (UN-HABITAT 2003). Improbably, as I write, the achievement of this goal, three years ahead of schedule, has just been announced. Even if accepted at face value, this achievement has had little to do directly with the actions of the UN or the World Bank. At most, the influence of the World Bank, an advocate for market reforms, has been felt indirectly. By adopting such reforms, the agency that has done the most to reduce poverty over the past generation has been the Chinese government, albeit at the price of a rapid increase in income inequality.

Overall, there are reasons for optimism. More than at any time, there is a consensus that global prosperity, and environmental security, depend on urbanization, and the better management of cities. Experts and policy-makers increasingly agree that housing must play a major role, for economic as well as social development (Angel 2000: 56–9; Pugh 2001; Tibaijuka 2009). This constructive trend will surely continue. The financial crisis of 2008–9 with its sub-prime mortgage debacle showed that Western nations are not the fount of wisdom on housing issues. In the colonial period, during the cold war, and indeed until the turn of the millennium, it seemed that the West knew best. No longer. We may already have entered an era when, in terms of policy, culture, and economy, there is a truly international dialog, one which links institutions at multiple scales.

NOTE

1 For this paper, "global South" and "developing world" are treated as synonymous. UN-HABITAT defines a slum as a dwelling with one or more of the following qualities: more than three persons per room; lacking sufficient, affordable clean water; lacking improved sanitation (private or public); not durable; insecure tenure. The criteria are necessarily arbitrary, measurement is difficult, and reliable data are often lacking, so I have rounded the published estimates. Treated with caution, they are a meaningful basis for broad historical-geographical comparisons.

Acknowledgements: I would like to thank Patricia Annez, Billy Cobbett, Charlotte Lemanski and the Editors of this volume for providing valuable comments and suggestions on earlier drafts of this essay. This research was supported by the Social Sciences and Humanities Research Council of Canada through funding from the Major Collaborative Research Initiative, "Global Suburbanisms. Governance, Land and Infrastructure in the 21st Century (2010–2017)."

REFERENCES

Angel, S. (2000) *Housing Policy Matters,* New York: Oxford University Press.

Benjamin, S. (2004) "Urban land transformation for pro-poor economies," *Geoforum,* 35: 177–87.

Boo, K. (2012) *Behind the Beautiful Forevers: Life, Death and Hope in a Mumbai Undercity,* New York: Random House.

Buckley, R. M. and Kalarickal, J. (eds.) (2006) *Thirty Years of World Bank Shelter Lending. What Have We Learned?* Washington, DC: World Bank.

Cohen, M. (2001) "Urban assistance and the material world: Learning by doing at the World Bank," *Environment and Urbanisation,* 13(1): 37–60.

Datta, K. and Jones, G. (1998) *Housing and Finance in Developing Countries,* New York: Routledge.

Davis, M. (2006) *Planet of Slums,* London: Verso.

De Soto, H. (2000) *The Mystery of Capital,* New York: Basic Books.

Gilbert, A. (2007) "The return of the slum: Does language matter?" *International Journal of Urban and Regional Research,* 31(4): 697–713.

Gilbert, A. (2011) *A Policy Guide to Rental Housing in Developing Countries. Quick Policy Guide Series Volume 1,* Nairobi: UN-HABITAT.

Hardoy, J. E. and Satterthwaite, D. (1981) *Shelter. Need and Response. Housing, Land and Settlement*

Policies in Seventeen Third World Nations, New York: Wiley.

Harris, R. (2010) "Meaningful types in a world of suburbs," in M. Clapson and R. Hutchinson (eds.), *Suburbanisation in a Global Society. Research in Urban Sociology No.10*, Bingley: Emerald Group Publishing.

Harris, R. (2012) *Building a Market: The Rise of the Home Improvement Industry, 1914–1960*, Chicago: University of Chicago Press.

Harris, R. and Giles, C. (2003) "A mixed message: The agents and forms of international housing policy, 1945–1973," *Habitat International*, 27(2): 167–91.

Keivani, R. and Werna, E. (2001) "Modes of housing provision in developing countries," *Progress in Planning*, 55(2): 65–118.

Kumar, S. (2008) "Urban housing policy and practice in the developing world," in I. C. Colby, K. Sowers, and C. N. Dulmus (eds.), *Comprehensive Handbook of Social Work and Social Welfare: Social Policy and Social Policy Practice*, New York: Wiley.

McBride, B. and French, M. (2011) *Affordable Land and Housing in Latin America and the Caribbean*, Nairobi: UN-HABITAT.

Majale, M, Tipple, G. and French, M. (2011) *Affordable Land and Housing in Asia*, Nairobi: UN-HABITAT.

Malpezzi, S. (1999) "Economic analysis of housing markets in developing and transition economies," in P. Cheshire and E. Mills (eds.), *Handbook of Regional and Urban Economics Vol.3. Applied Urban Economics*, Amsterdam: Elsevier.

Man, J. Y. (ed.) (2011) *China's Housing Reform and Outcomes*, Cambridge, MA: Lincoln Institute for Land Policy.

Miraftab, F. (1998) "Complexities of the margin: Housing decisions of female householders in Mexico," *Society and Space: Environment and Planning D*, 16: 289–310.

Mooya, M. M. (2011) "Making urban real estate markets work for the poor: Theory, policy and practice," *Cities*, 28: 238–44.

Moser, C. O. N. (1987) "Women, human settlements, and housing: A conceptual framework for analysis and policy-making," in C. O. N. Moser and L. Peake (eds.), *Women, Settlements and Housing*, London: Tavistock.

Mukhija, V. and Monkkonen, P. (2007) "What's in a name? A critique of 'Colonias' in the United States," *International Journal of Urban and Regional Research*, 31(2): 475–88.

Neuwirth, R. (2005) *Shadow Cities: A Billion Squatters, A New Urban World*, New York: Routledge.

Payne, G., Durand-Lasserve, A. and Rakodi, C. (2009) "The limits of land titling and home ownership," *Environment and Urbanization*, 21(2): 443–62.

Perlman, J. (1987) "Misconceptions about the urban poor and the dynamics of housing policy evolution," *Journal of Planning Education and Research*, 6: 187–98.

Pugh, C. (1997) "Poverty and progress? Reflections on housing and urban policies in developing countries, 1976–1996," *Urban Studies*, 34(10): 1547–95.

Pugh, C. (2001) "The theory and practice of housing sector development for developing countries, 1950–99," *Housing Studies*, 16(4): 399–423.

Renaud, B. (1999) "The financing of social housing in integrating financial markets. A view from the developing countries," *Urban Studies*, 36(4): 755–73.

Santos, D. de S. (1977) "The law of the oppressed: The construction and reproduction of legality in Pasagarda," *Law and Society Review*, 12(1): 5–126.

Saunders, D. (2010) *Arrival City: The Final Migration and Our Next World*, Toronto: Knopf.

Tibaijuka, A. K. (2009) *Building Prosperity: The Centrality of Housing in Economic Development*, London: Earthscan.

Tipple, G. (1993) "Shelter as workplace: A review of home-based enterprise in developing countries," *International Labour Review*, 132: 521–39.

Turner, J. F. C. (1968) "Uncontrolled urban settlements: Problems and policies," *International Social Development Review*, 1: 107–30. Reprinted in G. Breese (ed.) (1969) *The City in Newly Developing Countries*, New York: Prentice-Hall.

UNCHS (United National Centre for Human Settlements) (1989) *The New Agenda for Human Settlements*, Nairobi: UNCHS.

UN-HABITAT (2003) *The Challenge of Slums: Global Report on Human Settlements 2003*, London: Earthscan.

UN-HABITAT (2005) *Financing Urban Shelter: Global Report on Human Settlements*, Nairobi: UN-HABITAT.

UN-HABITAT (2008) *Housing the Poor in Asian Cities: Quick Guide for Policy Makers*, Bangkok, Thailand: UNESCAP.

UN-HABITAT (2010a) *State of the World's Cities 2010/2011: Bridging the Urban Divide*, London: Earthscan.

UN-HABITAT (with Global Land Tool Network) (2010b) *Count Me In: Surveying for Tenure Security and Urban Land Management,* Nairobi: UN-HABITAT and GLTN.

World Bank (1993) *Housing: Enabling Markets to Work,* Washington, DC: World Bank.

World Bank (with Development Research Center of the State Council, People's Republic of China) (2012) *China 2030. Building a Modern, Harmonious and Creative High-income Society,* Washington, DC: World Bank.

"Women and Self-Help Housing Projects: A Conceptual Framework for Analysis and Policy-Making"

Kosta Mathéy, *Beyond Self-Help Housing* (1992)

Caroline O. N. Moser

[. . .]

SELF-HELP HOUSING POLICY AND THE NEEDS OF WOMEN

To identify the extent to which women have particular housing needs on the basis of gender, it is important to start by understanding the assumptions underlying housing policy relating both to the structure of low-income families and to the division of labor within the family. Policy-makers, planners, architects and designers, within governments and international agencies, perceive themselves as planning for people. But regardless of the reality of the particular planning context there is an almost universal tendency to make three assumptions: firstly, that the household consists of a nuclear family of husband, wife and two or three children; secondly, that within the family there is a clear gender division of labor in which the man of the family as the "breadwinner" is primarily involved in productive paid work outside the home, while the woman, as housewife and "homemaker," takes overall responsibility for the reproductive and domestic work involved in the organization of the household; thirdly, that within the household there is an equal control over resources and power of decision-making between the man and the woman in matters affecting the household's livelihood, and therefore that the household can be identified as functioning as a unit. In most societies this gender division of labor is seen to reflect the "natural" order and is ideologically reinforced through such means as the legal and educational system, the media and family planning programs, without recognition that within it the woman's position is subordinate to the man's.

Housing policies based on these unacknowledged assumptions are well known and widespread. For instance, policy-makers tend to identify target groups in terms of the income of the male breadwinner; project authorities design eligibility criteria for participating in sites-and-services projects in terms of the income of men and evidence of regular employment; and architects design houses to meet the needs of nuclear families, in which all productive paid work is undertaken by men outside the home.

THE "TRIPLE ROLE" OF WOMEN

In most Third World contexts there are problems with this abstract stereotype model of society and the division of labor within it. First, it fails to recognize the triple role of women. In most low-income households "women's work" includes not only reproductive work (the childbearing and rearing responsibilities) required to guarantee the maintenance and reproduction of the labor force, but also productive work, as primary or secondary income earners, in rural areas in agriculture, in urban areas in informal sector enterprises located either in the home (in subcontracting or piece-rate work) or at the neighborhood level (Moser, 1981; Beneria and Roldan, 1987; Afshar, 1985).

In addition, women are involved in community managing work, undertaken at local community settlement level in both urban and rural contexts. With the increasingly inadequate state provision of housing and basic services such as water and health care, it is women who not only suffer most, but who also are forced to take responsibility for allocating limited basic resources to ensure the survival of their households. Where there is open confrontation between community-level organizations and local authorities in attempts to put direct pressure on the state or non-governmental organizations for infrastructure provision, again it is women who, as an extension of their domestic role, take primary responsibility for the formation, organization and success of local-level protest groups (Barrig and Fort, 1987; Moser, 1989b; Cole, 1987; Sharma et al., 1985). In their gender-ascribed roles as wives and mothers, women struggle to manage their neighborhoods. In performing this third role they implicitly accept the gender division of labor and the nature of their subordination.

[...]

In most Third World societies the stereotype of the man as breadwinner, i.e., the male as the productive worker, predominates, even when it is not borne out in reality. Invariably when men perceive themselves to have a role within the household it is the primary income earner. This occurs even in those contexts where male unemployment is high and women's productive work provides the primary income. In addition, generally men do not have a clearly defined reproductive role, although this does not mean empirically that they do not play with their children or assist their women partners with domestic activities.

Men are involved in community activities but in markedly different ways from women, reflecting a further gender division of labor. The spatial division between the public world of men and the private world of women (where the neighborhood is an extension of the domestic arena) means that men and women undertake different community work. While women have a community managing role based on the provision of items of collective consumption, men have a community politics role, in which they organize at the formal political level, generally within the framework of national politics. In organizations in which these two activities overlap, especially in societies where men and women can work alongside each other, women most frequently make up the rank and file voluntary membership, while men are only involved in positions of direct authority and work in a paid capacity.

Because the triple role of women is not recognized, so neither is the fact that women, unlike men, are severely constrained by the burden of simultaneously balancing the three roles of productive, reproductive and community managing work In addition, by virtue of its exchange value, only productive work is recognized as work. Reproductive and community managing work, because they are both seen as ""natural" and non-productive, are not valued. This has serious consequences for women. It means that most of their work fails to be recognized as such, either by men in the community or by those planners whose responsibility it is to assess different needs within low-income communities. While the tendency is to see the needs of women and men as similar, the reality of women's lives is very different.

WOMEN-HEADED HOUSEHOLDS

The second problem is that this model fails to recognize that low-income households are heterogeneous in structure. The most important type of household, apart from nuclear families, are de facto and *de jure* women-headed households in which the male partner is absent, either temporarily (for instance because of migration refugee status) or permanently (because of separation or death). It is estimated that one third of the world's households are now headed by women (Buvinic et al., 1978). In urban areas, especially in Latin America and parts of Africa, the figure exceeds 50 percent, while in the refugee camps of Central America it is nearer 90 percent. Globally this is a growing rather than declining phenomenon.

The economic condition of these households varies considerably, depending on the marital status of the woman, the social context of female leadership, her access to productive resources and income, and lastly the composition of the household. Frequently, women-headed households have a high dependency ratio and limited access to employment and basic services, and consequently all too often fall below the poverty line, and are disproportionately represented amongst the poorest of the poor (White et al., 1986). Although women who head households do not constitute a separate category, their problem of the triple burden is likely to be exacerbated, which may have specific policy implications.

HOUSEHOLD-LEVEL DECISION-MAKING

The third problem relates to the assumption that, for planning purposes, the household can be treated as a single decision-making unit, within which members make free and voluntary economic choices, motivated by the desire to maximize total family welfare. In this case the household, a residential unit, is conflated with the family, a social unit based on kinship, marriage and parenthood. Although in much policy-making marriage is conceived of as a partnership with a common set of interests, in reality this may not be the case. Women with competing time constraints may have differently prioritized needs from those of men. Equally there may not be an equal sharing of resources within the household; food distribution is often biased in favor of men, whether old or young; expenditure on education and health care is often given to sons before daughters. Therefore the concept of the household requires desegregation if the different roles in decision-making, diverse choices and needs of the various household members are to be identified.

IDENTIFYING PRACTICAL AND STRATEGIC GENDER NEEDS

When planners are blind to the triple role of women, and to the fact that women's needs are not always the same as men's, they fail to recognize the necessity of relating planning policy to women's specific requirements. Women, because of their triple role, have particular needs that differ from those of men, or low-income families generally. At the same time it is also important to identify how far planning for the needs of low-income women is necessarily "feminist" in content. The distinction between practical and strategic gender needs may be useful in the clarification of this issue.

Practical gender needs are the needs that women identify in their socially accepted roles in society. Practical gender needs do not challenge the gender divisions of labor or women's subordinate position in society, although arising out of them. They are a response to an immediate perceived necessity, identified within a specific context. They are often practical in nature and concerned with inadequacies in living conditions. Therefore, in many contexts needs such as adequate housing, clean water supply or community crèche facilities are identified as the practical gender needs of low-income women, both by planners and by women themselves. In reality practical needs such as these are required by all the family, especially children, and their identification as "women's needs" serves to preserve and reinforce the gender division of labor. At the same time it is important to recognize that such needs are in no sense "feminist" in content.

By contrast, strategic gender needs are the needs women identify because of their subordinate position to men in society. These vary according to particular contexts related to gender divisions of labor, power and control and may include such issues as legal rights, domestic violence, equal wages and women's control over their own bodies (Molyneux, 1985, 233). Meeting strategic gender needs assists women to achieve greater equality, changes existing roles and therefore challenges women's subordinate position. The strategic gender needs identified to overcome women's subordination will vary depending on the particular cultural and socio-political context within which they are formulated. Strategic gender needs are often identified as "feminist," as is the level of consciousness required to struggle effectively for them. In the examination of the self-help housing needs of low-income Third World women which follows, it is therefore important to identify how far different policy interventions will meet different needs.

GENDER AND PARTICIPATION IN SELF-HELP HOUSING PROJECTS

Because of the self-help nature of so many settlement and housing policies in the Third World, the discussion of stereotype housing target groups must also include the issue of women's participation in self-help housing projects. Again it is necessary to identify the implications of present assumptions relating to the structure of the family and division of labor within it, in terms of the planning of community participation in human settlement projects. The importance of community participation in human settlement projects is now widely recognized in terms of the role that intended beneficiaries and local organizations can and do play, not simply in the implementation and management phases of development projects, but also in decision-making about the design and allocation of resources.

Although participation is universally considered a "good thing," because the implication is that the community influences the direction and execution of development projects rather than merely receiving a share of project benefits, the motives and objectives behind it are not always clear. Paul, for instance, recently identified the objectives as ranging from empowerment and capacity building to increasing project effectiveness, improving project efficiency and project cost sharing (Paul, 1987). The important distinction is between projects that include an element of empowerment and those that do not. Another "measurement" of empowerment in projects has been through the distinction between participation as a means and participation as an end (Oakley and Marsden, 1984). Where participation is a means, it generally becomes a form of mobilization to get things done. This can equally be state-directed, top-down mobilization (sometimes enforced) to achieve specific objectives, or bottom-up "voluntary" community-based mobilization to obtain a larger share of resources. Where participation is identified as an end the objective is not a fixed quantifiable development goal but a process the outcome of which is an increasingly "meaningful" participation in the development process (UNRISD, 1979; Moser, 1983, 1989b).

Whatever the objectives of participation, ultimately it is a question of who is participating, and the accessibility of a project to the target population, which determines the extent to which real participation occurs. At the policy level the tendency has been to conceptualize "the community" in homogeneous terms, with further disaggregation, when necessary, done at the household level. Above all this means that the important role that women play in community participation is only rarely mentioned at the policy level, despite the constant comments from those working on the ground that "without the women the project would not have worked." The lack of disaggregation on the basis of gender can have serious implications. For despite cultural variations and regional differences with regard to the role of women in society and in the household, which obviously limits their capacity to participate, nevertheless there are particular reasons for recognizing the role all women can play in community participation.

Three reasons have been advocated for incorporating women's participation in human settlement projects, each of which has a different underlying objective.

(a) Women's participation is an end in itself. Women as much as men have the right and duty to participate in the execution of projects which profoundly affect their lives. Since women accept primary responsibility for childbearing and rearing, they are most affected by housing and settlement projects. They should, therefore, be involved in the planning and decision-making as well as the implementation and management of projects which relate particularly to their lives.

(b) Women's participation is a means to improve project results. Since women have particular responsibility for the welfare of the household, they are more aware than men of the needs for infrastructure and services and are also more committed to the success of a project that improves living conditions. The exclusion of women can negatively affect the outcomes of a project, while their active involvement can often help its success.

(c) Participation in housing activities stimulates women's participation in other spheres of life. Through active involvement in housing projects women may be encouraged to participate fully in the community. Participation in projects has been seen as an important mechanism to "overcome apathy" and "lack of confidence" and it can make women visible in the community. It can enable women to come out of their houses and show them the potential of self-help solutions. In so doing, it may raise awareness that women can play an important role in solving problems in the community (UNCHS, 1986: 2).

In the case of (a), the objective of women's participation identified here could be said to relate to the "top-down" end of empowerment; in the case of (b) . . . [women's participation is] a means to achieve efficiency, effectiveness and cost recovery, while in (c) . . . [participation is an end in itself, towards] capacity building.

[. . .]

CONCLUSION: MEETING GENDER NEEDS IN SELF-HELP HOUSING

The fact that women, because of their triple role, have particular needs in self-help housing that differ from those of men, or low-income families generally, indicates that women's needs will not be addressed by simply grafting on "women" as a category to existing self-help housing policy. Fundamental changes in attitude toward traditional stereotypes are required before a gender-aware approach can be reached that is an approach capable of acknowledging that women have different housing needs from those of men, by virtue of their engendered position, and that in many societies there are specific constraints that limit women's access to housing on the basis of gender, regardless of income.

The danger in current self-help housing projects is that women are being grafted onto projects without gender awareness. This is best illustrated by the increasing tendency to rely on women's participation in the implementation phases purely for project efficiency, but without recognition of women's triple role. The assumption that women have "free time" in reality means that projects, at both community and individual household level, often only succeed by forcing women to extend even further their working day. When women fail to participate it is not women who are the problem, as is frequently identified, but a lack of gender awareness on the part of authorities about the different roles of men and women in society and the fact that women have to balance three roles (Moser, 1989b).

At the same time it is important to recognize that the majority of interventions intended to meet the housing needs of low-income women are not necessarily "feminist" in content. The provision of housing and community-level infrastructure which meets family needs ipso facto meets the practical gender needs of women, who in their reproductive role are the principal users of housing. However, this does not mean that more strategic gender needs are met. For instance, in self-help housing projects women do not automatically become the equal or sole owners of either the house or land. In many housing projects tenure is the most significant issue, with rights to land generally given to men on the assumption that they are household heads. It is only where housing projects have been designed to provide for ownership regardless of the sex of the household head and women have been able to achieve equal right to own property that strategic gender needs are addressed. In such cases the project can certainly be evaluated as "feminist" in content, if not in design.

Although women's participation is increasingly becoming identified as a means to ensure the success of self-help housing projects, widespread constraints are still encountered by those who seek to challenge more fundamentally the existing policy and practice in human settlements and housing. In many contexts there is resistance to changing the existing gender division of labor within the household, and the division of responsibility between men and women in issues relating to housing, particularly over the issue of land tenure and home ownership.

This opposition may derive equally from two sources. First it may derive from "integrationalists" who want better conditions and the incorporation of women into policy frameworks but within the existing structures, accepting the triple role and endorsing what for many is an inequitable status quo. This may include not only policy-makers but often the majority of low-income women, whose power derives from their gender-ascriptive roles as wives and mothers. Secondly, opposition may derive from policy-makers who conveniently base their work on inappropriate abstract models of society.

Thus the specific circumstances by which conventional assumptions can be challenged and restructured may be dependent on the development of consciousness at two levels. They may depend as much on the "bottom-up" emergence of women's consciousness through the experience of participation in self-help housing projects as on the "top-down" opening up of political space. Ultimately it may be the conjuncture of the two that is necessary for any fundamental change in the nature of gender relations and housing to occur.

[. . .]

BIBLIOGRAPHY

Afshar, H. (ed.) (1985) *Women, Work and Ideology in the Third World*, London: Tavistock.

Barrig, M. and Fort, A. (1987) "La Ciudad de las Mujeres: Pobladores y Servicios: El Caso de El Augustino," in *Women, Low-Income Households and Urban Services*. Working Papers. Lima, Peru.

Beneria, L. and Roldan, M. (1987) *The Crossroads of Class and Gender*, Chicago and London: University of Chicago Press.

Buvinic, M. and Youssef, M. with Von Elm, B. (1978) *Women-Headed Households: The Ignored Factor in Development Planning,* Washington, DC: International Center for Research on Women.

Cole, J. (1987) *Crossroads: The Politics of Reform and Repression 1976–1986*, Johannesburg: Ravan Press.

Molyneux, M. (1985) "Mobilization without emancipation? Women's interests, state and revolution in Nicaragua," *Feminist Studies*, 11(2), 227–54.

Moser, C. (1981) "Surviving in the *Suburbios*," *Bulletin of the Institute of Development Studies*, 12(3), 19–29.

Moser, C. (1983) "The problem of evaluating community participation in urban development projects," in C. Moser (ed.), *Evaluating Community Participation in Urban Development Projects,* Development Planning Unit Working Paper, no. 14, London: University College London.

Moser, C. (1989a) "Gender planning in the Third World: Meeting practical and strategic gender needs," *World Development*, 17(11).

Moser, C. (1989b) "Community participation in urban projects in the Third World," *Progress in Planning*, 32(2).

Moser, C. and Chant, S. (1985) "The role of women in the execution of low income housing projects, Draft Training Manual," DPU Gender and Planning Working Paper, no 6, London: University College.

Oakley, P. and Marsden, D. (1984) *Approaches to Participation in Rural Development*, Geneva: ILO.

Paul, S. (1987) "Community participation in development projects: The World Bank experience," World Bank Discussion Papers, no 6, Washington, DC: The World Bank.

Rogers, B. (1980) *The Domestication of Women*, London: Kogan Page.

Sharma, K., Pandey, B. and Nantiyal, K. (1985) "The Chipko Movement in the Uttarkhand region, Uttar Pradesh, India," in S. Muntemba (ed.), *Rural Development and Women: Lessons from the Field,* Geneva: ILO/Danida.

UNCHS (1984) *Community Participation in the Execution of Low Income Housing Projects*, Nairobi: UNCHS-HABITAT.

UNCHS (1986) *The Role of Women in the Execution of Low-Income Housing Projects*, Nairobi: UNCHS-HABITAT.

UNRISD (1979) *Inquiry into Participation—a Research Approach*, Geneva: UNRISD.

White, K., Otero, M., Lycette, M. and Buvinic, M. (1986) *Integrating Women into Development Programs: A Guide for Implementation for Latin America and the Caribbean*, Washington, DC: International Center for Research on Women.

"The Suburbanization of Jakarta: A Concurrence of Economics and Ideology"

Third World Planning Review (1994)

Michael Leaf

THE SUBURBANIZATION OF THE GLOBE OR THE GLOBALIZATION OF THE SUBURB?

It is not an overstatement to claim that the Western suburb is quintessentially an imperialist form of human settlement. In recent years, the low density, residential suburb has come under increasing attack as being inherently wasteful of land and energy and detrimental to both the natural and social environments (Lowe, 1991). As the predominant form of settlement in the United States, the suburb is a significant factor contributing to the high per capita consumption of resources which characterizes American society (Brown and Jacobson, 1987). The suburb, more than any other component of the post-industrial Western city, is the locus of accumulation of imported surplus production.

The usual historical interpretation of Western suburbanization portrays this settlement form as arising from rapid economic growth and the concomitant expansion of the middle classes, coupled with unprecedented technological change, particularly the development of motor cars powered by plentiful (and therefore underpriced) fossil fuels.

Significantly, all of these forces came together at very specific points in the development of the world economy, when the resource-extracting processes of latter-day colonialism and then neo-colonialism were at their zeniths (King, 1990). The economic boom periods of the 1920s and the 1950s—the periods of major suburbanization in the West—must be seen in the context of a world economic system which facilitated the expansion of export markets for manufactured goods on the one hand and the extraction of resources from peripheral to core societies on the other. Unlike earlier stages of colonialism where the spoils of empire accrued only to the elites, the democratic tendencies of modern industrial society and the expanded market economy had, by the middle of the twentieth century, led to a greater distribution of the wealth within metropolitan societies. In the 1950s and 1960s, Americans learned that the democratic dream of "every man a king in his own castle" could be realized in suburbia.

From this interpretation of the suburb as a product of colonialist and neo-colonialist processes in the world economy, it is understandable how the automobile-dependent mass suburb could have come to be the predominant form of residential settlement in the cities of the core societies in the world. This perspective leads us to an interesting question: how are we to interpret the growth of low-density automobile-based residential districts in Third World cities? Should we consider it ironic that this settlement form is now being exported from the metropolitan societies of the West to the former colonies?

Here, it is worthwhile to make a distinction between two possible means by which the suburb might be spreading: the suburbanization of the globe, and the globalization of the suburb. The suburbanization of the globe is predicated upon the idea that the suburb is a settlement form which can expand beyond the cultural hearth of the West, where it now dominates the human landscape, to eventually house equally large proportions of urban populations throughout the world. Like the core precept of modernization theory, which envisages the possibility of all societies eventually attaining the standard of living of today's developed countries, this view assumes that the world economy has the potential for continuous growth.

The globalization of the suburb, on the other hand, refers to the spread of suburbia as one more facet of global material culture, like McDonald's hamburgers, blue jeans, and Michael Jackson's latest video. It is also of great significance to note that—like the Big Mac—the globalized suburb, although seemingly ubiquitous, is in fact only accessible to the very few. Global culture, although popular in its appeal, can be very exclusionary.

The processes of suburbanization in Jakarta, the capital and largest city in Indonesia, are examined here with this basic issue in mind. We will see that, in this case, the adoption of suburban enclave housing as the preferred form of settlement is an outcome of interactions between market forces and the ideological framework promoted by the Indonesian state. Policy in particular plays a great role in promoting the concurrence of economic and ideological factors which underlie the building of Jakarta's suburbs. To the extent which the phenomenon of suburbanization is linked to shifting relations within the world economy, lessons can be drawn for comparison with the cities of other developing countries.

THE SUBURBANIZATION OF JAKARTA

Although the emphasis here is upon suburban enclave housing, it should be understood that as a proportion of total housing stock in Jakarta, this form of settlement is still a relatively minor [...] component. Looking only at the periphery of the city, where such estates are concentrated, one would find that even there, suburban developments do not constitute the bulk of new construction. Indeed, real estate developments are most notable because they are exceptional. Located on the periphery of the city, they generally appear as walled compounds containing clusters or series of houses arrayed along standardized automobile thoroughfares. This is in sharp contrast to the extraordinarily mixed land uses, spatial patterns, and economic activities of the surrounding areas. A reasonable metaphor is of islands of homogeneity set in a sea of diversity; the perimeter walls of these enclaves act as dykes, put in place to maintain social and physical exclusivity.

[...]

There are two important areas of policy which have been instrumental in fostering the suburbanization of Jakarta.

[...]

These are a massive program of subsidized housing finance promulgated at the national level, and a municipal permit system for land development created specifically to feed urban lands into the corporate development sector.

The subsidized housing finance program was first established in 1974 and is administered principally through the State Savings Bank (BTN). Although this program is ostensibly aimed at fostering the construction of low-cost housing, its primary effect has been the stimulation of the housing industry overall, as fully three-quarters of all formal sector housing built in Indonesia since the program's inception have been with subsidized funds (Kantor Mentri Negara Perumahan Rakyat, 1990). This proportion has, in fact, increased over time, to include more than 85 percent of all private developer-built housing in the 1988–1989 fiscal year. These numbers include not only the subsidies through BTN, but those of a subsidiary program administered by the parastatal company PT Papan Sejahtera (PTPS), which specifically targets households with incomes above the eightieth percentile of the urban population. The BTN program, which is intended for households with incomes between the twentieth and eightieth percentiles, has often been criticized for failing to reach low-income households. For example, a 1988 random sample of privately developed houses built under this scheme found that the median income of current residents exceeds the eightieth percentile cut-off for the program (Struyk et al., 1990, 293–310). The typical beneficiaries of this subsidy program cannot be considered to be the urban poor.

The land development permit system, also put into place in the early 1970s, has further facilitated the activities of the corporate housing sector. The principal component of this system is the Indicative Land Use Permit (SIPPT), which is intended to be a mechanism for improving compliance with the city's land use plan. The granting of SIPPT permits also promotes the aggregation of numerous small, individually held plots of land into large parcels which can then be fed into the real estate development process. In practice, this amounts to a portioning off of large tracts of peripheral land for exclusive use by designated development companies.

[...]

The central point here is that, in the case of Jakarta, the expansion of suburban enclave housing as a new urban form is a direct outcome of policy, rather than merely being the natural result of market forces. The formulation of these policies must be considered within the political economic context of the New Order government. Those who have benefited the most from this policy regime are those members of the capitalist class, which has evolved from a mercantile base under the aegis of the New Order. The extraordinarily high concentration of capital ownership within the Indonesian economy in general (Robison, 1986) is paralleled within Jakarta's development industry. One indication of this is seen in the relative proportions of lands under permit by specific developers. By far the largest landholdings are those of the companies controlled by Ciputra, an Indonesian Chinese developer whose operations have long been financially linked to Liem Sioe Liong, Indonesia's pre-eminent industrialist.

[...]

It is insufficient, however, to maintain that the suburbanization of Jakarta is merely the result of collusion between capital and state, as this tells us very little about the particular form which this suburbanization is taking. To understand this, we need to look at both the broader ideological and economic contexts which have shaped Jakarta's suburbanization.

IDEOLOGICAL FACTORS IN JAKARTA'S SUBURBANIZATION

[...]

The first of these is the idea that the capital of the nation must necessarily express the power and centrality of the state. The importance of the symbolism of power for the Indonesian state has been emphasized before from a number of perspectives. Culturalist interpretations, for example, stress how the traditional Javanese aristocracy viewed power as an end unto itself, as virtually a palpable substance to be amassed for eventual spiritual benefit (Anderson, 1972). The physical expressions of this power in mandala-derived layouts for court complexes and royal towns are perceived as means by which traditional courts could enhance their existing stocks of power. Rather than being interpreted as a persistence of precolonial practices, the contemporary Indonesian emphasis upon the aggrandizement of the capital city can also be seen in pragmatic geo-political terms as part of the efforts of a post-colonial state to promote unity in a socially and culturally disparate nation.

[...]

The adoption of urban forms for the sake of conveying the power of the city—especially the capital city—is by no means unique to the Indonesian state. It is, in fact, part of a long standing historical trend. In the planned building or rebuilding of major cities, there is a general tendency to emulate the forms of the contemporary dominant culture of the region or the world, often with little regard for the ideological expression inherent in the adopted forms. The ironic contradiction which can result from this is seen in the case of Washington DC, the French baroque capital of the United States. Despite the egalitarian underpinnings of the American revolution, the founding fathers chose a royalist urban form in the layout of their capital. A similar irony can be seen in the comparison of post war Moscow and the modern Brazilian capital of Brasilia. Both of these cities drew upon international style conceptions of the modern city, thereby obtaining surprisingly similar results despite vast differences in economic structure, political ideology, social make up and climate.

In aspiring to attain the American way of life, the elites of Jakarta have begun to mold their city in a form reminiscent of US standards of urbanism, with a central business district (CBD) full of high rise office buildings, a network of freeways, a waterfront theme park, and incipient suburban sprawl. Unintended ironies abound: American cultural forms such as the drive-in theatre and the suburban flea market take on wholly different class connotations when they are adopted by the elites of Jakarta. But being able to present to the rest of Indonesia an image of Jakarta as a world-class city, on a par with the Dallases, the Las Vegases and the Los Angeleses which fill the nation's television screens, is important in reinforcing the idea that the power of the nation rests in the center.

Closely related to the question of how the city expresses the state's power is the issue of modernization itself. Here, the role of the city as the core of modern Indonesia goes beyond the merely symbolic. Jakarta truly is the center of economic growth, industrialization, consumerism, and, indeed, westernization for all of Indonesia (Castles, 1989). For better or for worse, it is Indonesia's prime point of connection with world culture and the world economy. As such, it is a key component of the government's effort to demonstrate that it is fulfilling an important promise to the people. This is the promise of development.

[. . .]

The ideological importance of Jakarta as a symbol of modernization is not unique to the New Order period. The rebuilding of the city to international standards of urbanism was also a central concern—in fact, nearly an obsession—of President Sukarno, whose internationalist ambitions for the city are clear in his statement that he envisioned Jakarta becoming "the beacon of the whole of mankind" (quoted in Abeyasekere, 1989, 168). A crucial distinction between Sukarnoist efforts to transform the city [. . .] and those of the later New Order lies in the fact that Sukarno, under the polemic of Indonesian socialism, was restricted to state resources, whereas the Suharto government is able to draw upon the now vastly expanded capital of the private sector. In fostering the role of the private sector in the current redevelopment of the city, it is apparent that New Order policy makers have also been willing to accept the specific forms of development favored by private capital. This is as true of suburban housing enclaves and multi-story shopping arcades as it is of the new "superblocks" now under construction in the midst of Jakarta's Golden Triangle (*Warta Ekonomi,* 1991). All of these forms of development are clearly New Order phenomena.

We see then that the ideological factors which underlie the attempt to transform Jakarta into a city of international standards have long preceded the attainment of the economic conditions necessary to make it happen. Perhaps it is ironic that some semblance of Sukarno's dream of Jakarta as a modern metropolis could only be achieved under the free market capitalist regime of the New Order. In the following section, we will consider the economic conditions underlying Jakarta's current development trends, with particular attention given to the growth of suburbs.

[. . .]

ECONOMIC FACTORS IN JAKARTA'S SUBURBANIZATION

The most significant factor affecting changes is the protracted movement of the Indonesian economy away from a purely colonial or neo-colonial role within the world economy to that of a diversifying economy with growing levels of indigenous ownership (Hill, 1989). Recent developments, such as increased local control of the agricultural and resource-extracting industries which were the basis of the colonial economy; the growth of indigenous middle management positions and a growing service economy in general as the country industrializes; and—at the high end of the social pyramid—the transformation of the so called "comprador" class into a growing body of industrial and finance capitalists (Yoshihara, 1988), have all contributed to a shift in patterns of accumulation which directly link to Indonesia's changing role within the world economy. As a broad generalization, there is an ongoing move in the locus of surplus accumulation in the post-colonial period from the metropoles to local centers in peripheral states. This shift has had significant local impacts, a number of which bear directly upon the issue of suburbanization.

The most visible of these impacts is the growth of Jakarta's middle class. For the time being, it may be more proper to use the term "consumer class," for in the Indonesian context, these people are by no means representative of the median strata of society. Nonetheless, their numbers are growing, and their purchasing power is growing even faster.

[. . .]

A related force which is influencing the behavior of Jakarta's consumer class is the growth of consumerist ethics. This is a trend again fostered in general by privatization and specifically by the growing size and sophistication of the advertising and communications industries. The impact of growing consumerism clearly favors the growth orientation of transnational capital, which must constantly look for new markets in order to maintain expansion. The growth of the consumer class is significant not only for the money they have in their pockets, but the aspirations which they hold, aspirations inspired largely by the promise of modernization. The consumerist ethic applies no less to the consumption of developer-built housing than it does to other market commodities; that this is understood by the developers themselves is evident from their sales brochures.

[. . .]

Due in large part to Indonesian property laws which sharply restrict foreign ownership, the real estate industry has emerged directly from indigenous capital. This is not to say that transnational capital has had no effect upon the industry, but only that this effect has been indirect.

[. . .]

A further indirect impact of foreign capital upon the industry has been through the functioning of the city's land market. The influx of foreign capital throughout the 1970s and 1980s has been highly concentrated in the commercial and office sectors at the city's core. By the late 1980s, the competition for land in the central city had resulted in the rapid escalation of land prices, with horror stories of commercial land prices doubling and tripling within a single year (*Tempo,* 1990). Ripple effects—including the impact of former inner city residents buying on the urban periphery—were felt in all land markets, with annual real increases in excess of ten percent for residential lands throughout the city (Dowall and Leaf, 1991). The long-term effect of this has been to continually push low-density land uses—such as suburban housing estates—outward to the edge of the city where lower land prices still allow a developer to build within BTN cost requirements. The outward expansion of urban infrastructure, particularly the freeway system which enables buyers to have direct access from peripheral estates to the central city, has accommodated this trend. These presumably natural market forces, propelled by land price leaps at the center, are working in concert with the regulatory and institutional milieu in effecting the rebuilding of Jakarta.

In summary, there are now three essential components in place for the growing suburbanization of Jakarta:

1 Demand, stimulated by rising real incomes within particular strata of society,
2 Supply, being created by a highly concentrated development industry with direct ties to the capitalist elites and through them to the political leadership, and
3 A policy regime which fosters low-density suburban development through subsidized housing finance; a permit system which empowers developers to aggregate atomized land holdings; the government-sponsored provision of trunk infrastructure; and a land use plan which promotes the further separation of residential and commercial functions in the City.

THE SUBURBAN PYRAMID

For analytical purposes, we have separated out the fundamental economic and ideological factors which underlie the emergence of the suburban settlement form in Jakarta. Yet the strong interlinkages between the ideology of modernization and the economic effects of capitalist expansion ensure that the two can never be fully separated. Market forces cannot properly be analyzed outside their particular ideological milieu, and here we are looking at processes of suburbanization within a context of growing consumerism and popular aspirations to Western standards of living. We have seen in Jakarta how the interconnectedness of economics and ideology is further reinforced through the regulatory functions of the government. Policy has steered market forces to obtain results—the suburban enclave and the modern city of which it is a part—which are in accordance with the state's ideological framework. This concurrence of economic and ideological forces is also supportive of the interests of developers, for they are operating from essentially the same ideological perspective, and are more than willing to meet the demand for "modern" suburban homes built in Western fashion. The suburban outcome of Jakarta's push to create a formal corporate housing sector is as much policy-driven as it is market-driven. If policy, working in concert with the market, has been able to achieve so much, what are the possibilities for the future?

[. . .]

It is now well understood that the earth's biosphere would not be able to support its current population if the average citizen of the planet consumed resources at the rate of the typical North American. By actively promoting a vision of Jakarta's future as a city which emulates the most profligate aspects of Western urbanization, Jakarta's policy makers are encouraging a decidedly unsustainable form of urban development.

In comparison with the many forms of human settlement of the past and present, the Western suburb appears to be more like a pyramid scheme than any other. There are a number of basic suburban characteristics, including low build densities, zoning for exclusively residential use, and a high dependency upon automobiles and other environmentally costly forms of infrastructure, which ensure that only a limited proportion of the world's population will ever be able to live the "classic" suburban lifestyle. In the recent past, under a neo-colonial world economic system which administered the transfer of wealth from resource- and labor-rich peripheral societies to those of the industrial and financial core, the mass of suburbanites were constituted in the countries of the core. Now, as North America undergoes a fundamental rethinking of suburbia as the preferred settlement form (and, not incidentally, the American middle class begins to decline), it is significant that this form of urbanization is arising elsewhere in the world. Can we interpret this as being a shift in the suburban pyramid scheme from the scale of the world economy to that of local societies?

REFERENCES

Abeyasekere, S. (1989). *Jakarta: A History* (revised edn), Singapore: Oxford University Press.

Anderson, B. R. (1972). "The idea of power in Javanese culture" in C. Holt (ed.), *Culture and Politics in Indonesia,* Ithaca, NY: Cornell University Press, 1–69.

Brown, L. B. and Jacobson, J. (1987). *The Future of Urbanization: Facing the Ecological and Economic Constraints* (Worldwatch Paper 77), Washington, DC: Worldwatch Institute.

Castles, L. (1989). "Jakarta the growing centre" in H. Hill (ed.), *Unity and Diversity: Regional Economic Development in Indonesia since 1970,* Singapore: Oxford University Press.

Dowall, D. and Leaf, M. (1991). "The price of land for housing in Jakarta," *Urban Studies,* 28: 707–22.

Hill, H. (1989). *Unity and Diversity: Regional Economic Development in Indonesia since 1970,* New York: Oxford University Press.

Kantor Mentri Negara Perumahan Rakyat (State Ministry for Housing) (1990). *Pembangunan Peru mahan Tahun 1990 (Housing Development in 1990),* Jakarta: State Ministry for Housing.

King, A. D. (1990). *Urbanism: Colonialism and the World Economy,* London: Routledge.

Lowe, M. (1991). *Shaping Cities: The Environmental and Human Dimensions* (Worldwatch Paper 105), Washington, DC: Worldwater Institute.

Robison, R. (1986). *Indonesia: The Rise of Capital, Asian Studies Association of Australia,* Southeast Asia Publication Series No. 13, Sydney: Allen and Unwin.

Struyk, R., Hoffman, M. L., and Katsura, H. M. (1990). *The Market for Shelter in Indonesian Cities,* Washington, DC: The Urban Institute Press.

Tempo (1990). "Laporan Utama: Gairah Membeli dan Arus Menguasi" (Lead Report: The lust to buy and the trend to control), XX: 84–90.

Warta Ekonomi (1991). "Superblock: Sebuah Kota de dalam Kota" (Superblock: A city within a city), III: 38.

Yoshihara, K. (1988). *The Rise of Ersatz Capitalism in Southeast Asia,* Singapore: Oxford University Press.

PART IV

The city environment

Figure 25 Baby Clothes, a woman sells clothes outside a maternity hospital located in the heart of old Karachi.
Source: © Naila Mahmood, 2012.

SECTION 6
Basic Urban Services

Editors' Introduction

What are urban services? They are the services that residents use every day: water, sanitation, and electricity that are fundamental to a healthy life; parks and other open spaces for recreation; streets and roads for mobility; community spaces to come together; schools to educate children; and police and emergency services that can put out fires, control floods, and respond to other disasters, small and large, everyday and episodic. This diverse array of services has differing levels of complexity in terms of production and consumption. They are produced by various entities from private non-state actors to governments and they are consumed at different scales from the individual to various collectivities. Allocation of basic services relies on a range of political and economic mechanisms.

There are several theoretical approaches to understanding the provision and impact of urban services. In economic terms basic services possess the characteristics of collective public goods, serve a social welfare function, and have externalities that include negative spillover effects (Weisbrod 1964). In neo-Marxist analysis the focus is on how the collective consumption of urban services facilitates the reproduction of labor. The sociologist Manuel Castells (1977, 1985) was the first to theorize the urban as the sphere of collective consumption and to examine its effects. He argued that the increasing role of the state in service provision did not resolve the contradictions of capitalism. Instead it led to collective action around consumption and produced social movements that challenged the state around the provision and management of urban services.

Across colonial cities in the global South, despite unequal distribution between the white settler and local population, the provision of urban services such as water and sanitation became an important project (McFarlane 2008). Service provision drew on ideas developed to remedy the ills of Victorian cities such as London, as King describes in his contribution to this volume. However, the idea of "basic" urban services (water, sanitation, roads, electricity) targeted at poorer city residents emerged around the late 1960s and 1970s with the introduction of the paradigm of "basic needs" in development (Goldman, this volume; Moon 1991). At this time, policies that promoted "sites and services" also began to dominate solutions to the low-income housing crisis in the megacities of Latin America, Asia, and Africa (Harris, this volume). This brought renewed attention to the provision of basic urban services.

In the early and mid-twentieth century, local governments dominated urban service provision. In addition, large portions of tax revenue went toward subsidizing these services. Today, this scenario has changed with services increasingly paid for by user charges and fees, and provided not only by local governments, but also by private firms, through public–private partnerships or by private firms contracted through and regulated by the government. Sometimes, another territorial level of government may be involved in service provision as in school districts, water utility and power utility districts or in subsidized food distribution systems. Urban services depend on extensively supporting infrastructure systems and often rely on interrelated technologies (for example, water and sewerage systems use electricity for pumping). How a service provision system works is important for analysis and policy planning. Nevertheless, the bulk of research and academic work on urban services has focused on the issues of allocation and access. As William Baer (1985: 883) has noted in reviewing the literature of urban service provision in the United States, the question of "who gets what, when, [and] how?"

remains central. Issues related to inequality of coverage and various methods of service provision dominate the study of basic urban service delivery in the cities of the global South as well.

In this introduction we start with Baer's question of "who gets what, when?" Highly inequitable levels of basic urban service provision are the norm across cities in the global South, with slums and low-income localities often having little or no access to services as compared to the more affluent middle- and upper-class areas. These produce inequitable outcomes in terms of health, quality of life, and economic productivity indicators. Even though larger cities in countries like China, Argentina, Indonesia, or Thailand have made considerable progress in improving coverage and narrowing the basic service provision gap, most lower-income urban residents across the South, and a disproportionate number of residents of smaller cities, rely on a patchwork of urban services often provided through private operatives working in the informal economy. The changing perspectives on basic service delivery are shaped by larger debates on development, urbanization, and economic growth, which currently emphasize working markets for particular services.

KEY ISSUES

"Who gets what, when?"

Huge inequities exist in the provision of basic urban services across global regions, between urban and rural areas in any given country, and within individual cities. A good example is access to safe drinking water and adequate sanitation. Since the Millennium Development Goal (MDG) targets were established by the UN in 2000, progress in drinking water and sanitation provision has been tracked more carefully than before. Despite acknowledged problems with standardized data collection and collation across countries, some trends are clear: there have been impressive gains in coverage (of which half is driven by China and India's growth). Yet striking disparities persist. Globally, drinking water provision by piped water on premises is highest in Latin America (86 percent) and lowest in sub-Saharan Africa at 16 percent. Sanitation coverage figures indicate that North Africa and the Caribbean are at 96 percent with Latin America at 94 percent and sub-Saharan Africa again the lowest at 30 percent. While overall global coverage for drinking water provision through all sources at 94 percent in 2010 is not much higher than in 1990, the numbers of people receiving coverage has gone up in those two decades by about 1 billion people (UNICEF and WHO 2012).

The inequality of drinking water and sanitation services between urban and rural residents is striking. Globally 80 percent of urbanites had access to piped water in 2010 while only 29 percent of rural residents enjoyed the same service. For sanitation, the figures stood at 79 percent and 47 percent with about a third of the rural population still practicing open defecation. However, regardless of where one lives, the chances of having access to sanitation or drinking water through pipes on premises are highest if one belongs to the top wealth quintile (the top 20 percent) of the population. In urban areas in Sierra Leone, in West Africa, for instance, the top wealth quintile had 97 percent coverage for drinking water. By contrast, the lowest wealth quintile (the bottom 20 percent) had only 56 percent coverage in urban areas and 10 percent in rural areas. While the percentage difference elsewhere may be a little less dramatic than in Sierra Leone, the disparities according to urban-rural residency and wealth remain significant even in China, which has seen the largest growth in terms of urbanites gaining access to drinking water provision. (All data from UNICEF and WHO 2012.)

The access and coverage data highlights inequities but does not reveal problems with quality of water or frequency in delivery. Water sources in many parts of the world are heavily dependent on climatic conditions. In China and India, for instance, water bodies and groundwater rely on monsoonal patterns for recharging. As a result, severe water shortages are not uncommon in one part of the country even as there is flooding in another. Climate change is expected to further intensify problems related to water availability, and the impacts on city supplies are a crucial topic for research. The duration and frequency of water in the pipes is also problematic, currently uncovered only through case studies and anecdotal evidence. Tales of taps spluttering water for an hour or so at two o'clock in the morning are not unusual. The extent to which piped water is contaminated by fecal or industrial waste and garbage that is dumped, unprocessed, into water bodies is also not fully recorded or understood, though several horrifying anecdotal and journalistic accounts exist.

Increasing coverage represents a considerable challenge and differs for each type of basic urban service. Coverage varies not just because of physical terrain but also due to the ways culture, governance institutions, power relations, financing and maintenance mechanisms shape outcomes. The provision of basic urban services continues to be a major focus of development aid. However, despite substantial evidence that governance reform and administrative issues are the most problematic issues in providing services, most aid emphasizes financing mechanisms (Rondinelli 1990; Bakker and Kooy 2008).

Impact of lack of access to basic urban services

The impacts of the lack of access and inequitable distribution of services are substantial. Purchased water in low-income neighborhoods (from tankers, for instance) is much dirtier than piped water delivered through municipalities. In addition, purchased water is more expensive. It can cost up to eight times more per unit than piped water (IRIN 2007).

Unsafe water, poor sanitation and hygiene are the cause of 4–8 percent of the overall disease burden across the global South, and nine-tenths of all diarrheal diseases that disproportionately affect young people and children under the age of five (Fay, Leipziger, Wodon and Yepes 2005: 1270). In addition, the poor lack access to public health education that assists prevention, and often cannot afford the medical care that could lower their health risks (Sverdlik 2011). By contrast, in some instances, the middle and upper classes lower risk because of their monopoly over whatever basic services the state provides (Chaplin 1999).

Difficulties in accessing water and sanitation place considerable burdens on women and female children. It is in areas where access to water is most difficult that these burdens are most onerous, whether through the extra time and energy spent in bringing water to the home or in finding space for safe storage. Women and girls have to often wait till after dark to relieve themselves, leading to various health issues. Finding a safe place to defecate is another problem, with instances of rape and violence not uncommon as women search for privacy or make their way to ill-lit public toilets at a distance from home (Brewster, Herrmann, Bleisch and Pearl 2006; World Bank 2011).

Access to electricity has another set of implications. Electrification can influence children's schooling outcomes by enabling pupils to do homework in the evenings under a light. In addition, children, especially girls, gain study time when they do not have to help forage for cooking or heating fuel. Electric power can also reduce incidents of tuberculosis and acute respiratory problems resulting from cooking fuels that produce smoke. This most seriously affects those responsible for most household cooking and chores, primarily young people and women (Hardoy and Satterthwaite 1991, this volume). The benefits of electricity extend into other areas of life as well, such as providing more leisure time and increased security in open places and streets. Unfortunately, comprehensive studies of the impact of service provision are in short supply. Further research is needed, for example, to assess the impact of the absence of parks and open spaces for children's play, the limited options for adult leisure, and the lack of access to safe streets.

Provision of and access to basic services remain among the biggest issues of our time in the cities of the global South. Interventions in basic urban services produce multi-sectoral, linked outcomes with intergenerational effects. For example, reducing ill-health and chronic poverty increases community and city resilience and enhances people's ability to cope with and adapt to a range of risks–from disasters and violence to climate change. (See Section 8 Cities at Risk, this volume.)

So far, we have focused on the issue of "who gets what, when?" We will now turn to the second part of the question: how basic urban services are provided.

How are basic urban services provided?

Typically enormous civil-engineered infrastructure systems bring services like water, sanitation and electricity to city dweller's homes. These systems have historically been created, maintained, and regulated by local government entities, often in response to organized citizen pressure, as the early periods of rapid urbanization in the global North have demonstrated (Monkkonen 1988; Chaplin 2011). While cities across the global South share some of this experience of state-provided infrastructure, the informal or community-based provision of basic services dominates and continues to expand. It is linked both to longstanding, traditional systems of service provision (such as the garbage system described by Assaad, this volume) and to entrepreneurial systems produced under modern conditions of scarcity and crisis (Batley and Moran 2004; Moretto 2006; see also Editors' Introduction to Urban Economy, this volume).

In the developmentalist approaches that continue to dominate planning for urban areas across the South (Robinson, this volume), basic service delivery of water and sanitation, streets and occasionally electricity were often tied to low-income housing provision. A site typically came with a core bundle of services: a connection to a water main, a sewerage system, and an electrical grid as well as access to streets. In the late 1970s, with the advent of slum upgrading policies, attention was paid to providing infrastructure systems within existing settlements, a process that proved to be expensive and difficult to implement. The questions of whether water flowed regularly and predictably through the pipes, or whether roads connected settlements to places of employment, and whether there was enough electricity being generated to meet increasing demand were left unanswered.

The shift to neoliberal development strategies under structural adjustment starting in the 1980s resulted in a push for decentralization, privatization, and an increasing role for markets in the provision of public services (Rivera 1996; World Bank 2000/2001; see Editors' Introduction to Governance, this volume). The alleged benefits of privatization included reductions in bloated and corrupt state apparatus, reduced subsidies and raised tariffs to pay for increased coverage, and more efficient service provision through non-state actors (see Editors' Introduction to Governance and Mascarenhas, this volume). There is a large literature on the experience of decentralized, privatized service delivery across the global South, which we discuss in the next section of the *Reader*. Here we will focus on the ways in which the role of citizens and residents in service delivery is articulated.

The role of citizens and residents has been depicted in a number of ways. Are city residents *victims* of government inaction or manipulation by populist politicians? Are these residents *villains* who opportunistically take advantage of state-provided subsidized services? Alternatively should we view them as entrepreneurial *fixers* who solve the problem of lack of access by self-provisioning–creating and managing urban services independent of the authorities (see Beall, Crankshaw and Parnell, this volume)? Harriss (2005) stresses this last point, emphasizing that the urban poor differ from the middle classes and the elite in the ways in which they "fix" their service provision problems, relying much more on informal mechanisms of self-provisioning than political claim-making through patron-client relations with politicians or, even, aid agencies.

Castells's insights into collective consumption as helping forge movements against the state also continue to find purchase across cities in the global South. Examples include a range of activist politics. For example, the Mumbai-based NGOs, SPARC (Society for the Promotion of Area Resource Centers), Mahila Milan, and the National Slum Dwellers Federation, organized *sandas melas* (toilet fairs) to raise awareness, organize communities and advocate with the state and international funders for better services that take poor people into account–particularly the needs of women (Appadurai, this volume; Patel 2013). In Porto Alegre, Brazil, the people's councils successfully pressed for more equitable capital expenditure on basic services through a process known as participatory budgeting (Baiocchi, this volume). Other important instances of such collective activism include the organized resistance of residents associations in China (Read 2012) and the activist struggles of summer 2013 in Istanbul, Turkey, against the state's takeover of Gezi Park. In Brazil, protests erupted against corruption and state spending on mega-events like the World Cup and the Summer Olympics as opposed to addressing issues of basic services in poor communities. As Patel (2013) reminds us in her assessment of India's long-term and heavy involvement in the provision of basic services to the poor, governance arrangements that are not inclusive of the poor are rarely effective.

ARTICLE SELECTIONS

The article selections touch on three of the issues this introduction has framed. The first selection comes from the seminal paradigm-changing article, "Environmental Problems of Third World Cities: A Global Issue Ignored?"

by urbanists and planners Jorge Hardoy and David Satterthwaite (1991). The problems that Hardoy and Satterthwaite wrote about over two decades ago persist. They were among the first to argue that the most serious urban environmental problems are the result of a lack of urban services. The authors highlighted the impacts of issues such as inadequate and unsafe drinking water, insufficient provision for sanitation and solid waste disposal, ineffective pollution control, crowded conditions, and hazardous working conditions. They also noted that associated environmental burdens land disproportionately on poorer groups. Hardoy and Satterthwaite's work, and that of many others since, connects environmental issues to poor basic urban services. While other issues like biodiversity and climate change do impact local environments, lack of access to services continues to be a substantial environment problem that shapes people's life chances across the cities of the global South.

The second selection is from an article co-authored by a well-known development economist specializing in urban issues, Jo Beall, sociologist Owen Crankshaw, and geographer Susan Parnell (2000). In this excerpt, they describe the inequitable distribution of services given apartheid's spatial and political legacy in South Africa, where the white nationalist government systematically segregated Africans and Colored (mixed-race) people and denied them certain basic services. During apartheid, rent boycotts of state-owned housing and struggles around urban services played an important role in political mobilization and protest that led to the overthrow of the regime. Yet, these same groups became villains in the post-apartheid regime when they protested privatization and cost-recovery for services, which had become the new norm. Beall, Crankshaw, and Parnell focus on the different ways in which citizens have been conceptualized with regard to basic service provision by the developmentalist state. The implications of these positions have been discussed in the introduction.

The third selection comprises an excerpt from Egyptian planner Ragui Assaad's path-breaking article on the Zabbaleen and the Wahiya communities in Cairo. It was one of the first works to describe an informal but highly organized system of garbage collection in two communities and their symbiotic relations. More recently, the literature on service provision through informal systems in water provision, garbage collection, sanitation and housing in several other contexts has exploded. Assaad also describes the difficulty of trying to formalize the informal. He is interested in what gets lost in this process of translation in one of the first municipal experiments at engaging and even sub-contracting garbage collection to informal service providers.

REFERENCES

Assaad, R. (1996) "Formalizing the informal? The transformation of Cairo's refuse collection system," *Journal of Planning Education and Research*, 16(2): 115–126.

Baer, W. C. (1985) "Just what is an urban service, anyway?" *Journal of Politics*, 47(3): 881–898.

Bakker, K. and Kooy, M. (2008) "Governance failure: Rethinking the institutional dimensions of urban water supply to poor households," *World Development*, 36(10): 1891–1915.

Batley, R. and Moran, D. (2004) *Literature Review of Non-State Provision of Basic Services*, Birmingham: International Development Department, School of Public Policy, the University of Birmingham.

Beall, J., Crankshaw, O., and Parnell, S. (2000) "Victims, villains and fixers: The urban environment and Johannesburg's poor," *Journal of Southern African Studies*, 26(4): 833–855.

Brewster, M., Herrmann, T. M., Bleisch, B., and Pearl, R. (2006) "A gender perspective on water resources and sanitation," *Wagadu: A Journal of Transnational Women's and Gender Studies*, 3 (special issue). Available online at http://appweb.cortland.edu/ojs/index.php/Wagadu/article/viewArticle/249/462 (accessed May 16, 2013).

Castells, M. (1977) *The Urban Question: A Marxist Approach,* Cambridge, MA: The MIT Press.

Castells, M. (1985) *The City and the Grassroots,* Berkeley: University of California Press.

Chaplin, S. (1999) "Cities, sewers and poverty: India's politics of sanitation," *Environment and Urbanization*, 11(1): 145–158.

Chaplin, S. (2011) "Indian cities, sanitation and the state: The politics of the failure to provide," *Environment and Urbanization*, 23(1): 57–70.

Fay, M., Leipziger, D., Wodon, Q., and Yepes, T. (2005) "Achieving child-health-related Millenium Development Goals: The role of infrastructure," *World Development*, 33(8): 1267–1284.

Hardoy, J. and Satterthwaite, D. (1991) "Environmental problems of Third World cities: A global issue ignored?" *Public Administration and Development*, 11: 341–361.

Harriss, J. (2005) "Political participation, representation and the urban poor: Findings from research in Delhi," *Economic and Political Weekly*, 40(11): 1041–1054.

IRIN. (2007) "Kenya–the fight for water, a valuable slum commodity," *IRIN Humanitarian News and Analysis*. A Service of the UN Office for the Coordination of Humanitarian Affairs. Available online at http://www.irinnews.org/report/73718/kenya-the-fight-for-water-a-valuable-slum-commodity (accessed July 17, 2013).

McFarlane, C. (2008) "Governing the contaminated city: Infrastructure and sanitation in colonial and post-colonial Bombay," *International Journal of Urban and Regional Research*, 32(2): 415–435.

Monkkonen, E. (1988) *America Becomes Urban: The Development of US Cities & Towns 1780–1980*, Berkeley, London: University of California Press.

Moon, B. (1991) *The Political Economy of Basic Human Needs,* Ithaca, NY: Cornell University Press.

Moretto, L. (2006) "Urban governance and multilateral aid organizations: The case of informal water supply systems," *The Review of International Organizations*, 1(4): 345–370.

Patel, S. (2013) "Upgrade, rehouse or resettle? An assessment of the Indian government's basic services for the urban poor (BSUP) program," *Environment and Urbanization*, 25(1): 177–188.

Read, B. (2012) *Roots of the State: Neighborhood Organizations and Social Networks in Beijing and Taipei*, Palo Alto: Stanford University Press.

Rivera, D. (1996) *Private Sector Participation in the Water Supply and Wastewater Sector: Lessons from Six Developing Countries*, Washington, DC: World Bank.

Rondinelli, D. (1990) "Financing the decentralization of urban services in developing countries: Administrative requirements for fiscal improvements," *Studies in Comparative International Development*, 25(2): 48.

Sverdlik, A. (2011) "Ill-health and poverty: A literature review on health in informal settlements," *Environment and Urbanization*, 23(1): 123–155.

UNICEF and WHO. (2012) *Progress on Drinking Water and Sanitation, 2012 Update.* Available online at http://www.unicef.org/media/files/JMPreport2012.pdf (accessed July 23, 2012).

Weisbrod, B. (1964) "Collective-consumption of individual-consumption goods," *The Quarterly Journal of Economics*, 78(3): 471–477.

World Bank. (2001) "Making markets work better for poor people" in *World Development Report 2000/2001: Attacking Poverty,* Washington, DC: World Bank, pp. 61–76.

World Bank. (2011) *Violence in the City: Understanding and Supporting Community Responses to Urban Violence*, Washington, DC: World Bank.

"Environmental Problems of Third World Cities: A Global Issue Ignored?"

***Public Administration and Development* (1991)**

Jorge E. Hardoy and David Satterthwaite

SUMMARY

This article describes the massive scale and range of environmental problems in Third World cities, considered in terms of the impact mainly on human health. The first half of the article is an overview of these problems at different geographic scales, ranging from the home and workplace to the city region. It also discusses the interaction between city-based production/consumption and environmental degradation in the wider region. The main problems identified include unsafe and inadequate water supplies, inadequate provision for sanitation and solid waste disposal (including toxic waste), overcrowding, hazardous working conditions and ineffective pollution control. The second half presents some conclusions. The poorer groups in cities suffer most of the environmental burden. Governments and aid agencies allocate little to addressing the most serious environmental problems; local government is weak and ineffective in most Third World nations and citizen groups and NGOs that might offer some redress are often repressed. But without representative local government, and without NGOs and citizen group action, these problems are unlikely to be solved. Finally, different perceptions as to what constitute the World's major environmental problems threaten to divide North from South. If the North wants the South's cooperation in addressing global problems, it must help the South address those environmental problems which impact most on the health and livelihoods of its poorer citizens.

CONCLUSIONS

[. . .]

Perhaps the least surprising conclusion of this review is that it is the poor who bear most of the ill-health and other costs of environmental problems. Their houses and neighborhoods are the worst served with water, sanitation, garbage collection, paved roads and drains. When figures for average per capita water consumptions for whole cities are in the range of 20–40 liters per person per day (well below the minimum required for good health), most poorer groups' consumption is likely to be less than half the average. (See, for example, Schteingart, 1989.) Many studies of intraurban differentials in health status or infant mortality show rates in poorer areas being several times the city average (Harpham, Lusty and Vaughan, 1988; Basta, 1977; Rohde, 1983; Guimaraes and Fischmann, 1985). Poorer groups have the least possibility of medical treatment if they are injured at work or fall ill from pollution. They will also have the least possibility of taking time off to recover from sickness or injury, because the loss of income from doing so would press heavily on their survival.

It is almost always the poorer groups who live in the places where the pollution levels are worst; as noted earlier, poorer groups often choose such places because these are the only locations where they can find land for their housing close to sources of employment without fear of eviction. It was the high concentration of low income people around the Union Carbide Factory in Bhopal which caused so many people to be killed or permanently injured by the leak of methyl isocynate. In Mexico City, the highest concentrations of dust particles in the air are found in the south-east and north-east areas, which are areas where lower income groups live (Schteingart, 1989). In Manila, some 20,000 people live around a garbage dump known as Smokey Mountain, where the decomposition of organic wastes produces a permanent haze and a powerful, rank smell affecting the whole area. Some of these people have lived there 40 years or more. Moving to a cleaner, safer location is beyond their means, and many of them make a living scavenging on the dump, often sorting it with their bare hands (Jimenez and Velasquez, 1989).

It is also the poorest groups who suffer most from the floods, landslides or other "natural" disasters which have become increasingly common in cities of the Third World. Once again, poorer groups occupy the sites most at risk, since these are often the only centrally located sites where poorer groups can find (or build) housing. Each time a disaster is reported, most of those who die or are injured, and most of those evacuated and lodged in transitory shelters, are from lower income groups. They lose their housing (often their only capital asset) because no provision was made by their society to allow them a safer site; they lose their source of income as they are moved by some public agency to a different ("safer") site which is too far from their previous source of work. Moving to a new site often damages their contacts with family, friends and others who have helped them find work or survive periods of no income (Hardoy, 1988).

In the regions around cities, it is generally the poorer households who suffer, as their cheap or free sources of wood and land for grazing are pre-emptied by commercial concerns for use to meet urban demands, while their air and water suffer from wastes originating in the cities. It is possible to envisage governments in the more prosperous nations or cities taking action to meet the demands of richer households and businesses for more water, better drains and a cleaner environment, and still ignoring the more pressing problems of poorer groups, both within the city and within the surrounding region. A second conclusion is that very few governments or aid agencies give much attention to cities' environmental problems—especially the problems which impact most on the health and livelihoods of poorer groups.

[. . .]

Yet a lack of piped water and sanitation, of adequate housing on safe sites with paved roads and drains, of regular collection of household wastes, and of health care systems oriented towards prevention and the provision of health care and emergency services, are the main factors in the very unhealthy housing and living environments which affect close to half of the Third World's urban population. Deficiencies in water supply, sanitation and housing and the provision of health care are responsible for tens of millions of preventable deaths each year, and they contribute to serious ill-health or disablement for hundreds of millions (Cairncross *et al.*, 1990a). As such, they should rank as environmental problems which deserve a high priority in cities and in smaller towns and rural areas.

Among the other interventions which could help improve city environments, few aid agencies give much support to city or municipal governments to improve garbage collection and disposal, or to implement more effective pollution control. Public transport tends to receive support only where this means large civil construction contracts for companies from the donor nation. A few agencies have, in recent years, given more support to developing the institutional capacity, of city and municipal governments, to address such problems—perhaps most notably the World Bank and, in technical assistance, the joint Urban Management Program of the World Bank, UNDP and the United Nations Centre for Human Settlements; however, the scale of such support remains small in relation to needs (Satterthwaite, 1990; UNCHS, 1989, 1991; Guarda, 1990).

A third conclusion is that major improvements can be made to city environments at relatively modest per capita costs through six interventions (Cairncross *et al.*, 1990b):

- water piped into or near to the home of each inhabitant;
- systems installed to remove and dispose of human excreta;
- a higher priority to the installation and maintenance of drains and services to collect garbage;

- primary health care systems available to all (including a strong health prevention component which is so central in any primary health care system);
- house sites which are not on land prone to flooding, landslides, mudslides or any other site-related hazard made available to poorer groups;
- the implementation of existing environmental legislation.

The cost constraints for applying these measures appear to be overstated; work on alternatives to conventional water, sanitation and garbage collection systems have shown a range of cheap options (Kalbermatten *et al.*, 1980; Cairncross, 1990; Sinnatamby, 1990; Furedy, 1990; Cuentro and Gadji, 1990), while poorer groups' willingness and ability to pay for improved services—if these match their own priorities—appears to have been underestimated (Cairncross *et al.*, 1990b). In many cities, it is not so much a lack of demand for water, sanitation, health care and garbage collection that is the problem, but an institutional incapacity to deliver cheap and effective services.

A fourth conclusion is linked to this—that the long-term solution to any city's environmental problems depends on the development within that city of a competent representative local government. Lessons drawn from 40 years' experience of national or international agencies demonstrate that most local problems need local institutions. Outside agencies may bring knowledge, expertise, capital and advice—but they cannot solve most environment problems without effective local institutions.

[. . .]

Basic infrastructure and services—water, sanitation, garbage disposal, health care, drains—cannot be adequately provided to poorer groups without effective local government. There may be potential for private sector enterprises to provide some of these, but most of the literature on privatization overstates this potential and may even forget that effective privatization needs strong, competent and representative local government to oversee the quality and control prices charged, especially in those services which are a natural monopoly (Cairncross *et al.*, 1990a).

A fifth conclusion is that most Third World nations cannot follow the path taken by European or North American governments towards environmental policies. One obvious difference is the level of wealth and prosperity. But three other differences need stressing.

(1) Most Third World economies, employment patterns and foreign exchange earnings are more dependent on natural resource exploitation. In the North, a very small proportion of the labor force depend on farming or forestry, so a decision to protect the beauty of a rural landscape by putting controls on farming or forestry hardly affects the balance of payments, the scale of economic activities or employment patterns. The impact of comparable controls is much greater in nations where a substantial proportion of foreign exchange earnings and jobs depend on farming and forestry.
(2) Virtually all Third World nations lack the institutions and the infrastructure on which to base effective actions to address urban environmental problems. In Europe and North America, when environmental awareness began to grow in the early 1960s, the basic infrastructure was already in place—the water pipes, drains, sewers and treatment plans, the landfills sites—even if more investment was needed to improve the quality of sewage treatment and solid waste disposal. Perhaps more importantly, so too was the institutional structure, especially the local government structure and an established system to fund local government investments and expenditures. Most Third World governments cannot implement comparable environmental policies because of weak institutional structures (especially at local government level), very inadequate capital budgets, large backlogs in basic infrastructure and an economy far less able to generate the capital needed to address the backlog.
(3) There are also important differences in the nature of government. In western Europe and North America, action on environmental problems was almost always citizen-led; citizen groups and NGOs provided the lobby which eventually produced changes in government policy. This required representative government structures. Many Third World nations do not have representative governments; many of those which have returned to democratic rule at national level still retain relatively undemocratic structures at local level. These are unlikely to implement environmental policies unless there are strong

democratic pressures pushing them in these directions. There has been no shortage of citizen groups pressing strongly for water, sanitation, health care, and the right to live in cities without the constant fear of eviction, but this does not appear to have had much impact on the housing and health policies of most Third World governments.

This link between democracy and action on the environment is illustrated by the literature on environmental problems in Third World cities. Our review of this literature produced numerous examples from India and Malaysia. This might be taken to imply that problems were more serious there. We suspect a different reason—the fact that in both nations, there are national and local NGOs active on environmental issues. They include some, such as the Centre for Science and Environment in Delhi, the Consumers' Association of Penang and Sahabat Alam Malaysia, which have international reputations. If this is the reason, it gives a clue about how to mobilize action in the Third World. Outside agencies wanting to stimulate more action on environmental problems in cities might consider providing support direct to citizen groups and NGOs within these cities to document the problems and organize lobbies.

A sixth conclusion is that the environmental issues which dominate discussions in Europe and North America about the Third World are not the environmental issues which pose the most serious threats to most Third World citizens' health and well-being. For instance, the concentration in the North on the issue of global warming—especially in the lead-up to the 1992 United Nations Conference on Environment and Development—risks exacerbating the divisions between North and South apparent 20 years ago, when a concern for "environment" was seen by many as a way through which the North could limit the competitiveness of Southern economies in the world market. In addition, the continued stress by groups in the North on the links between environmental degradation and "over-population," where the South is seen as the guilty party, is also divisive. Such a stress is also of questionable accuracy when it is the high consumption life-styles of the relatively well-off—most of them in the North—which accounts for most non-renewable resource consumption, most of the generation of toxic and hazardous wastes, most of the greenhouse gases currently in the atmosphere, and a considerable proportion of the pressure on soils and forests. In addition, it can be argued that it is not over-population which underlies most of the deforestation and soil erosion caused by poorer groups, but the inequitable land-owning structures which prevent poorer groups obtaining sufficient land to allow sustainable exploitation; in most nations with significant amounts of rural poverty, there is no absolute shortage of good quality agricultural land.[1]

Most Third World citizens find it difficult to share the concerns of the North on global warming. Questions of survival 20 or more years into the future have little relevance to those concerned with survival today. Attempts by the governments of the North (or the United Nations system) to obtain agreement on controls of greenhouse gas emissions should have, as a precondition, support for Third World societies to develop the capacity to tackle the environmental problems which daily affect the health and livelihoods of poorer groups. The North should not hope to promote important long-term perspectives on maintaining the integrity of global systems in the Third World, when such a high proportion of Third World citizens are suffering from such enormous short-term environmental health problems.

A joint program to address environmental problems must have the long-term goal of building the capacity within each Third World society to identify, analyse and act on their own environmental problems. Building such a capacity in turn demands action on such issues as impossible debt repayment levels and the removal of protectionist barriers around the North's own markets. More prosperous, stable economies in the Third World, with citizens no longer suffering constantly from preventable diseases, disablement and premature deaths arising from environmental problems, should be preconditions for global agreements. They also provide a more realistic basis from which to advance other important development goals—not least among these being stronger democracies and more effective, accountable local governments.

NOTE

1 If the degree of a nation's "over-population" is measured by land area per inhabitant, cultivable land area per inhabitant, non-renewable resource consumption per

inhabitant, or contribution to greenhouse gases now in the atmosphere per inhabitant, most Third World nations and virtually all the poorest Third World nations are among the least "over-populated" nations in the world.

REFERENCES

Basta, S. S. (1977) "Nutrition and health in low income urban areas of the Third World," *Ecology of Food and Nutrition,* 6, 113–124.

Cairncross, S. (1990) "Water supply and the urban poor," in S. Cairncross, J. E. Hardoy and D. Satterthwaite (eds), *The Poor Die Young: Housing and Health in the Third World,* Earthscan: London, 109–126.

Cairncross, S. *et al.* (1990a) "The urban context," in S. Cairncross., J. E. Hardoy and D. Satterthwaite (eds), *The Poor Die Young: Housing and Health in the Third World,* Earthscan: London, 1–24.

Cairncross, S. *et al.* (1990b) "New partnerships for healthy cities," in S. Cairncross., J. E. Hardoy and D. Satterthwaite (eds), *The Poor Die Young: Housing and Health in the Third World,* Earthscan: London, 245–268.

Cuentro, S. C. and Gadji, D. M. (1990) "The collection and management of household garbage," in S. Cairncross, J. E. Hardoy and D. Satterthwaite (eds), *The Poor Die Young: Housing and Health in Third World Cities,* Earthscan: London, 169–188.

Furedy, C. (1990) "Social aspects of solid waste recovery in Asian cities," *Environmental Sanitation Reviews,* no. 30, ENSIC, Asian Institute of Technology, Bangkok, 2–52.

Guarda, G. C. (1990) "A new direction in World Bank urban lending in Latin American countries," *Review of Urban and Regional Development Studies,* 2(2), 116–124.

Guimaraes, J. J. and Fischmann, A. (1985) "Inequalities in 1980 infant mortality among shantytown residents and non-shantytown residents in the municipality of Porto Alegre, Rio Grande do Sul, Brazil," *Bulletin of the Pan American Health Organization,* 19, 235–251.

Hardoy, J. E. (1988) "Natural disasters and the human costs in urban areas of Latin America," (Mimeo), IIED Human Settlements Programme, London.

Harpham, T., Lusty, T. and Vaughan, P. (eds) (1988) *The Shadow of the City: Community Health and the Urban Poor,* Oxford University Press: Oxford and New York.

Jimenez, R. D. and Velasquez, A. (1989) "Metropolitan Manila: A framework for its sustained development," *Environment and Urbanization,* 1(1), 51–58.

Kalbermatten, J. M., De Anne, J. D. and Gunnerson, C. C. (1980) *Appropriate Technology for Water Supply and Sanitation: Technical and Economic Options,* World Bank: Washington.

Rohde, J. E. (1983) "Why the other half dies: The science and politics of child mortality in the Third World," *Assignment Children,* 61–62, 35–67.

Satterthwaite, D. (1990) "La ayuda internactional," in Clichevsky, N. *et al.* (ed.), *Construccion y Administracion de la Ciudad Latinoamericana,* IIED-AL y Grupo Editor Latinoamericano: Buenos Aires, 435–492.

Schteingart, M. (1989) "The environmental problems associated with urban development in Mexico City," *Environment and Urbanization,* 1(1), 40–49.

Sinnatamby, G. (1990) "Low cost sanitation," in S. Cairncross, J. E. Hardoy and D. Satterthwaite (eds), *The Poor Die Young: Housing and Health in the Third World,* Earthscan: London, 127–167.

UNCHS (Habitat) (1989) "Financial and other assistance provided to and among developing countries for human settlements—report of the Executive Director," Biennial Report presented to the Commission on Human Settlements at their meeting in Cartagena, April.

UNCHS (Habitat) (1991) "Financial and other assistance provided to and among developing countries for human settlements—report of the Executive Director," Biennial Report presented to the Commission on Human Settlements at their meeting in Harare, April.

FOUR

"Victims, Villains and Fixers: The Urban Environment and Johannesburg's Poor"

Journal of Southern African Studies (2000)

Jo Beall, Owen Crankshaw and Susan Parnell

INTRODUCTION

In their relationship with the urban environment, poor urban dwellers are variously characterized as victims, villains or fixers (Rocheleau 1990). As victims they are seen to suffer from poor services and environmental conditions, a situation highlighted in the South African context by apartheid's spatial legacy and racial inequalities in the segregation of urban residential areas and the provision of public goods and services. As villains, urban people in poverty are seen as perpetrators of environmental degradation through illegal, wasteful and polluting practices.

In Johannesburg their role as villains has been additionally underscored by past and on-going practices of boycotting of rents and service charges and illegal tapping of municipal services.

As fixers, poor people are often called upon to participate in community-based responses to environmental management, cost sharing or payment of user charges for service provision and maintenance. Two critical issues are being debated and explored in Johannesburg in the present context. The first is, which citizens are to participate in community-based responses to environmental management, given that in the past little participation was expected of better-off (predominantly white) residents, while poorer (predominantly African) urban dwellers often had to rely almost entirely on their own resources or initiatives? Second, and equally crucial is how service provision and maintenance is to be paid for in the longer term. Both issues constitute major social, economic political and environmental challenges for the Greater Johannesburg Metropolitan Council (GJMC), the new metropolitan government structure for the city.

This article explores the relationship of Johannesburg's poor to the urban environment and more specifically to three key urban services, water supply, sanitation and electricity. [Ed. note: electricity is not included in this excerpt.] This focus is justified because these services are of central importance to environmental health, urban economic growth and social relations.

The provision of basic services and the construction of infrastructure to meet the basic needs of the poor are the widely accepted priority of both the post-apartheid government and of the GJMC, a priority that far outweighs any other urban environmental focus. The metropolitan council is equally determined to establish the commercial viability of service delivery, both to improve efficiency and in order to facilitate its commitment to a level of cross-subsidization across the city. However, while these issues have been politically and legislatively resolved, they have not yet stood the test of implementation.

[...]

VICTIMS: APARTHEID'S LEGACY

Under apartheid, South Africa was a country exhibiting levels of inequality in wealth and access to services among the highest in the world (Wilson and Ramphele 1989: 4). A combination of policies and legislation dating from the early 20th century consistently denied Africans vital components of well-being and a secure base in the cities where, in principle at least, they were not allowed to live permanently. This gave rise to racial imbalances in the provision of housing, infrastructure and services, which were inherited by post-apartheid local governments.

The legacy of apartheid impacted specifically on the provision of services in Johannesburg in two ways. First, the well-known policy of providing inferior quality services for Africans meant that standards of social and physical infrastructure were intentionally set lower than they were for whites. In public education, health, housing and transport, racially defined standards of construction and service gave tangible expression to the political and economic hierarchy on which white supremacy was based. The second explanation relates to the decision taken by the apartheid government in 1968 to stop the development of African areas in "white" South African cities. The metropolitan outcome of the policy of separate development, which insisted that African development be restricted to rural settlements or small towns in racially defined homelands, was the massive backlog of housing and infrastructure development in the old township areas of Johannesburg (Beall, Crankshaw and Parnell 2000).

[. . .]

It is not just apartheid South Africa that has provided adequate and reliable services to only a minority of its urban citizens. Across many cities of the South, mains water and sewerage connections are concentrated in better off areas, while new investment has tended to be in existing serviced areas. Thus it is common for local governments to subsidize elites heavily in terms of urban services (Jarman 1997; Black 1994) and this is compounded by the fact that cost sharing or community participation increasingly and commonly characterizes new investment in low-income areas. Referring specifically to inequalities in access to urban water and sanitation facilities, Caroline Stephens identifies the health inequalities that can arise:

> the urban poor often have least access to piped water and are forced to pay more than the wealthy for poor quality and limited quantities of water from vendors. This becomes a doubly regressive taxation in which one group is doubly disbenefited (in health and economic terms) while another doubly gains. Put bluntly, the poor pay more for their cholera. (Stephens 1996: 16)

What follows in this section is the presentation of some statistical data drawn from our own analysis of the October Household Survey (OHS) on provision of water supply [and] sanitation in Johannesburg. The data provide a picture of the conditions of some of apartheid's victims when a democratic government took office in 1994.

One thing that emerges from the tables below is that when access to such services is used as an indicator of poverty, then Johannesburg's poor are better off than many other urban dwellers across the continent. In Africa, 36 percent of the urban population are thought to be without an adequate water supply and 45 percent are not covered by sanitation (Nunan and Satterthwaite 1999). It should also be pointed out that the situation of Johannesburg's poor also compares well with national figures. For example, it has been estimated that for the country as a whole, in the immediate post-apartheid period, only 21 percent of households had access to piped water and only 28 percent to sanitation facilities. Over 80 percent of poor rural households had no access to either (May 1998: 138). Nevertheless, intra-urban inequalities exist and this is undoubtedly the most startling picture that emerges within Johannesburg. Although almost all the residents of backyard shacks and informal settlements are African, there is nonetheless considerable differentiation within the African population. [. . .] Whereas coloured, Indian and white households live almost exclusively in formal houses or flats, African households are distributed across a much wider range of informal and formal types of accommodation.

WATER

In Johannesburg, the metropolitan government is acutely aware that the extensive and generally very effective network of water coverage is not universal across the city.

[. . .] Moreover, in a pattern that is repeated across many services, standards of water delivery vary enormously according to both the original racial occupation of the suburb and the type or generation

Source of water	African	Coloured	Indian	White	All races
Tap in house/flat	67	100	100	97	80
Tap on the stand	29	0	0	3	18
Public tap/kiosk/borehole	4	0	0	0	2
Total	100	100	100	100	100

Table 1 Main source of domestic water in Johannesburg by race (%)

Source: Own analysis of the 1995 October Household Survey

Note: The following analyses are for the magisterial districts of Johannesburg, Randburg, Roodepoort and Soweto, the boundaries of which approximate the present boundary of the Greater Johannesburg Metropolitan Council. The OHS data are weighted to reflect the actual size of the provincial populations and not the population of the metropolitan areas. However, because the population of the GJMC contributes a relatively large proportion to the province of Gauteng, the weighting of the data for the districts that make up the GJMC is fairly accurate. The total number of households reflected in the weighted results is 396,809, which represents a population of roughly 2.4 million. This figure is only slightly higher than the 1996 Population Census estimate of 2.3 million. Another problem with using the OHS data at the level of only a few districts is that the sample size is only 711 households (339 African, 95 coloured, 80 Indian and 197 white households). This means that in this and most of the following tables, the figures are accurate to within 10 percent in 95 percent of cases.

Type of sanitation	African	Coloured	Indian	White	All races
Flush toilet in dwelling	50	89	94	99	70
Flush toilet on site	38	11	6	1	23
Toilet off site (all types)	4	0	0	0	3
Other toilet on site (chemical and bucket)	5	0	0	0	3
Pit latrine on site	2	0	0	0	1
Total	100	100	100	100	100

Table 2 Type of sanitation provision in Johannesburg by race (%)

Source: Own analysis of the 1995 October Household Survey Column totals may not add up to 100 percent due to rounding up/down.

of housing in which a household lives. In practice and as illustrated in Table 3, the worst supplied are the more recently urbanized African households or children of established urban dwellers who have moved from formal housing stock in townships into informal settlements. There are limited cases where water is trucked into informal settlements and a small fraction of the population depends on water from rivers and streams. However, taps in houses or on the stand provide most of the population with safe and drinkable water.

[. . .]

Compared to other cities of the South, purchase of water from private vendors is relatively uncommon in Johannesburg. However, in townships, rental arrangements mean that sub-tenants pay landlords for water and in informal settlements shacklords can control access to water sources. Consequently, a large proportion of those paying for water in the Province of Gauteng, of which Johannesburg forms a significant part, are not necessarily paying the service provider but an intermediary. What this signals is that the politics of water supply are not confined to who gets what at the level of the city. They also involve vested interests at the local level that will oppose improvements in delivery because they are making a profit from the current unregulated supply of water.

Source of water	Formal dwelling (house or flat)	Formal dwelling in backyard	Informal dwelling in backyard	Informal dwelling not in backyard*	Hostel	Other	All housing types
Tap in house/flat	90	65	0	4	100	40	80
Tap on the stand	10	34	55	78	0	60	18
Public tap (incl. kiosk/borehole)	0	2	45	18	0	0	2
Total	100	100	100	100	100	100	100

Table 3 Main source of domestic water by type of housing in Johannesburg (%)

Source: Own analysis of the 1995 October Household Survey

*These include free-standing squatter settlements and informal settlements set up or sanctioned by the state.

Column totals may not add up to 100 percent due to rounding up/down.

It is frequently argued in the international literature that because poor urban residents pay informally for water supply that they are therefore willing to pay for a more efficient formal water supply. This assumption is reinforced by the affirmative results from contingent valuation and other methods for assessing demand for services. However, it is also increasingly recognized that willingness to pay is simply that, willingness to pay and does not necessarily say anything about affordability or the proportion of household income that poor households have to spend on water. Moreover, as Michael Goldblatt points out in the context of Johannesburg, "it does not provide answers to the moral questions concerning the welfare role of government, and whether water supplies, and other urban services, are a good vehicle for pursuing welfare strategies" (Goldblatt 1997).

SANITATION

Planning for urban water supply and sanitation is often neglectful of sanitation (DfID 1998). If the official, albeit preliminary report, *State of the Environment in Gauteng* (Gauteng Provincial Government, n.d.), is anything to go by, it appears that Gauteng Province is no exception. The report does not even have a chapter on sanitation and to the extent that it is covered it is only in respect of the effect of waste sources on water quality. Having said this, as illustrated in Tables 2 and 4, sanitation coverage in Johannesburg is fairly extensive. Nevertheless the same patterns of service delivery as for water are reflected. Formal housing has a high level of service, while newer informal areas are poorly supplied.

The standard of formal sanitary services in Johannesburg is very high, with flush toilets being the norm. The GJMC is the sole agent responsible for bulk sewerage and waterborne sewerage and wastewater treatments dominate, covering almost 80 percent of the metropolitan area. Historically, areas without waterborne sewerage had the bucket system, with local authorities being responsible for ultimate disposal. Today, chemical toilets are used extensively in informal settlements but it is shack owners who often have access to chemical toilets and pit latrines, while many tenants who rent shacks are still forced to use buckets.

Within the townships there are also important differences in standards of sanitation that emerge more readily from micro studies. Results from the *Survey of Soweto* [an historically black township] *in the Late 1990s* (Morris and Bozzoli 1999) show very clearly that it is only in the elite new private sector developments that flush toilets in the house are the norm. So levels of service differ dramatically across different settlement and housing types but there are also important variations within settlements.

The issue of service standards also has as much to do with the micro-politics of the number of people per toilet as with the technical quality of the service. For example, the flush toilets of Soweto's hostels [Ed. note: single sex dormitories for workers] are essentially communal facilities associated with the

Type of sanitation	Formal dwelling (house or flat)	Formal dwelling in backyard	Informal dwelling in backyard	Informal dwelling not in backyard	Hostel	Other	All housing types
Flush toilet in dwelling	79	54	0	0	98	0	70
Flush toilet on site	20	43	55	13	2	100	23
Toilet off site	0	3	40	19	0	0	3
Other toilet on site (chemical and bucket)	0	0	0	50	0	0	3
Pit latrine on site	0	1	4	18	0	0	1
Total	100	100	100	100	100	100	100

Table 4 Type of sanitation by housing in Johannesburg (%)

Source: Own analysis of the 1995 October Household Survey
Column totals may not add up to 100 percent due to rounding up/down.

worst humiliations of abject living conditions (Pirie 1988). Backyard renters usually have to share a toilet on site and in Alexandra [Ed. note: an historically black township] the proportion of households who share toilets with other households on the site is as high as 87 percent (CASE 1998). In the same study Alexandra backyard shack dwellers complained that not only did flush toilets not work but access was difficult because landlords locked working toilets at night to stop them being used by squatters and other non-tenants in the area.

Their position is less desirable than that of informal settlements, where although technically lower standards prevail, there is better control over the maintenance of facilities. For example, in the official informal settlement of Diepsloot, which falls under the Eastern Municipal sub-structure, private operators under contract to the municipality provide chemical toilets and tank water. They are also strictly managed and monitored by the Diepsloot Community Development Forum (Beall, Crankshaw and Parnell 2000). Tensions over sanitation are one of the issues that some squatters voiced as the reasons for their move away from the backyard shacks into the informal settlements (Crankshaw 1996). However, the perception that sanitation is less contested in informal settlements may not be true either. A large proportion (40 percent) of residents in informal settlements in Gauteng also reported having to share their toilets with other households (DCD 1997).

No separate charges are levied for sanitation in Johannesburg and the costs form part of the rates. In poor areas, sanitation charges are part of the flat rate system. The key debate in the post-apartheid city, therefore, relates to the issue of standards and technology choice; waterborne sewerage versus pit latrines or chemical toilets. [. . .] Despite their adverse impact on the environment, chemical toilets have been the technology of choice over pit latrines because of political imperatives (they appear more modern and temporary) and the speed at which they can be erected in the face of popular demand.

[. . .]

FROM HEROES TO VILLAINS AND THE STRUGGLE FOR URBAN SERVICES

There is no doubt that apartheid had its victims and it is equally true that since apartheid's demise, poor conditions have persisted for many among the historically disadvantaged populations. However, it would be wrong simply to characterize Johannesburg's poor as passive victims or to see their condition in static terms. As described by Crankshaw and Parnell the townships of Johannesburg no longer bear the homogenizing hallmarks of high apartheid (Crankshaw and Parnell 1999). Alongside the matchbox houses and hostels are backyard shacks, informal settlements and the incremental growth of shantytowns. Equally there are signs of upward mobility

Type of toilet	Council houses (forced removals)*	Council houses (pre-apartheid)	Council houses (apartheid period)	Private sector	Backyard shacks	Informal settlement	Hostel	Site and service	%
Flush toilet in backyard	64.3	77.6	86.7	0.8	95.7	0	0.3	0.3	41.0
Flush toilet inside	35.7	22.4	13.0	98.3	3.7	0	0	2.4	21.7
Communal flush toilet	0	0	0	0	0	14.1	99.7	0.3	14
Pit latrine	0	0	0	0	0	11.4	0	7.7	2.4
Chemical toilet	0	0	0.3	0.6	0	74.5	0	0.3	9.4
No toilet	0	0	0	0	0.5	0	0	0	0.1
Septic tank	0	0	0	0	0	0	0	88.9	11.4
Total	100	100	100	100	100	100	100	100	100

Table 5 Types of toilet in Soweto by housing type (%)

Source: Morris and Bozzoli (1999)

*State-owned houses for rental by officially recognised African urban dwellers.

Column totals may not add up to 100 percent due to rounding up/down.

with the growth of middle class neighborhoods and as more modest houses are extended, renovated and adorned with various accoutrements of status.

During the early decades of apartheid the uniformity of the township landscape was matched by little class differentiation within urban African society. The changes evident in the built environment reflected not only increasing social differentiation among township residents but also the political dynamics of reform and resistance. In the 1980s in particular, the structural conditions of inequality were actively challenged. Collective action contributed not only to the struggle for political transformation but also to struggles around goods of collective consumption in organizational forms said to be akin to social movements (Swilling and Shubane 1991). As a result access to services such as water supply, sanitation and electricity in Johannesburg became highly politicized.

During the 1970s, the nationally appointed administration boards ruled the townships with an iron fist. However, with the abolition of the administration boards and the creation of elected black local authorities (BLAs), which were given augmented powers in the early 1980s, white officialdom withdrew from the volatile pit-face of urban governance (Bekker and Humphries 1985).

[...]

The unpopularity of the BLAs was compounded by their circumscribed mandates, which were matched by even more limited resources. Having responsibility for both spending and raising revenue, a style of patronage politics emerged within a context of local fiscal constraints of crisis proportions. This situation was accompanied by popular uprisings during the mid-1980s and before long the community councils lost control of the townships. Rents were a major source of local income and in 1979 the Greater Soweto Council raised rents by over 80 percent in an effort to balance their books. Soweto was the first township to protest against rents although the rent boycott subsequently spread across the country and in particular, through the townships in Johannesburg, Pretoria and the Witwatersrand (Seekings 1988). In reaction to these widespread protests, the apartheid government began to initiate a number of reforms, both to woo support from a black elite and to win African hearts and minds more broadly. For the former, urban property rights were returned to those African residents who could afford to pay for them and for a broader urban African constituency costly development plans focusing on electrification and sewerage systems were proposed "with the cost, unsurprisingly, being passed on to residents" (Seekings 1988). Further rent increases were announced in 1984, justified in terms of the cost of installations of service provisions. Combined with recession, inflation and the introduction of General Sales Tax in 1983, the fact that residents saw little progress led to an escalation of the rent boycott in Gauteng, which has persisted to the present day.

That the struggle around housing and urban services resulted in increasingly widespread political mobilization and protest, has led some to argue that the rent boycotts in Johannesburg during the 1980s "finally brought home the unviability of urban apartheid" (Swilling and Shubane 1991: 225). Certainly calls for installations or improved services were critical to the demands of the rent boycott, both in Soweto and Alexandra. For example, in response to a survey on why Sowetans supported the boycott, over half said it was because people thought housing and services were inadequate and around a third said it was because people did not have the money (ibid: 234). What is more, these concerns did serve to fuel urban-based political action towards fundamental social change in South Africa (Mayekiso 1996). However, once political change was wrought, problems of poverty, inequality and inadequate urban services persisted, as did the rent boycott. It was this that turned heroes of the struggle against apartheid into villains of the post-apartheid peace

WHO ARE THE FIXERS? IT CANNOT BE BUSINESS AS USUAL

A key problem facing the new Metropolitan Council is the tension between maintaining established service levels (in historically white areas) and extending services to new and historically under-serviced (mainly black) areas. [. . .] The GJMC is crucially aware that the introduction of rates and service charges, no matter how minimal, may represent a significant increase in living costs for the poor (Beall, Crankshaw and Parnell 2000). Equally, they cannot ignore the view that under apartheid, Johannesburg's strong black-supported tax base allowed the city to subsidize white ratepayers, a variation on the "sweat equity" argument. For reasons of social justice the argument goes, a system of cross-subsidization at the city level is warranted, in order to right history and to address the issue of sub-standard services in Johannesburg's townships and informal settlements (Bond 2000). However, an opposing view of the problem is that the rates and services boycotts of the last 15 years in Johannesburg's townships has meant that the state has effectively been subsidizing the consumption of water, sewerage and electricity in black residential areas.

[. . .]

As laid out in the city's new policy document, *iGoli 2002* (GJMC 1999), the plan for improved service delivery involves the establishment of public utilities for water supply and sanitation, electricity and solid waste management. These are considered to be "trading services" and the utilities will operate as independent companies with the metropolitan council as the single or major shareholder in each utility and playing a regulatory role. The advantages envisaged from this model include "the introduction of sound management practices, improve the quality and coverage of service delivery (particularly through raising capital to meet backlogs), and enhance the financial stability of Greater Johannesburg (through generating additional revenue)" (ibid.).

On paper, *iGoli 2002* has strong neo-liberal overtones and appears designed to tackle head-on the critical financial and institutional problems faced by the GJMC. There is scant mention of social or environmental issues bar a section on the impact on labor relations and the need for careful management and partnership. Nevertheless, Kenny Fihla, Chairperson of the Transformation *Kgotla* of the GJMC, has emphasized the importance of all citizens of Johannesburg coming to terms with redistribution at a citywide level (Beall and Lawson 1999).

Addressing the poor through urban services delivery will include, according to Gordhan [Chief Executive Officer of the GJMC], offering free of charge sixty liters per capita per day (l/c/d) of free water, 60 kilowatt hours of free electricity, free waste management in the form of two bin bags per week and no rates charges for anybody who owns property of less than R10,000 [Ed. note: about US$600 at the time] in value. He goes on to say:

> I think that this is more than any city in South Africa has been able to offer the marginalized poor in its environment. And this is a plan that

> tries simultaneously to address two very diverging constituencies. One is providing that basic level of service to the poor while upgrading systematically the services tc the rich northern suburbs I live in the northern suburbs and I would like the city to take basic responsibility for keeping the place clean and tidy and ensure that I get water and electricity and the lights work, and traffic lights operate and other basic things. The northern suburbs residents are paying a huge premium for a very poor quality of service at the moment, and I think we owe them value for money.

So that's the story of Johannesburg. That's the story of South Africa (Beall and Lawson 1999).

What Johannesburg intends to do then is address universal service provision by replacing a lifeline tariff with a lifeline service. In this way it is calculated that the better off who will use more in the way of services and resources will pay more, effectively cross-subsidizing those who cannot afford it.

Targeting of any sort is costly and administratively complex but it is believed that by opting for a universal approach to lifeline services along with having the technical capacity in place, difficulties in administering the system will be minimized, the problems of fluid populations and changing occupancy notwithstanding. However, there are those who object strongly to the basic standards approach being adopted by the GJMC. It is advocated both in the MIIF [Municipal Infrastructure Investment Framework] and is influenced by the World Bank's neo-liberal urban policy advice to South Africa. Opponents argue that excessively low standards are inappropriate for a country that the World Bank classifies as "upper middle-income" (Bond and Swilling 1992). Not only is this considered an inequitable solution but an inefficient one as well, as Patrick Bond has argued:

> The basic service levels contemplated in the *MIIF* are not merely emergency services (piped water or portable toilets in slum settlements that are without water or hygienic facilities at present), but represent more fundamentally, development policy that will be in place for at least a decade. It is extremely difficult to incrementally upgrade infrastructure—particularly sanitation systems—from pit latrines to waterborne sewage, resulting in permanently segregated low-income ghettoes. (Bond 2000)

Nevertheless, given the financial crisis in which Johannesburg finds itself, where according to Ketso Gordhan "we are spending zero on capital expenditure, we have zero in our reserve funds and we are basically a cash-based organization" (interview in 1996), the GJMC appears to have come up with a reasonably creative response to the question of who are to be the fixers. It is addressing issues of housing, living environments and more specifically urban services in a pro-poor way. The issue of inequality is being tackled more coyly. The question at stake here is whether a metropolitan government such as that of Johannesburg can do any more in the current political climate. To extend service delivery and infrastructure further at the local level requires investment. This is clearly beyond the capacity of metropolitan councils and municipalities facing financial stringency. There is a need for much stronger commitment by national government to restructuring national tariffs to ensure cross-subsidization, for example between industrial and domestic consumers. This is particularly the case given that it is difficult for low-income householders to share in the recurrent costs of service delivery, let alone to participate significantly in cost recovery of capital expenditure. The GJMC has begun to grapple with these concerns but pursuit of equality and social justice can begin at local government level but cannot end there. Indeed national government cannot simply pass the buck.

REFERENCES

Beall, J. and Lawson, S. (1999) *Smart Johannesburg, Leading the African Renaissance?* Audiocassette course material for Open University graduate program "Third World Development" (Milton Keynes).

Beall, J., Crankshaw, O. and Parnell, S. (2000) *Urban Poverty and Urban Governance in Johannesburg: Phase Two Report*, ESCOR commissioned research on urban development.

Bekker, S. and Humphries, R. (1985) *From Control to Confusion: The Changing Role of Administration Boards in South Africa, 1971–1983*, Pietermaritzburg: Shuter and Shooter.

Black, M. (1994) *Mega-Slums: The Coming Sanitary Crisis*, London: Water Aid.

Bond, P. (2000) *Cities of Gold, Townships of Coal: Essays on South Africa's New Urban Crisis*, Trenton, NJ: Africa World Press.

Bond, P. and Swilling, M. (1992) "World Bank financing for urban development: Issues and options for South Africa," *Urban Forum*, 3(2).

CASE (1998) *Determining Our Own Development: A Community-based Socio-economic Profile of Alexandra*, Johannesburg: CASE.

Crankshaw, O. (1996) "Social differentiation, conflict and development in a South African township," *Urban Forum*, 7: 53–68.

Crankshaw, O. and Parnell, S. (1999) "Interpreting the 1994 African township landscape," in H. Judin and I. Vladislavić (eds), *Architecture, Apartheid and After*, Cape Town: David Philip, 439–443.

Department of Constitutional Development (DCD) (1997) *Municipal Infrastructure Investment Framework*, Pretoria: Government Printer.

Department for International Development (DfID) (1998) *Guidance Manual on Water Supply and Sanitation*, Leicestershire: WEDC.

Gauteng Provincial Government (n.d.) *State of the Environment in Gauteng, a Preliminary Report*, Johannesburg: GPC.

Goldblatt, M. (1997) "The provision, pricing and procurement of water: A willingness to pay survey in two informal settlements in greater Johannesburg," MSc Dissertation, Johannesburg: University of the Witwatersrand.

Greater Johannesburg Metropolitan Council (GJMC) (1999) *iGoli 2002*, Johannesburg: GJMC.

Jarman, J. (1997) "Water supply and sanitation," in J. Beall (ed.), *A City for All, Valuing Difference and Working with Diversity*, London: Zed, 182–193.

May, J. (ed.) (1998) *Poverty and Inequality in South Africa*, Report prepared for the Office of the Executive Deputy President and the Inter-Ministerial Committee for Poverty and Inequality, Pretoria: Government Printer.

Mayekiso, M. (1996) *Township Politics, Civic Struggles for a New South Africa*, New York: Monthly Review.

Morris, A. and Bozzoli, B. (1999) *Change and Continuity: A Survey of Soweto in the Late 1990s*, Johannesburg: Sociology Department, University of the Witwatersrand.

Nunan, F. and. Satterthwaite, D. (1999) "The urban environment," Theme Paper 6, ESCOR commissioned research project on urban development, University of Birmingham, International Institute for Environment and Development, London School of Economics and University of Wales, Cardiff.

Pirie, G. (1988) "Housing essential service workers in Johannesburg: Locational constraint and conflict," *Urban Geography*, 9: 568–583.

Rocheleau, D. (1990) "Gender, complementarity and conflict in sustainable forestry development: A multiple user approach," Paper presented to the IUFRO World Congress, August 5–11: Montreal.

Seekings, J. (1988) "The origins of political mobilisation in the PWV townships, 1980–84," in W. Cobbett and R. Cohen (eds.), *Popular Struggles in South Africa*, Trenton, NJ: Africa World Press.

Stephens, C. (1996) "Healthy cities or unhealthy islands: The health and social implications of urban inequality," *Environment and Urbanization*, 8(2): 16–25.

Swilling, M. and Shubane, K. (1991) "Negotiating urban transition," in R. Lee and L. Schlemmer (eds.), *Transition to Democracy: Policy Perspectives 1991*, Boston: Oxford University Press.

Wilson, F. and Ramphele, M. (1989) *Uprooting Poverty: The South African Challenge*, Cape Town: David Philip.

“Formalizing the Informal? The Transformation of Cairo’s Refuse Collection System”

The Journal of Planning Education and Research (1996)

Ragui Assaad

INTRODUCTION

I examine in this paper the process that led to the transformation of Cairo’s household refuse collection system in the late 1980s. The process began with a decision by Cairo’s municipal authorities to replace the city’s informal household waste collection system with a more “modern” system that would be subject to the state’s regulatory authority. In attempting to implement this decision, the municipal authorities set the stage for a confrontation between the state’s legal order and the decades-old system of customary rights and practices that governed household waste collection in Cairo. The *zabbaleen* and the *wahiya*, two groups identified by close-knit ethnic, kinship and residential ties, had hitherto provided waste collection services with minimal involvement by the municipality. Despite the vast imbalance in power between the state, on the one hand, and the *zabbaleen* and *wahiya*, on the other, the outcome of the confrontation was not a foregone conclusion. A drawn-out process that involved various forms of resistance, mediation, negotiation, and compromise finally resulted in an accommodation that partially satisfied the needs of the municipal authorities for regulation and oversight without jeopardizing the social norms, practices, and rights that were the mainstay of the informal system.

The purpose of this case study is to shed light on the design of public interventions in the area of urban service delivery in Third World cities. The main question I address is how to build on existing informally organized private sector arrangements, while ensuring the provision of adequate, affordable, and more comprehensive service levels. The case study demonstrates that planning goals can be achieved without either supplanting informal arrangements with “modern,” centralized, hierarchical, and bureaucratic systems or fully formalizing them. Instead, the study shows that attempts to formalize these informal arrangements can give rise to a dynamic, openended process that leads to partial formalization of the *zabbaleen* system, as well as greater tolerance of informality on the part of municipal authorities.

From a planning perspective, another key issue addressed by the study is the role of intermediary actors and institutions in mediating conflicts and providing channels of communication between two parties with unequal power and different interests and who operate according to different sets of rules and norms. In this case, the main intermediary role was played by Environmental Quality International (EQI), an Egyptian consulting firm, that expended considerable time and effort in understanding the intricacies of the *zabbaleen* system.[1] EQI was able to act as an effective mediator because of years of prior ground work with all the parties involved. Its history of involvement with community development work among the *zabbaleen* had earned it their confidence. As an internationally recognized consulting firm, it had credibility with the municipal authorities.

Because EQI was not in a position to play such an interim-diary role on a sustained basis, it helped create the Environmental Protection Company (EPC), a legally registered company owned by members of the *zabbaleen* and *wahiya* communities, to bridge the institutional gap between the municipal authorities, on the one hand, and the *zabbaleen* and *wahiya*, on the other. The experience of the EPC provides valuable lessons on how to structure such interim-diary institutions.

The implications of this research for the design of planning interventions in Third World cities go well beyond the case at hand. As is well documented in the Third World planning and urban studies literature, many activities that are crucial to the economic and social vitality of Third World cities take place predominantly outside the control and regulation of the state.

[. . .]

States invariably attempt to formalize these activities by imposing on them legal and bureaucratic standards and regulations that clearly favor centralized, hierarchical, and standardized modes of organization. Threatened with expulsion or, at the very least, marginalization, participants in these informal activities and systems often respond with what James C. Scott has termed the *weapons of the weak*: everyday forms of resistance like false compliance, clandestine acts of sabotage, and other acts of hidden resistance aimed at undermining and subverting the system that the state seeks to impose (Scott 1985). Cognizant of their relative powerlessness vis-à-vis the state, the protagonists, be they squatters, street peddlers, or scavengers, rarely engage in open acts of rebellion or defiance. Their acts of resistance are akin to guerrilla warfare rather than the frontal assaults of conventional battlefields (Scott 1990: 192–193). Generally, these confrontations between the state's modernizing legal and bureaucratic order and the day-to-day practices of ordinary people generate disastrous consequences, not only for the people involved but also for the state's own objective of instituting "rational" and "efficient" policies.[2] The importance of this case is that it documents a process that successfully averted such a planning failure and provides valuable lessons that can be applied to similar situations elsewhere.

This case is also relevant to the design of interventions aimed at fostering greater private sector participation in urban service delivery as a solution to shrinking fiscal budgets (Roth 1987). Most of these privatization efforts implicitly assume a model of the private sector that does not correspond to the reality of the private sector in Third World cities.

[. . .]

Government agencies, with their reliance on bureaucratic and routine procedures, are often ill-equipped to deal with the informal modes of transacting in this non-corporate private sector. As Peattie points out, however, if government planners were so inclined, they could have resorted to appropriate intermediary institutions adept at working with the "myriad of little local actors" (Peattie 1987: 78).

When the Cairo municipal authorities sought to assert control over the city's solid waste collection system, a publicly financed and operated system was simply not an option because of the severe financial limitations under which the municipal government was operating. Because the *zabbaleen* and *wahiya*, which had been providing the service for decades, did not fit the corporate private sector model that the authorities had in mind, they were not deemed legitimate private counterparts to the authorities. Through EQI's intervention, the *zabbaleen* and *wahiya* were able to create a corporate entity that the authorities could recognize. Even this entity, which was nominally owned by the two groups, proved to be too alien to their informal mode of transacting to succeed. Taking their cue from this experience, the *zabbaleen* and *wahiya* registered a number of small companies that, as formal entities, could obtain municipal refuse collection licenses and pay the licensing fees. These companies were minimally involved in either the financing of capital equipment or the management of day-to-day operations, which continued to be done through informal ties between the *zabbaleen* and *wahiya*. These companies were in effect no more than shells that allowed this informal sphere to connect to the municipal authorities on their own terms.

[. . .]

THE *ZABBALEEN* SYSTEM OF WASTE COLLECTION

The literature on the urban informal sector has long been concerned with groups, often referred to as scavengers, that make a living by extracting useful products from solid wastes (Birkbeck 1978; Blincow 1986;

Furedy 1984; Sicular 1991). A distinction is usually made in this literature between refuse workers who treat waste as waste by being involved in its collection, haulage, and disposal, and scavengers who treat waste as ore, that is, as a source from which valuable materials can be extracted (Sicular 1991: 138). The uniqueness of the *zabbaleen* is that they perform both functions. They undertake the haulage and disposal of Cairo's household refuse, as well as material recovery from it. The confluence of these two roles has led to a complex system of customary rights and practices that is discussed in some detail later.

[. . .]

THE ORIGINS OF THE SYSTEM

Cairo has had groups associated with the collection of household wastes for hundreds of years. A list of guilds dating back to 1670 includes a guild of *zabbaleen*, a word whose literal meaning is garbage men (Evliya Celebi, quoted in Baer 1964: 35). Until the mid-1930s, the refuse collection trade was tightly linked to the operation of public baths and bean cookeries. The refuse was dried and used simultaneously as fuel for the baths and the cookeries, which were typically combined in a single establishment. Toward the end of the nineteenth century, migrants from three villages in the Dakhla Oasis in the Western Desert took control of the trade. The *wahiya* (singular: *waht*), or people of the Oasis, as these migrants came to be known, charged Cairo residents a small fee for refuse collection services and continued to use the refuse itself as a source of energy. In the 1930s, as fuel oil replaced dried organic wastes as the preferred fuel for public baths and bean cookeries, the *wahiya* began selling the refuse to a more recent wave of migrants who settled in squatter settlements on the outskirts of the city. These were poor peasants from villages around Assiut in Upper Egypt, who used the organic components of the wastes to raise pigs.[3] Because of their need to obtain waste food before it rots, the newcomers, who became known as the *zarraba* (singular: *zarrab*), or pig pen operators, preferred to collect the refuse themselves using their own donkey carts. Although the *wahiya* continued to be the party that was responsible to building owners and tenants for refuse collection, the task of haulage and disposal was turned to the *zarraba*.

[. . .]

The *zabbaleen* are organized as household enterprises where all members of the household participate in waste collection and recovery in one way or another. Children of both sexes assist their father or older brothers in the collection rounds by driving the donkey cart and guarding it while their older male relative travels from unit to unit to collect the refuse.[4] Once the cart returns to the household compound or *zeriba*,[5] the adult women and teenage girls take over. They sort the refuse by hand, separating the edible materials, which they feed to the pigs in the adjoining pig pen. They also sort the recyclable materials, such as paper, cloth, glass, bone, tin cans, plastics, and the usually small quantities of non-ferrous metals into separate piles. When the family has accumulated sufficient quantities of recovered materials, the materials are sold to primary materials dealers located in the family's settlement for further processing. Food remnants and pig manure are periodically cleaned out from the pig pen and sold to specialized dealers (and more recently to a community composting plant) for composting. Finally unusable materials are disposed of by burning on community streets or in nearby open spaces and crevices.

THE *ZABBALEEN* SYSTEM AS A SEMI-AUTONOMOUS SOCIAL FIELD[6]

As the system evolved, a body of rights and obligations developed to regulate its operation. The *wahiya* refer to the collection of these rules as *'urf al-mihna*, which can roughly be translated as the custom of the trade.[7] According to this normative order, the *wahiya*, as a group, have exclusive rights to organize refuse collection services in Cairo. Upon the completion of a new building, an individual *wahi* "purchases" the right to service the building in perpetuity by making a one-time payment to the building owner.[8] The *wahiya* are able to block effectively anyone outside their community from participating in the ostensibly competitive bidding process leading to the purchase. In fact, they can restrict competition even further by allocating different zones of the city to particular family groups among them, thus reducing the payments to building owners. The rights thus obtained, although not legal in any formal sense, can be inherited or transferred to others, so long as they remain in the *wahiya* community.

Once the *wahi* obtains collection rights to a building, he arranges with a *zabbal* for the haulage of the refuse. The *wahiya* used to charge the *zabbaleen* a monthly fee to give them access to the refuse, but this practice had virtually disappeared by the mid-1970s.[9] A group of buildings containing a total of 250 to 300 residential units makes up a *zabbal*'s route. With collection services six days a week, the refuse generated from these units fills a standard donkey cart with a 2.1 cubic meter container. There is no necessary correspondence between a *zabbal*'s route and the buildings controlled by a given *wahi*. A route may include buildings controlled by several *wahiya*, and, conversely, a *wahi* may control buildings falling in a variety of routes. While the *wahiya*'s informal property rights give them exclusive access to the monthly fees paid by the residents, the *zabbaleen* acquire the rights to the refuse itself. Rights to a "route" can also be passed on to an heir and transferred to other members of the *zabbaleen* community, thus constituting another set of extralegal property rights. Just as a *zabbal* who attempts to collect fees from residents on his own account or who accepts to work with a contractor from outside the trade would be breaching the rules of conduct, a *wahi* who gives the wastes to someone other than their rightful "owner" or otherwise deprives a *zabbal* of access to them would also be breaking the rules of the game.

When asked, the *wahiya* justify their exclusive rights to collect service fees by referring to the ancestral rules of the trade, which specify a strict division of labor between them and the *zabbaleen*. The *wahiya* assume overall responsibility for regular removal of the wastes vis-à-vis residents and building owners and are responsible for collection of the wastes from individual dwelling units to the street level. The *zabbaleen*, on the other hand, are responsible for the haulage and disposal of the wastes. In practice, a *zabbal* can participate in door-to-door collection and may even collect the user fees and remit them to the *wahi*, but he would be doing so as an agent of the *wahi* and under his supervision.

An important question that arises is why do the *zabbaleen* abide by rules that clearly give the *wahiya* a distinct advantage? Moreover, why aren't there more frequent challenges to the *wahiya*'s monopoly on waste collection contracts by outsiders? The *wahiya-zabbaleen* social field contains a number of mechanisms to induce or coerce compliance with its rules by both insiders and outsiders. The *wahiya* block entry to outside contractors by forging mutually beneficial relationships with building doormen and by developing long-term personalized relationships with clients. But if these fail to stem outside encroachment, they sometimes resort to acts of vandalism against the intruders' property and, in rare instances, to physical assaults on their representatives.

The *zabbaleen* do not challenge the *wahiya*'s monopoly over fee collection because their own livelihood depends on the stability of the system and their continued access to the waste stream. A challenge to the *wahiya* may jeopardize that access, threatening a *zabbal*'s investment in live pigs by disrupting the supply of fresh garbage to his *zeriba*. The asymmetry of the relationship between the *wahiya* and *zabbaleen* is a source of frequent tensions. Conflicts sometimes occur over the allocation of new buildings and the distribution of revenues when it is the *zabbal* rather than the *wahi* who collects the user fees. When these conflicts do arise, however, they are usually resolved through the informal mediation and arbitration of community leaders from both sides. Although both the *zabbaleen* and the *wahiya* have their respective formally registered organizations to represent them in their dealings with government institutions and other formal entities, these organizations play a minimal role when it comes to enforcing the rules of the trade and resolving internal disputes. Thus the relationship between the two groups combines both rivalry and cooperation. Despite their sometime opposing interests the *wahiya* and *zabbaleen* realize that, by respecting the other party's rights, they also uphold them against "outsiders."

THE TRANSFORMATION OF THE TRADITIONAL SYSTEM

The forces leading to the transformation of the system

Although the traditional system provided good incentives for waste recovery and recycling, it did not provide adequate incentives for comprehensive coverage of the city. The *zabbaleen*, who provide the haulage capacity, having no access to the service fees, could not keep up with expanding demand for waste collection service.

[. . .]

At the same time, the municipal authorities were becoming increasingly intolerant of the *zabbaleen*'s donkey carts, which they considered an eyesore and a traffic hazard, precisely in the neighborhoods that were being more fully served.

[...]

The traditional division of labor between the *zabbaleen* and the *wahiya* was getting in the way of making the system respond more effectively to increasing demand for waste collection and disposal services.

[...]

In effect, the combination of the "rules of the trade" and the individualized arrangements between the *wahiya* and their clients prevented a socially efficient solution from emerging on its own. Some kind of public intervention was necessary to break the logjam.

[...]

The new CCBA (Cairo Cleansing and Beautification Authority), set up by presidential decree in 1983 to oversee municipal sanitation in all 12 of Cairo's districts, leadership's first instinct was to scrap the entire system and replace it with a "modern" mechanized refuse collection system. Instituting a publicly run and financed system was not feasible because the municipality was already burdened with the collection and disposal of street wastes and was severely limited in its capacity to generate new revenues to finance a household waste collection system. The CCBA planned to rely instead on private contractors to whom it would issue licenses to provide fee-based services in specified locations under municipal supervision. The fees charged by the contractor would be fixed by the CCBA and would include a monthly license fee payable to the CCBA to cover supervision and administrative costs. The contractor would also post a security bond with the agency to guarantee adequate service levels. There appeared to be no place for the *zabbaleen* and the *wahiya* in this vision.

[...]

THE GENESIS OF THE NEW ORDER

After lengthy negotiations between the CCBA and the leadership of the *wahiya* and *zabbaleen*, which were brokered by EQI, the CCBA agreed to grant the *wahiya* and *zabbaleen* a license to provide waste collection services. This system was tested in Zamalek and Manial, two upper income neighborhoods, on condition that the two groups establish a legally registered company that the CCBA could contract with and that they used motorized vehicles for haulage. These two neighborhoods were selected because they are highly visible and, as two islands on the Nile, vehicular access to them could be easily monitored. With a total of about 25,000 residential units, the two neighborhoods were being served by 40 donkey carts under the existing system.

[...]

With technical assistance from EQI and a substantial grant from a European-based charitable organization, the *wahiya* and *zabbaleen* founded the Environmental Protection Company (EPC) as a private for-profit company in September 1987.

[...]

CONFRONTATION, NEGOTIATION, AND COMPROMISE

[...]

The EPC's financial difficulties and its failure to expand beyond the borders of Zamalek and Manial made it vulnerable to pressures from large, well-connected, private companies that were becoming increasingly interested in the waste collection business. Some top administrators at the CCBA, who were reluctant to grant a *zabbaleen*-based company the license to serve Zamalek and Manial in the first place, were sympathetic to these companies. The CCBA was also faced with insistent demands by powerful neighborhood organizations for improvements in the cleanliness of streets and public spaces. Because it was limited in its ability to impose new taxes, the CCBA was looking for ways to finance the cleaning of streets and public spaces from user charges for commercial and residential waste collection. One way to achieve this was to force the EPC out of Zamalek and grant the license to Misr Service, a large private firm, which would also undertake street cleaning. In February 1993, the CCBA signed a three-year contract with Misr Service for the provision of street cleaning and commercial waste collection services in exchange for a generous monthly payment of £E109,000. The CCBA planned to undertake fee collection itself to recover this amount. In April, the CCBA informed the EPC that it would not renew its license for residential waste collection, thus setting the stage for extending the services of

Misr Service to residential units. This was the final blow to the EPC, which continued operations for two or three more months in Manial and then declared its bankruptcy and liquidation.

Faced with the threat of being completely pushed out of Zamalek, the *wahiya* and *zabbaleen* took it upon themselves to provide uninterrupted service using their own trucks with or without the approval of the CCBA. They undertook a series of concerted actions to block Misr Service from collecting household wastes. They redoubled their efforts to provide regular collection service early in the morning, so that when Misr Service crews arrived they would find that the refuse had already been collected. Using their long-term relationships with building doormen, they agreed to transfer temporarily part of the monthly proceeds to the doormen in return for preventing entry into the premises by Misr Service or CCBA employees. They also made use of their personalized relationships with their clients to encourage them not to pay the CCBA fees, which would have been more than three times higher than those charged by the EPC. Unlike Madinet Nasr, where some *zabbaleen* agreed to cooperate with the "outsider" by purchasing the refuse from it, the Zamalek *zabbaleen* formed a united front with the *wahiya*. Finally, in one instance, there was a violent confrontation pitting the *wahiya* and the *zabbaleen* against CCBA and Misr Service employees, when the latter attempted to confiscate the former's trucks.

In addition to the troubles the CCBA was having with the *wahiya* on the residential front, it faced severe problems in its attempt to collect the newly imposed fees from commercial establishments. Shop owners and other businesses balked at paying the fees, which were much higher than anything they had previously paid. The fees for shops were set at 30 Egyptian pounds per month, a tenfold increase over those charged by the EPC. Fees for other commercial establishments such as restaurants, hotels, banks, and private clubs were set much higher. In five months, the CCBA managed to collect less than 4% of what it owed Misr Service.

As these developments were unfolding, the *wahiya* and *zabbaleen* sought the intermediation of EQI, which had been instrumental in creating the EPC. EQI proposed to the CCBA that it grant a new *zabbaleen*/*wahiya*-owned company a license to provide residential collection service in Zamalek, but that the user charge be increased by two thirds and the difference remitted to the CCBA to cover partially the cost of street cleaning. The role of Misr Service in residential waste collection would be reduced to that of monitoring. Taking a purely legalistic approach to the matter, the CCBA initially turned down the EQI proposal, arguing that the abrogation of the EPC's license meant the end of the *zabbaleen*'s role in Zamalek. With continuing trouble on the ground, however, the CCBA finally agreed to consider EQI's proposal on condition that the *wahiya* pay retroactively for the four months since the EPC's license was revoked. After three months of negotiation, in which the CCBA had clearly lost the upper hand, a contract was signed between the newly formed *wahiya* company and the CCBA, stipulating a higher user charge from which a street cleaning fee would be remitted to the CCBA. Discussion of the retroactive payments was postponed indefinitely.

The preceding account shows that despite being the embodiment of the regulatory power of the state, the CCBA could not impose its preferred solution. From a purely legalistic point of view, it had the power to revoke the service license and therefore deny the *wahiya* access to Zamalek. The *wahiya* started at a diametrically different position: Their rights of access were based on decades of presence in the area and a multitude of extralegal contractual relationships with building owners and residents that, in their view, could not be revoked by a stroke of a pen. The opposing positions inevitably led to a confrontation between the two parties.

Although such a confrontation did take the form of an open violent clash in one instance, such an occurrence is usually very rare. As Scott (1985) pointed out, open rebellion by subordinate groups is dangerous and even outright suicidal at times. The *wahiya* used forms of resistance that were, for the most part, hidden and indirect, akin to guerrilla warfare tactics. They made use of their long-standing relationships with building doormen and the residents themselves to disrupt the ability of the Misr Service to provide the service and the CCBA's ability to recover its cost. Occasionally, they also engaged in small acts of sabotage that made it difficult for Misr Service to undertake its task effectively. Although such low-profile techniques are rarely successful in allowing subordinate groups to maintain their way of doing things, they can succeed in thwarting the changes that the more powerful actors, like the CCBA, are trying to impose. This kind of stalemate is a common outcome of confrontations between the state and subordinate groups. What led to a resolution in this case is the successful mediation role played by EQI, which was able to exploit the stalemate to reach a compromise solution.

[. . .]

NOTES

1 Even though the system involves two distinct groups, the *zabbaleen* and the *wahiya*, it is commonly referred to as the *zabbaleen* system.
2 Prominent examples of such development or planning failures are the Ujamaa village collectivization project of the late 1960s and 1970s in Tanzania and the Ciudad Guayanna Project in Venezuela.
3 Such a strict, seemingly arbitrary, division of labor was a central feature of guild organization in Egypt. There were for instance five guilds for butchers, each specializing in a different animal (Raymond 1973). Raymond argued that such division of labor was not designed to rationalize the technical process but to limit competition.
4 The recent changes in the system described in this paper have resulted in the substitution of most donkey carts with motorized vehicles.
5 The word *zeriba*, which means pig pen in Arabic, is used to refer to the entire household compound, which also includes a living area and an area for sorting the waste.
6 The information in this section is derived from my experience working with the *zabbaleen*, from interviews I conducted with leaders of the *wahiya*, and from EQI reports. See in particular Neamatalla *et al.* (1981) and Neamatalla *et al.* (1984).
7 The Hans Wehr Dictionary of Modern Arabic provides the following definitions for the word *urf*: beneficence, kindness; custom, usage, practice, convention, tradition, habit; legal practice; custom, customary law. In its definition of the adjective *urfi*, the dictionary describes it as implying "private, unofficial" as opposed to *rasmi*, which means official or formal.
8 Most residential building in Cairo comprise multiple rental units.
9 The recent switch to motorized collection vehicles has given rise to complicated financial transactions between the two parties, which I discuss later.

REFERENCES

Baer, G. (1964). *Egyptian guilds in modern times*, Jerusalem: The Israel Oriental Society.

Birkbeck, C. (1978). Self-employed proletarians in an informal factory: The case of Cali's garbage dump, *World Development* 6(9–10): 1173–1185.

Blincow, M. (1986). Scavengers and recycling: A neglected domain of production, *Labour, Capital, and Society* 19(1): 94–116.

Furedy, C. (1984). Socio-political aspects of the recovery and recycling of urban wastes in Asia, *Conservation and Recycling* 7(2–4): 167–173.

Neamatalla, M. S., R. Assaad, L. Oldham and T. Etreby (1981). *The people of the Gabbai: Life and work among the Zabbaleen of Manshiet Nasser*, Report No. 3. Solid Waste Component, First Egypt Urban Development Project. Submitted to Governorate of Cairo/Joint Housing Projects Committee/International Development Association. Cairo, Egypt: Environmental Quality International.

Neamatalla, M. S., R. Assaad, L. Oldham, A. Soueni and F. Gohary (1984). *Solid waste collection and recycling in Cairo: A system in transition*, Cairo, Egypt: Environmental Quality International.

Peattie, L. (1987). *Planning: Rethinking Ciudad Guyana*, Ann Arbor: University of Michigan Press.

Raymond, A. (1973). *Artisans et Commercants au Caire au XVIII Siecle*, Damascus, Syria: Institut Francais de Damas.

Roth, G. (1987). *The private provision of public services in developing countries*, New York: Published for the World Bank by Oxford University Press.

Scott, J. C. (1985). *Weapons of the weak: Everyday forms of peasant resistance*, New Haven: Yale University Press.

Scott, J. C. (1990). *Domination and the arts of resistance: Hidden transcripts*, New Haven: Yale University Press.

Sicular, D. T. (1991). Pockets of peasants in Indonesian cities: The case of scavengers, *World Development* 19(2): 137–161.

SECTION 7
Urban Infrastructure

Editors' Introduction

Infrastructure networks are the physical structures that form the basis for city growth and development. In the contemporary world, place-bound infrastructure networks facilitate the circulation of information, money, people, and commodities while anchoring capital in place to create the physical city. As geographer and planner Stephen Graham (2000: 114) noted, infrastructure networks "represent long-term accumulations of finance, technology and organizational and geopolitical power" and "offer up a powerful and dynamic way of seeing contemporary cities and urban regions."

In the previous section we focused on the differentiated nature of urban service provision across cities in the global South. We paid special attention to the informal systems through which some level of services are provided to the majority poor who remain largely invisible to governments and elites. Here, we turn our focus to the technologies and undergirding systems of these services, the infrastructure networks. While some infrastructure networks are clearly visible, others such as sewage pipes or cellular signals remain largely invisible, even buried. Regardless of visibility, infrastructure networks remain central to growth and development, and in the current patterns of wealth creation, benefits only flow to those who can afford to pay for services. David Harvey's (2005) concept of "accumulation by dispossession" captures this process of immiseration. For Harvey, "[a]ccumulation by dispossession is about plundering, robbing people of their rights" in order to build wealth. The patterns that emerge deepen inequality and poverty.

The past decade has seen a surge of work on infrastructure systems and networks (Graham and Marvin 2001; Coutard 2008; McFarlane and Rutherford 2008; Legg and McFarlane 2008). Much of this comes from the fields of sociology and geography and has focused on water and sanitation. Of these, Graham and Marvin's (2001) "splintering urbanism" offers a valuable framework that suggests complex interactions between urban dynamics, socio-spatial configurations, and urban functioning. Though largely drawing on the experience of the global North, this widely cited and influential framework is worth examining in some detail. Coutard (2008) has summarized its four main elements. First, a *modern infrastructural ideal* supports the notion of monopolistic, integrated, standardized provision of urban services or collective goods. Second, a combination of neoliberal ideals of privatization and market efficiency combines with technological advances that facilitate new systems of metering and allow powerful actors to *unbundle or splinter infrastructures* into different elements and service packages. These unbundled systems allow for *bypass strategies* that connect valued and powerful users and bypass those deemed less important. Third, bypass strategies produce *premium networked spaces* of higher income groups who are linked to each other through global enclaves and yet are withdrawn from their immediate urban fabric. Finally, Graham and Marvin suggest that communities who face splintered urban processes develop forms of resistance, recalling Castells's notion of collective consumption effects discussed in the Editors' Introduction to Basic Urban Services in this volume.

While much in the splintering urbanism framework is useful, the experience of infrastructure production in the cities of the global South presents a somewhat different picture. We will address some of these differences through examining patterns of segregation and inequality as well as presenting a section on the infrastructure of transportation.

KEY ISSUES

Infrastructure ideals versus urban realities

Case studies of the colonial and postcolonial experience of infrastructure provision as well as historically grounded research (McFarlane and Rutherford 2008; Legg and McFarlane 2008) have demonstrated that the modern infrastructural ideal that emerged from the global North was always a fiction in the cities of the global South. Standardized, integrated systems that served the entire city were absent in the colonial period. In addition, provision differed for the colonizer and the colonized as well as along lines of class, race, ethnicity, or caste, an issue that has been noted several times in this *Reader*.

Bypass strategies across the global South are not a recent phenomenon. While denying entire groups of city residents services on the basis of affordability and solvency is not acceptable in Northern cities, it is not uncommon in cities like Cape Town where poor households may be permanently disconnected from electricity for lack of payments. The majority of the poor in slums across a range of large and small African and Asian cities are bypassed when it comes to water or sanitation infrastructure.

The assumed relationship where universal, integrated services produced urban integration is also ambivalent. As Pflieger and Matthieussent (2008) demonstrated in their study of Santiago, Chile, despite near universal connections to services, fragmentation of the city in socio-spatial terms has continued since the 1980s. Similar processes took place in Los Angeles in the United States where essential infrastructure design and service provision (of water and electricity) were tied to local oligarchs who controlled local government and owned desert land they wished to develop and urbanize. Moreover, the high levels of self-segregation in LA have led over time to the splintered urbanism that Graham and Marvin (2001) discuss. Yet infrastructure systems can produce certain levels of integration as witnessed by recent experiments in "slum networking" in cities like Indore in India (Parikh 2012) or the establishment of gondola transportation structures over slums in Medellin and Caracas (Dale 2010). The gondola system is designed to meet the twin needs of access to transportation and community-based services such as sports halls and music centers located at transport hubs while linking low-income neighborhoods into city-wide transportation networks.

Let us turn to the implications of long-term bypassing and more recent unbundling strategies that have been promoted under privatization by neoliberal regimes. Both strategies intensify socio-spatial segregation. The multiple interactions of different sorts of deprivation only lead to further segregation as explored in several section introductions in this *Reader* (see Editors' Introductions to Basic Urban Services, Cities at Risk, and Housing, this volume, for example). In such conditions of segregation and manufactured scarcity, both the poor and the wealthy seem to opt out of the system. While the former are forced to turn to small-scale informal infrastructure providers, the latter pay to self-segregate into hypermodern enclaves and premium networked spaces. These premium networked spaces include the new downtowns and technology parks that Mitra describes (this volume), gated communities and enclaves (see Leaf, this volume; Baviskar, this volume), and the "city doubles" that Murray describes (forthcoming; see also Editors' Introduction to Housing, this volume). They also include various new forms of city-building being carried out to prove arrival on a global stage, from Dubai's architectural fantasy worlds (see Kanna, this volume) to a range of Asian city experiments (Roy and Ong 2011).

One important type of infrastructure development project that has not been covered yet are the international mega-events of recent years where national, state and metropolitan governments spend billions of dollars to create sporting, tourist, transportation and security infrastructures (Roche 2000). As studies of recent events like the Beijing 2008 Summer Olympics (Broudehoux 2007) or the 2007 Pan American Games and preparations for the 2014 World Cup (Gaffney 2010) have shown, these mega-events are accompanied by large scale clearances of low-income communities for new infrastructure into which massive amounts of public money are diverted. The driving force for governments is the idea of "'accelerated development"–a mantra that uncritically places public money in the service of private profit creating "neo-liberal dreamworlds" (Davis and Monk 2007) "wherein democratic processes are suspended, public space militarized, and urban space restructured in the image of global capital" (Gaffney 2010: 8). The Summer 2013 protests across Brazilian cities that drew millions of residents into the streets are evidence of widespread public outrage against such models of accelerated development. While price hikes in bus fares served as the tipping point, the protests grew to include a range of other issues, including government corruption and lack of funding for basic services when the government was spending large amounts of public money on the World Cup and the Olympics.

Infrastructure provision in the neoliberal era

Equally as important as the unbundling generated by the experience of decentralization is the privatized service delivery that accompanies neoliberal development policies. This has generated a large literature. Many of the issues privatization critics raised are analyzed in Loftus and McDonald's (2001) incisive analysis of Aguas Argentina's celebrated water privatization project in Buenos Aires. Promised reductions in tariffs never materialized and agreed-upon targets for coverage increases were not met. The failure of Aguas Argentinas also emerged as a result of conflicts in state regulatory institutions and the role of international financial institutions. While the Aguas Argentinas case questioned the outcomes of privatization, Patrick Bond (1999: 43) raised a more basic question: how do investment decisions in basic infrastructure provision incorporate socio-economic cost-benefit analysis that covers the "variety of direct, indirect, developmental, ecological and geographical factors" that service outcomes influence? Both supporters of privatization reforms and critics emphasize the inherently political nature of infrastructure provision and the difficulties involved in design and implementation. Embedded in these development debates on infrastructure provision are perspectives on the role of communities and individuals. As Coutard (2008: 1817) pointed out "this politics of infrastructure reforms does not boil down to a simple dialectic between regressive neoliberal reform and socially progressive "resistance." Neoliberalization is not a generalized phenomenon and resistance can be sporadic and splintered along lines of social grouping and power relations in communities.

The contextual diversity in the case studies of infrastructure provision suggests that ideals, interests, motivations of actors and communities need to be taken into consideration. Specific local conditions such as property ownership patterns, the legal status of services, and available subscription options warrant attention as well (Choguill 1996).

We have addressed various issues related to visible infrastructure systems. But what happens with hidden infrastructures that support people on the margins, that sustain marginal life-worlds? In an insightful essay on *People as Infrastructure*, AbdouMaliq Simone (2008) points to the ways in which people improvise, form connections, communicate and exchange what is needed to maximize life-chances, all of which form a support logic (or infrastructure) that does not rely on the territoriality and physical structure of infrastructure as it is commonly understood. The excerpt from Filip de Boeck and Marie-Françoise Plissart's book on Kinshasa, which is a selection for this section, echoes this logic.

We next turn to one specific infrastructure system, transportation, to better understand the range of issues that structure design, provision, and outcomes in any infrastructure system.

Understanding transportation as an infrastructure system

Discussions on the issues raised in the infrastructure literature are largely based on water and sanitation systems with an occasional reference to electricity. Mobility and access to transportation options within the city are rarely included. They remain the purview of transportation engineers and planners, a key infrastructure that remains, with few exceptions, seriously under resourced in the global South.

Like all infrastructure, transportation has always been crucial to the development of cities and to urban form. We know that the scale and spread of the city kept changing as the speed with which we move increased from walking, to riding in a horse drawn carriage, to pedaling a bicycle, and finally, driving in a private automobile. The suburb would never have been possible without the street-car, the commuter train, and then the automobile. Cities like Los Angeles–indeed most contemporary cities perhaps–are impossible to imagine without the car. In the cities of the global South, many different modes remain in active use–people walk, ride bicycles, or cycle rickshaws, or move about in buses, trains, BMWs, large SUVs, and a huge range of small cars like the Indian TATA-Nano. Goods are moved by truck but also by carts pulled by people, oxen, donkeys, or camels. We have thus a spread in terms of technologies, modalities and forms of energy people use to get around the city (Cervero 2000; Cervero and Golub 2007; Kumar and Barrett 2009).

The urban form itself often determines the types of transportation. The medieval European city or the crowded bazaars of Asian and Middle Eastern cities, for example, with their narrow streets and alleys make the movement of buses and large cars difficult. This is true in many of the lower income settlements in the cities of the global South as well. The sprawling form of Los Angeles makes the design of effective public transportation systems a challenge. Similarly the relative ease with which a freeway system was designed and built in the US would be very different today in heavily populated Asian countries where displacement costs of road construction would be substantial.

Understanding historical processes that shape transportation options is as important as following transportation trends. In every country in the world, the preferred mode of transportation is overwhelmingly the car. Regardless of urban form or people's values, automobile ownership is rising faster than GDP. These trends are produced by a number of factors. They include urbanization, smaller household sizes, higher incomes, the development of credit markets for car ownership, changing tariffs on cars through the WTO regime, and sociocultural factors such as the status accompanying car ownership. The differential costs of these trends, particularly increasing automobile ownership, on various groups have been extensively documented (Vasconcellos 2001; Gwilliam 2002; Badami, this volume).

Access to different forms of private and public, motorized and non-motorized modes of transport is vital to poverty alleviation and equity, enabling access to jobs, housing, and a range of services central to increasing opportunities for residents. Research consistently demonstrates that wealth is a crucial factor in transportation access. Wealthier households tend to own multiple cars and other forms of motorized transport, and car use shifts from a peak hour phenomena to more trips that do not fit the peak traffic pattern. In contrast, people who have to use public transit systems, which are often bad or non-existent, are forced to walk or use modes of transportation like rickshaws. Such users are overwhelmingly lower income residents with an overrepresentation of the old, the very young, and women.

Gender is particularly important in understanding access and mobility. The gender-based division of labor within households and communities is reflected in travel patterns. Men tend to be more task- and work-related in their travel patterns while women's travel involves work, shopping, and care-giving. Women are also more likely to carry out "chained or linked trips," often using multiple modes. In addition, evidence shows that women often use cheaper/slower/more local modes of transportation (Turner and Grieco 1998). The use of public transport, especially by women, also relates to cultural norms. Women-only buses, or special compartments for women on trains, are common in cities across the global South.

In certain cities of the global South, important advances have been made in transportation systems. For example, Curitiba in Brazil implemented a highly publicized and innovative Bus Rapid Transit (BRT) system, which has been replicated with notable success across Latin America, China and some cities elsewhere (Cervero 1998). BRTs involve dedicated bus lanes with platforms where tickets can be bought beforehand and special buses that allow large numbers of people to move in and out very quickly. Nonetheless, the inequities of access and the impacts on mobility still contribute to the poverty and inequality of residents of most cities in the global South (Vasconcellos 2001). It is this inequity and the lopsided priorities of government agencies and elites whose transportation policies are largely private automobile focused that Madhav Badami addresses in the article excerpt we have selected for this section.

ARTICLE SELECTIONS

Madhav Badami is an engineer who spent many years in diesel engine development before getting degrees in environmental science and planning. He now teaches in the School of Urban Planning and in the School of the

Environment at McGill University in Montreal, Canada. His research interests include environmental policy, urban infrastructure and services, urban transport and alternative fuels. For Badami, the active destruction of pedestrian accessibility is at the core of the transportation crisis. The loss of access to pedestrians, who make up the majority of people moving across the Indian city, supersedes environmental considerations and resource–use impacts. Current policy privileges cars and car-owners, a position that Badami strongly opposes in this article excerpt.

The second selection comes from French anthropologist Filip de Boeck's stories and exploration of the urbanscape of the capital of the Democratic Republic of the Congo (CRC), Kinshasa. The text is accompanied by the photographs of Marie-Françoise Plissart (not includes in this selection). Together they provide an evocative account of the physical, visible, and invisible structures that emerge and support city-life in Kinshasa. Based on many years of field work, the excerpt focuses on the imaginative ways people invent to deal with the lack of urban infrastructure–a condition that is not uncommon, especially in small cities across the global South.

The third and final selection is by Ahmed Kanna, an Iraqi-American anthropologist currently teaching in California. The excerpt is from the second chapter of his book, *Dubai: the City as Corporation* (Kanna 2011), which is an exploration of the interactions between the built environment and sociocultural processes in the context of a Gulf State. For Kanna, Dubai represents, a quasi-utopia, an architectural and urban experiment and a version of Arab modernity that outsiders seem to understand and find acceptable. The chapter excerpted here is a critique of representations and spatial formations in the premium networked spaces of the city, in the language of Graham and Marvin (2001). Written as a friendly provocation, the chapter nevertheless highlights the problematic manner in which architectural innovation and practice dominated by a small number of celebrity architects dominates the hypermodernity of Dubai.

REFERENCES

Badami, M. (2009) "Urban transport policy as if people and the environment mattered: Pedestrian accessibility the first step," *Economic and Political Weekly*, xliv(33): 43–51.

Bond, P. (1999) "Basic infrastructure for socio-economic development, environmental protection and geographical desegregation: South Africa's unmet challenge," *Geoforum*, 30: 43–59.

Broudehoux, A.-M. (2007) "Spectacular Beijing: The conspicuous construction of an Olympic metropolis," *Journal of Urban Affairs*, 29(4): 383–399.

Cervero, R. (1998) "Creating a linear city with a surface metro: Curitiba, Brazil," in *The Transit Metropolis: A Global Inquiry*, Washington DC: Island Press, pp. 265–293.

Cervero, R. (2000) *Informal Transport in the Developing World*, Nairobi, Kenya: UN-HABITAT.

Cervero, R. and Golub, A. (2007) "Informal transport: A global perspective," *Transport Policy*, 14(6): 445–457.

Choguill, C. (1996) "Ten steps to sustainable infrastructure," *Habitat International*, 20(3): 389–404.

Coutard, O. (2008) "Placing splintering urbanism: Introduction," *Geoforum*, 39(6): 1815–1820.

Dale, S. (2010, March 25) *Medellin/Caracas, Part 7.* Available online at http://gondolaproject.com/2010/03/25/medellincaracas-part-7/ (accessed June 2013).

Davis, M. and Monk, D. B. (2007) *Evil Paradises,* New York: The New Press.

De Boeck, F. and Plissart, M.-F. (2005) "Kinshasa and its (im)material infrastructure." In *Kinshasha: Tales of the Invisible City*, Antwerp, Belguim: Ludion.

Gaffney, C. (2010) "Mega-events and socio-spatial dynamics in Rio de Janeiro, 1919–2016," *Journal of Latin American Geography*, 9(1): 7–29.

Graham, S. (2000) "Introduction: Cities and infrastructure networks," *International Journal of Urban and Regional Research*, 24(1): 114–119.

Graham, S. and Marvin, S. (2001) *Splintering Urbanism: Networked Infrastructures, Technological Mobilities and the Urban Condition,* London and New York: Routledge.

Gwilliam, K. (2002) "Urban transport and poverty reduction," in *Cities on the Move: A World Bank Urban Transport Strategy Review,* Washington, DC: World Bank, pp. 25–38.

Harvey, D. (2005) *The New Imperialism,* Oxford, New York: Oxford University Press.

Kanna, A. (2011) *Dubai: The City as Corporation*, Minneapolis: University of Minnesota Press.

Kumar, A. and Barrett, F. (2009) *Stuck in Traffic: Urban Transport in Africa,* Washington, DC: The World Bank.

Legg, S. and McFarlane, C. (Eds.) (2008) "Ordinary urban spaces: between postcolonialism and development," Guest Editorial, *Environment and Planning A*, 40: 6–14.

Loftus, A. and McDonald, D. (2001) "Of liquid dreams: A political ecology of water privatization in Buenos Aires," *Environment and Urbanization*, 13(2): 179–199.

McFarlane, C. and Rutherford, J. (Eds.) (2008) "Political infrastructures: Governing and experiencing the fabric of the city," Symposium, *International Journal of Urban and Regional Research*, 32(2): 363–374.

Murray, M. (forthcoming) "'City-doubles': Re-urbanism in Africa," in F. Miraftab, D. Wilson, and K. Salo (Eds.), *Cities and Inequalities in a Global and Neoliberal World,* New York: Routledge.

Parikh, P. (2012) "Slum networking," Paper presented at *Design Tactics and the Informalized City,* International Conference, Cornell University, Ithaca, NY, April 2012.

Pflieger, G. and Matthieussent, S. (2008) "Water and power in Santiago de Chile: Socio-spatial segregation through network integration," *Geoforum*, 39(6): 1907–1921.

Roche, M. (2000) *Mega-Events and Modernity: Olympics and Expos in the Growth of Global Culture,* New York: Routledge.

Roy, A. and Ong, A. (Eds.) (2011) *Worlding Cities: Asian Experiments and the Art of Being Global,* Chichester: Wiley-Blackwell.

Simone, A. (2008) "People as infrastructure: Intersecting fragments in Johannesburg," in S. Nuttal and A. Mbembe (Eds.), *Johannesburg: The Elusive Metropolis*, Durham, NC: Duke University Press.

Turner, J. and Grieco, M. (1998) "Gender and time poverty: The neglected social policy implications of gendered time, transport and travel," *International Conference on Time Use,* University of Luneberg, Germany.

Vasconcellos, E. A. (2001) *Urban Transport, Environment and Equity–The Case for Developing Countries,* Sterling, VA: Earthscan Publications.

"Urban Transport Policy as if People *and* the Environment Mattered: Pedestrian Accessibility is the First Step"

Economic and Political Weekly (2009)

Madhav G. Badami

URBAN TRANSPORT IN INDIA: A RAPIDLY WORSENING PROBLEM

The rapid growth in motor vehicle ownership and activity in India is causing a wide range of serious health, environmental, socio-economic, and resource use impacts, even as it provides mobility to millions, and contributes to employment and the economy.

Perhaps the most serious of these impacts, in health and welfare terms, result from road traffic accidents. Road traffic deaths, which stood at 15,000 in 1971, increased to around 93,000 in 2004. Pedestrians and cyclists, the most vulnerable road users, and two-wheeled motor vehicle users account respectively for 50–67% and a quarter of road fatalities, while car users do so for only around 5% (CIRT 2007; Sundar et al. 2007; Mohan 2004). The sad irony is that the road users and modes that are the least responsible for traffic fatalities (and other urban transport impacts) are the most adversely affected.

[. . .]

But of all the impacts due to rapidly growing motor vehicle activity, the loss of accessibility, in particular for pedestrians, is likely the most important, in terms of its implications for the overall urban transport situation. The pedestrian environment in Indian cities is so severely vitiated, that walking, the most natural of human activities, has become an extremely unpleasant, if not a hazardous, activity. Indeed, it may be said that in a nation of pedestrians, the pedestrian has been rendered a third-class citizen.

[. . .]

ROAD CAPACITY ADDITION: THE TRIUMPH OF HOPE OVER EXPERIENCE

Policy making related to urban transport has focused predominantly on road infrastructure development and transport system management to accommodate and improve the traffic characteristics of motor vehicles, along with technological measures to mitigate the impacts of motor vehicle activity per vehicle-kilometer, with a particular focus on congestion and air pollution. The former set of measures has comprised road widening, grade-separated intersections (known commonly as flyovers), limited access expressways, synchronized signals, and area traffic control systems (Tiwari 2002). For many governments, the construction of such infrastructure is proof of their commitment to development and modernization.

[. . .]

Urban road infrastructure projects are being implemented at great public expense. Finally, very significant investments are also being devoted to rail-based metro systems. Delhi's metro system, and the one currently being built in Bangalore, cost roughly $40–45 million (around 200 crore rupees) per kilometer to build (Delhi Metro Rail Corporation 2008; Bangalore Metro Rail Corporation Limited 2008). It is not clear to what extent these massive investments will be cost-effective in relieving congestion and other urban transport impacts over the long term. Meanwhile, budgets for the provision of infrastructure and facilities for pedestrians and cyclists have been miniscule.

While both the public and policy makers set great store by road infrastructure development as the principal response to traffic congestion, which is seen as the primary urban transport problem, building our way out of this problem is not only very expensive, it is, even worse, an exercise in futility, even in resource-rich contexts.

[...]

This trend in congestion despite continuous road-building is not surprising—as international experience has shown, while road-building may improve speeds for motor vehicles and ease congestion in the short term, these benefits tend to be neutralized over the longer term. This becomes a vicious spiral over time, leading to more motor vehicle activity and congestion, and the need to build more roads. The net result is that road-building as a means of addressing congestion is not only futile, it is counter-productive, since it worsens, indeed contributes to, congestion and the other urban transport impacts that it is intended to alleviate.

THE NEGLECT AND LOSS OF PEDESTRIAN ACCESSIBILITY

On many roads in Indian cities, there are no sidewalks, and where they do exist, they are largely unusable, on account of, among other problems, poor design and maintenance, vehicles being parked on them, electrical transformers and junction boxes, uncollected garbage, or encroachment by local businesses and hawkers. Worse, what little existed by way of footpaths, are being lost due to road widening and flyovers. Further, there are few if any facilities for pedestrians to cross roads safely and conveniently; where such facilities do exist, they are spaced too far apart, motorists show no concern whatsoever for pedestrians, and the crossing times are often inadequate. This situation is further exacerbated by the long blocks that characterize urban Indian roads, coupled with the hard, often barricaded medians that are increasingly implemented on these roads, for the purpose of ensuring smooth motor vehicle traffic flow.

The cumulative effect of these conditions is to severely compromise accessibility and safety for walking and other non-motorized modes, rendering their use both extremely inconvenient and hazardous. Travel distances and times are greatly increased for pedestrians and cyclists, and since controlled pedestrian crossings are few and far between, they are forced to cross roads wherever they can, often at mid-block, where motor vehicle speeds are very high. It is precisely because of the lack of pedestrian (and cyclist) infrastructure and facilities, and therefore the inability to walk and cycle safely, that such an overwhelming proportion of traffic fatalities are accounted for by these two modes.

[...]

The lack of pedestrian accessibility affects all, since everyone, including motor vehicle users, is a pedestrian at some stage of their travel, but groups such as young children, the elderly, and the physically disabled, are particularly disadvantaged, and at serious risk of being hurt or killed in road accidents. [T]he group most seriously affected by the lack of pedestrian (and cycling) infrastructure and facilities, in terms of time and productivity losses and road safety, may well be the poor, who for the most part have no choice but to walk or cycle, regardless of how arduous it might be to do so. The poor benefit the least from urban transport infrastructure, but are affected the most severely by motor vehicle activity, and the least able to cope with its impacts, which further exacerbate their poverty.

Therefore, while there is a range of negative externalities associated with motor vehicle activity, the loss of pedestrian (and cyclist) accessibility is particularly pernicious, because unlike vehicular traffic congestion, it is a user on non-user externality, which is rendered all the more serious by being caused as a result of discriminatory transport policy and planning. It is pernicious also because it is not merely a matter of time and productivity losses, and increased risk of fatalities and injuries, for pedestrians (and cyclists), but is directly linked, and is a major contributor to, other serious urban transport impacts.

It is because it is so time consuming, if not unsafe, because of the lack of adequate pedestrian facilities, for people to walk even over short distances, that many trips over these distances are by force of circumstance—and needlessly—conducted by motor vehicles, and often converted into longer motorized trips. [. . .] The largely avoidable use of motor vehicles for short-distance trips, which account for a significant proportion of all urban trips, exacerbates congestion, which in turn increases vehicle emissions and energy consumption.

[. . .]

The increased congestion that results from these avoidable short distance motor vehicle trips renders walking, cycling, and public transit even more compromised and unviable than before, further increasing the need for motor vehicle ownership and use, and forcing motor vehicle owners to needlessly drive even for short distances. In short, motor vehicle activity and planning for it to the exclusion of other modes lead to ever more motor vehicle activity and congestion, as people use motor vehicles to protect themselves from other motor vehicle users.

[. . .]

PEDESTRIAN ACCESSIBILITY: THE FOUNDATION OF URBAN TRANSPORT POLICY

An urban transport policy focused predominantly on road capacity addition to prioritize and improve traffic characteristics for motor vehicles is likely not merely to be infeasible in the Indian context, given inadequate resources to accommodate even present levels of motor vehicle activity and impacts, and ever growing multiple demands on those resources, but is also highly undesirable, given the high urban densities and poverty levels.

While incomes and motor vehicle ownership and use are growing rapidly in Indian cities, large sections of their population have low incomes or are poor and cannot afford even the least expensive motor vehicles, and indeed, even public transit fares (Tiwari 2002). [. . .] Crucially from the urban transport standpoint, the poor in our cities often live cheek by jowl with, and are therefore affected by the travel of, wealthier groups. The confluence of rapid urbanization, and growing incomes and motorization on the one hand, and poverty (and consequently, low motor vehicle ownership and use) among a significant proportion of the population on the other, is an important factor contributing to the severity of urban transport impacts.

Under these circumstances, large scale road infrastructure building to address congestion on an on-going basis will likely cause considerable social disruption. [. . .] Worse, an urban transport policy that not merely accommodates motor vehicles, but actively discriminates against other modes (increasingly the case in Indian cities), will only exacerbate the already serious access and time loss, and road safety impacts for the users of these modes, and for the urban poor in particular.

While high population densities, intensive mixed use, and low income levels make large scale road building undesirable, these characteristics lead to short and medium distance trips forming a significant share of urban travel, which in turn make walking and other non-motorized modes both possible and necessary. Indeed, the vast majority of trips in Indian cities are made by non-motorized modes and public transit, even as motorization increases, and despite the natural advantages of urban form having been lost due to rapidly growing motor vehicle activity, and the poor quality of the pedestrian environment and public transit service (Wilbur Smith Associates 2008).

[. . .]

The urban transport challenge in India is how to cater for rapidly growing mass mobility needs, while minimizing environmental, health and welfare, and socio-economic impacts, and being sensitive to resource constraints and other contextual realities. Given this imperative, it would be desirable for urban transport policy and planning to achieve synergies, by simultaneously addressing the wide range of urban transport impacts, and focus on problem avoidance or prevention, by minimizing motor vehicle activity and the need for it. There is a particular need to pay attention to the needs of low-income groups and the modes on which they depend, even as we plan for motor vehicle activity and apply technological measures to mitigate its impacts.

The achievement of these objectives will call for a wide range of measures, including public transit that is reliable, convenient, affordable, and widespread; pricing of road use that internalizes, to the greatest extent possible, the social costs of urban transport, and provides incentives for minimizing motor vehicle activity; and restoring accessibility for all, in particular pedestrian accessibility, as the foundation of urban transport policy and planning. It is important to stress that accessibility, which was the natural advantage of Indian cities, needs to be *restored* more than *created*, since it has been destroyed by motor vehicle activity and planning to accommodate it.

[...]

Restoring pedestrian accessibility should be the very foundation of urban transport policy and planning if we are to effectively address the urban transport challenge in India. Providing infrastructure and facilities for pedestrians (and cyclists) is only logical and fair, given that the majority does not own personal motor vehicles, that a significant proportion of trips are conducted by these modes, and that pedestrians and cyclists, and the urban poor in particular, bear the brunt of road traffic fatalities and injuries, among other urban transport impacts. But the benefits of this measure would go well beyond those for these groups, and in terms of traffic accidents—making walking and cycling safer and easier [it] would help reduce short distance motor vehicle trips, which are both the most avoidable and energy consuming and polluting on a per-kilometer basis, thereby contributing to reductions in congestion, energy consumption and emissions with high cost-effectiveness, and to that extent obviating the need for expensive end-of-pipeline technological cures.

Providing segregated facilities for pedestrians and cyclists would not cost much, but would allow all modes, including personal motor vehicles, to operate more efficiently, render bus service more attractive and effective (by improving access to it, *and* helping improve its operational efficiency), and potentially allow the movement of a significantly higher number of people per hour overall (Tiwari 2002). In this regard, note that pedestrians (and cyclists) use space far more efficiently than personal motor vehicles, even without accounting for parking; a 3.5-meter lane can carry more pedestrians per hour than car users, even when the cars are travelling on an arterial road, at a speed considerably higher than the current peak-hour average (Indian Roads Congress 1989 and 1990; Wilbur Smith Associates 2008).

Walking and cycling shares are high despite adverse circumstances, but could be higher still. Apart from the large share of short- and medium-distance trips, to which they are naturally suited, and their highly efficient use of space, these modes are competitive in terms of door to door journey times with motor vehicles and public transit over these distances, if adequate infrastructure and facilities are provided for pedestrians and cyclists, as the European experience shows (Whitelegg 1993). The weather is undoubtedly a factor, but people tend to walk more as well over longer distances when the quality of the pedestrian environment is improved. "Build it and they will come" is as true of pedestrians (and cyclists) as it is of motor vehicles.

Enhancing public transit service, rational pricing of road use, and pedestrian accessibility comprise a three-legged stool, with each measure depending on the other two. Improving the attractiveness and effectiveness of public transit, as discussed, depends importantly on ensuring safe and convenient pedestrian access, and curbing personal motor vehicle use, through pricing to increase its marginal cost relative to transit. At the same time, measures to curb personal motor vehicle use would be politically unacceptable without the provision of convenient and affordable transit options, and safe and easy pedestrian accessibility. Finally, increasing the attractiveness of walking and cycling depends not only on the provision of infrastructure and facilities for these modes, but also on reducing motor vehicle congestion. Because this in turn depends on reducing motor vehicle use through pricing, and more effective public transit, one can see how these measures rely on, and reinforce, each other.

PEDESTRIAN ACCESSIBILITY: OVERCOMING THE BARRIERS

Given the serious and worsening urban transport situation, and the need for and desirability of pedestrian accessibility, it is tragic that the state of pedestrian infrastructure and facilities is so very poor, so very little attention is devoted to pedestrian accessibility, and indeed, that non-motorized modes are discriminated against, even as we cater for motor vehicles.

To what may one attribute this state of affairs—a lack of awareness of the benefits of pedestrian accessibility, or a lack of political will, or that we do not care about pedestrians and cyclists? It may be argued that decision makers have what might be called a *car windshield view*, in part because, while those who walk and cycle have no say, those who have a say do not walk.

[...]

Sadly, and perhaps most importantly, the bias in favor of personal motor vehicles and a sense of inevitability in motor vehicle growth among decision makers, the public, and the media, is mirrored in (and contributed to by) urban transport planning in India, which focuses mainly on relieving traffic congestion for motor vehicles, while assigning lower importance to, if not ignoring, the non-motorized modes, and urban transport impacts such as accessibility and safety.

[...]

But hopefully, things are changing—the National Urban Transport Policy (Government of India, 2006) stresses the importance of putting people before motor vehicles in urban transport policy and planning, and commits to prioritizing non-motorized transport. At the same time, large amounts of funds are being made available for urban infrastructure. One hopes that these funds, and those forthcoming from international funding agencies, will be used strategically to ensure that infrastructure and facilities for pedestrians and cycling are incorporated in urban transport projects.

[...]

The extent to which investments that are intended to promote pedestrian accessibility in fact do so will depend importantly on what is understood to constitute, and how investment outcomes are measured in terms of, "pedestrian accessibility." Pedestrian accessibility is achieved not merely by means of good quality footpaths and crosswalks, important as they are, but more generally by enabling people to walk safely, conveniently, and seamlessly, from wherever they might be to wherever they might want to go, at a time of their choosing, at low cost. When understood in this manner, pedestrian accessibility involves a wide range of issues, including street lighting, road drainage, tree cover, modal segregation (by way of, for example, bus bays), traffic and parking management and control, garbage collection and disposal, the design and integration of electric and other utilities, and the provision of public toilets. Beyond this, urban roads need to be designed as public spaces for multiple groups, including pedestrians and cyclists, motor vehicle and transit users, the elderly, the young and the handicapped, and local businesses, including street hawkers—in short, as "complete streets."

BREAKING HEGEMONIC DOMINANCE

But perhaps the most important task in confronting the urban transport challenge is to expose and counter the conceptual underpinnings—and thereby break the hegemonic dominance—of conventional urban transport planning, which assumes ever increasing motor vehicle activity, and prioritizes motor vehicles, by assigning high value to small time savings for them, which in turn is used to justify highways designed for high speeds. The prioritization of motor vehicles is based on their perceived superiority relative to other modes, in part because overcoming the "friction of distance" is considered to be the essence of the urban transport problem.

In providing for high speeds for motor vehicles, access for other modes becomes severely compromised, because high-speed travel requires infrastructure that consumes vast amounts of space, and can be ensured only by limiting access (Whitelegg 1993). As Ivan Illich observed in his 1974 classic *Energy and Equity*, beyond a certain speed, "motorized vehicles create remoteness which they alone can shrink. They create distances for all and shrink them for only a few." Automobile passengers become "consumers of other people's time" (Illich 1974); indeed, as Whitelegg (1993) stresses, transport and spatial planning for motor vehicles "*steal time from (poor and other disadvantaged) groups and reallocate it to (usually) richer groups.*" But the substantial time and productivity losses for the disadvantaged groups, including the very large number of pedestrians and cyclists, are not accounted for. [...] Further, there is no reason why time savings for motor vehicles should always be counted as a societal benefit; indeed, there are many instances—for example, in improving road safety or neighborhood livability, or conserving fuel—in which time *losses* may be beneficial (Whitelegg 1993; Goodwin 2004). Lastly, the urban transport problem in India is at least as much about the *friction of motor vehicle activity*, which impedes (and often endangers) the vast majority who travel on foot, by bicycle, or public transit, as it is about the *friction of distance*, the overcoming of which necessitates high speeds.

By assuming (and accommodating) ever increasing motor vehicle use, urban transport planning only serves to make it more inevitable, as Brown and Jacobson (1987) observed, thus becoming a self-fulfilling prophecy. As planning for personal motor vehicles leads to their increased use, the system is designed more and more to suit them; what is worse, the fact that the other modes are becoming increasingly unviable is used as an excuse not to provide for them, which in turn makes them even less viable, and makes those who rely on these modes more vulnerable. And the more impacts they create, the more personal motor vehicles are seen as the solution to those impacts, thus becoming self-perpetuating.

As motor vehicles become more and more inevitable, the assumption is that their growing dominance (and the increasing unpopularity of other modes) reflects traveler preferences, which transport policy and planning must cater to. But the growing dominance of motor vehicles (and their assumed superiority) results *precisely* from the enclosure of the commons to adapt our cities to motor vehicles, *as a consequence of which* other modes are prevented from performing effectively, and people are forced to use motor vehicles. And the *iatrogenic* nature of transportation planning, that is, its tendency to exacerbate the very problems it purports to address, leads to the need for more expertise, and transportation planning (and planners), like personal motor vehicles themselves, become self-perpetuating. It is in this light that Illich's claim that the transportation industry exercises *a radical monopoly* by creating and shaping the need which it alone can satisfy, is profoundly important.

In the end, urban transport planning is fundamentally about moral and political choices—about what kind of cities we want for ourselves and our future generations, whether urban space is primarily for people or motor vehicles, and what we owe each other, especially the disadvantaged. While motor vehicles play a vitally important role, as do planning and infrastructure for them, and technological measures to mitigate their impacts, an urban transport policy that focuses on these measures to the exclusion of planning, infrastructure and facilities for walking and other non-motorized modes will likely prove futile, even counter-productive, despite great public expense. There is an urgent need, in order to effectively address our urban transport challenge, for an integrated approach that addresses multiple impacts (access loss, road safety, congestion, air pollution, energy consumption, and climate change), caters to multiple modes and road users, and is sensitive to the needs, capabilities and constraints of the Indian context. Such an approach, comprising pedestrian accessibility as the very foundation of urban transport policy, along with quality public transit, pricing of motor vehicle use, and land use-transport integration, would minimize the need for, and curb rapid growth in, motor vehicle activity, allow all modes (including personal motor vehicles) to operate more efficiently, enhance the effectiveness of mass transit, and help achieve an urban transport system that is cost-effective, health promoting, resource conserving, environmentally benign, and socially equitable.

REFERENCES

Bangalore Metro Rail Corporation Limited (2008) Available HTTP http://www.bmrc.co.in/, Accessed June 2008.

Brown, L. R. and Jacobson, J. L. (1987) *The Future of Urbanization: Facing the Ecological and Economic Constraints*, Worldwatch Institute, Washington, DC.

CIRT (2007) *State Transport Undertakings—Profile and Performance 2005–2006*, Central Institute of Road Transport, Pune.

Delhi Metro Rail Corporation (2008) Available HTTP: http://www.delhimetrorail.com/index.htm.

Goodwin, P. (2004) *Valuing the Small: Counting the Benefits*, A paper prepared for CPRE, CTC, Living Streets, Slower Speeds Initiative, Sustrans and Transport 2000, London, UK.

Government of India (2006) *National Urban Transport Policy*, Ministry of Urban Development, Government of India, New Delhi.

Illich, I. (1974) *Energy and Equity*, Marion Boyars, London.

Indian Roads Congress (1989) *Guidelines for Pedestrian Facilities*, IRC: 103–1988, Indian Roads Congress, New Delhi.

Indian Roads Congress (1990) *Guidelines for Capacity of Urban Roads in Plain Areas*, IRC: 106–1990, Indian Roads Congress, New Delhi.

Mohan, D. (2004) *The Road Ahead: Traffic Injuries and Fatalities in India*, Indian Institute of Technology, New Delhi.

Sundar, S. et al. (2007) *Report of the Committee on Road Safety and Traffic Management*, Ministry of Road Transport and Highways, Government of India, New Delhi.

Tiwari, G. (2002) "Urban transport priorities—Meeting the challenge of socio-economic diversity in cities: A case study of Delhi, India," *Cities*, 19: 95–103.

Whitelegg, J. (1993) *Transport for a Sustainable Future—The Case for Europe*, Belhaven Press, London.

Wilbur Smith Associates (2008) *Study on Traffic and Transportation Policies and Strategies in Urban Areas in India—Final Report*, Wilbur Smith Associates, Bangalore.

"Kinshasa and Its (Im)material Infrastructure"

Kinshasa: Tales of the Invisible City (2005)

Filip De Boeck and Marie-Françoise Plissart

SIMULACRA OF INFRASTRUCTURE

In ongoing discussions concerning the nature of the African city architects, urban planners, sociologists, anthropologists, demographers and others devote a lot of attention to the built form, and more generally, to the city's material infrastructure. Architecture has become a central issue in western discourses and reflections on how to plan, engineer, sanitize and transform the urban site and its public spaces. Mirroring that discourse, architecture has also started to occupy an increasingly important place in attempts to come to terms with the specificities of the African urbanscape and to imagine new urban paradigms for the African city of the future. Indeed, one can hardly underestimate the importance of the built form and of the material, physical infrastructure if one wants to understand the ways the urban space unfolds and designs itself. For example, studying the process of the "bunkerization" of the city, as it is called by its inhabitants, that is the fact that one of Kinshasa's crucial spaces, the compound, has evolved from an open space lined by flowers and shrubs in the 1940s and 1950s to today's closed *parcelles*, surrounded by high walls that make the inside invisible to the street, would certainly contribute to a better understanding of the city's history of unraveling social relationships, its altered sense of security and its changing attitude towards the qualities of public and private.

However, as Kinshasa's ports reveal, the city's infrastructure is of a very specific kind. Its functioning is punctuated by constant breakdown. The qualities of failing often give the urban infrastructure the character of a simulacrum. For example, the television set, a status symbol, occupies a central place in the living room, but often it just sits there without functioning at all because there is nothing to plug it into, or because it broke down a long time ago, or because there is no electricity. Often, while in Kinshasa, I am reminded of Mary Douglas' "*The Hotel Kwilu: A Model of Models.*" In this text she describes how she visits the Hotel Kwilu, located in the town of Kikwit overlooking the Kwilu river:

> The Hotel Kwilu looks like a modest version of the Sheraton or the Marriott or any of a number of well-standardized airport hotels: modest by comparison, but grandiose in its setting. As I remember, it is a handsome building made of solid stone, with broad steps up to the front entry, a reception desk on the right, a big glass-roofed atrium in front, potted palm trees around, a bar to the left, and a restaurant beyond that, all calm, cool, and inviting. Before looking in I asked to see the bedroom. It was still in the accepted Sheraton style: clean, big, huge mirror, air-conditioning, twin beds, twin pictures on the wall, the telephone, the reading lamp, well carpeted, the bathroom en suite. Inside the bathroom, again, perfectly in style, the bath, the gleaming fittings on the hand basin, shower, hairwashing spray, the lavatory. Everything was there, not forgetting the bottle of drinking water. The only thing I thought was odd was that the bath was full of cold water. I wondered if the last guest had not left them time to clean it, but no, I was told this was to economize water. The candle and

> matches by the bed I took for an extra courtesy in case of emergency.
>
> [...]
>
> The receptionist asked me to pay in advance so that they could procure the diesel fuel needed for refrigeration and electricity. He also said that the electric lighting went out at 8 o'clock, to save diesel. [...] However, when I got upstairs I found, with the help of the candle, that the taps did not run, the lavatory did not flush, and neither the phone nor the air-conditioning was connected. But I rejoiced in the huge bath full of water, and a dipper for carrying water to the hand-basin and the lavatory. (Douglas 1989)

If Douglas had arrived at her hotel some hours before, she would have witnessed two women, with large plastic buckets on their heads, walking back and forth from a nearby public water tap all the way up to her room to fill the bathtub.

As with the Hotel Kwilu, Kinshasa is full of such disconnected figments, reminders, and echoes of a modernity that exists as form but no longer has the content that originally went with it. The fragments themselves are embedded in other rhythms and temporalities, in totally different layers of infrastructure and social networks. Failing infrastructure and an economy of scarcity therefore constantly delineate the limits of the possible, although they also generate often surprising possibilities, through a specific aesthetics of repair, by means of which breakdown is bypassed or overcome.

THE UNFINISHED CITY

Along the bypass, the main road which coils around Kinshasa's southern and western parts, a dusty sand road leads to the commune of Mont Ngafula. This neighborhood emerged in the 1970s as a semi-residential area for executives, functionaries and upcoming politicians. Many compounds in this neighborhood are spacious, with lots of trees and green. But many houses were never finished. With the generalized breakdown that characterized the end of Mobutu's reign [Ed. note: Mobutu was leader of the DRC, then called Zaire from 1965–97], the emerging middle class that bought building plots in this neighborhood was gradually cut off from its income. The (often spectacular) houses they had dreamt of building for themselves were left in various stages of unfinished abandon, impressing upon one the image of the city as a never ending, perpetual building site, a characteristic Kinshasa shares with many other African towns. Today, people live in the skeletons of their frozen dreams of progress and grandeur, in constructions of concrete and cement without doors, windows, roofs. Only the ground plan betrays the original aspirations.

Other less fortunate inhabitants of this neighborhood witnessed how their houses disappeared overnight. During the rainy season, erosion is a constant threat in many parts of the city. Overnight, the erosion, which often finds its origin in deficient drainage, cuts through the sandy soil of Kinshasa's hills, leaving behind spectacular abysses in which houses, roads and other infrastructure disappear. Here Kinshasa becomes a cannibalistic city, literally devouring its own urban tissue. Today in Kinshasa, erosion threatens whole neighborhoods in at least 400 different spots.

Together with the dust roads, the generally spacious and green compounds, though usually surrounded by high walls, give the commune of Mont Ngafula a rather rural character. As in most neighborhoods of Kinshasa, water and electricity reach this part of the city only sparingly. Water, for example, usually comes between 2 and 4 am, whereas electricity is made available according to a system of what Kinois refer to as *delestage*: in different sectors of the network, SNEL or REGIDESO, the national electricity and water companies, switch electricity and water off at certain times in order to feed other sectors. It is totally unclear which criteria determine the distribution over the various communes and neighborhoods. Some receive water and electricity during certain hours of the day (but unfortunately these hours often vary from one day to the next). Other parts of the city are supplied for weeks and then cut off for weeks. Some areas are not served for months while many neighborhoods are not even connected. Each of these cases sets in motion a carrousel of people. Girls and boys are sent out with buckets, tiles and cans to fetch water in nearby or more distant neighborhoods, fathers visit their friends to charge the batteries of their cell phones, and whole neighborhoods move elsewhere to watch soccer games in those compounds where there is a television set that works. When technologies remain silent or break down, and thereby give form to yet another level of invisibility that shapes the city, these lacks and absences generate new spheres of social interaction and different coping strategies and regimes of knowledge and power.

POSSIBILITIES OF INFRASTRUCTURE

Infrastructural fragments thus also enable the creation of new social spaces. A couple of years ago, on the corner between the main Bypass road and the entrance to the commune of Mont Ngafula, someone started building a FINA gas station. It took several years to complete, but the gas station finally opened in 2002. A couple of months later the owner placed a huge lamppost on the premise. Since the gas station used its own generator and therefore did not depend on the city for its electricity supply, the lamp kept burning. In no time at all, the lamppost gave birth to a large numbers of bars, a cyber cafe and a telephone shop around the station, while buses and taxis began to use this place as the terminus of their trajectory, thereby bringing even more people to the bars. Business at the near by Fwakin Hotel started picking up again after many years. With fascination I observed how one lamppost transformed what was a quiet corner with little movement after nightfall into an important meeting point bustling with life until midnight. The process of random occupation of space also reveals the organic approach Kinois [Ed. note: residents of Kinshasa] have to the production of the city. Space, in a way, belongs to whomever uses it, despite the halfhearted attempts of the city authorities to control the slow but unstoppable occupation and the progressive denser use of that space.

Simple material infrastructures and technologies, as well as their dysfunctioning and breakdown, thus create, define, and transform new sites of transportation, new configurations of entangled spatialities, new public spaces of work and relaxation, new itineraries and clusters of relations, new social interactions.

The *phonie* [Ed. note: a form of cyber café] is a good example of how material infrastructure and technology create new forms of sociality and new topographies of propinquity, how they can bring people into physical proximity with each other, how they generate new (trans)urban public spheres, or enable, maintain and carry forward existing social landscapes, networks and affiliations under changed circumstances. *Phonies* connect Kinshasa to the rural hinterland. Every *phonie* transmits messages to specific towns. As a result, people from the same ethnic or regional background meet at these *phonies*, where they often spend several days before getting in touch with the person they want to talk to at the other end. The *phonie*, therefore, provides a place, a social island within the city to maintain, strengthen and reactivate different, often pre-urban, ties of locality and belonging.

However, such islands of communication, association, collaboration and proximity, with their possibilities of reimagining a different ethics for the current urban life, remain very dependent on the materiality of their infrastructure and are therefore very vulnerable. In the case of the *phonie* or the cyber cafe there is the constant dependence on hardware that is costly, electricity that is interrupted all the time, radio transmitters that risk being damaged because of unstable voltage, solar panels that are hard to get by and easily break down, batteries of poor quality that have to be constantly recharged, computer viruses that infect all the city's PCs and are as difficult to get rid of as the viruses that attack the people of this city in real life. Beyond that, the existence of the *phonie*, for example, is strongly dependent on the absence of other, newer technologies. These may be more sophisticated and efficient but also more demanding, necessitating a larger investment or a higher degree of technical knowhow, which inevitably turns its users into mere consumers rather than producers or controllers of that technology. For example, as I write, the introduction of sophisticated cell phone technology by international communication multinationals in Congo is already turning the *phonie* meeting points and the social geographies they engender into archeological sites. In 2003, seven different telecommunication networks, each with its own international (American, South African, French, Belgian, Chinese) affiliations, were competing with each other to control the potentially vast Congolese telecommunications market (VODACOM, CELTEL, TELCEL, OASIS, AFRITEL, CCT [*Congo-ChineTeliphone*], COMCEL). In 2003, also, large parts of the interior were opened up through the implementation of cell phone technology in which these international companies invested a lot of money. For the first time ever, people in remote villages can call Kinshasa, Brussels, Paris or New York and reach beyond their own horizon within seconds. Wonderful as this is in itself, it also means that established forms of cooperation, communication and collective responsibility, with their specific social capital and the particular levels of trust they summon, will become obsolete and disappear in the very near future (no doubt to be replaced by something else).

INVISIBLE ARCHITECTURE

In spite of the fact that an analysis of the different physical sites through which the city exists and invents itself helps us to better understand the specific ways in which the materiality of the infrastructure generates particular sets of relations in the city, I would submit that in the end, in a city like Kinshasa, it is not, or not primarily, the material infrastructure or the built form that makes the city a city. The city, in a way, exists beyond its architecture. In Kinshasa, the built form is not, or is no longer, the product of a careful planning or engineering of the urban space. It is, rather, produced randomly in human sites as living space. Constantly banalized and reduced to its most basic function, that of a shelter, the built form is generated by this much more real, living city which exists as a heterogeneous conglomeration of truncated urban forms, fragments and reminders of material and mental urban "elsewheres" (a lamppost, a radio antenna, a television screen, dreams of life in the diaspora). These are embedded into autochthonous dynamics and into an urban life that is itself produced through the entanglement of a wide variety of rhizomatic trajectories, relations and mirroring realities. They enjoin, merge, include, fracture, fragment and reorder the urban space through the practices and discourses of its inhabitants. Within these local dynamics, within these syncretic multiplicities, the cultural status of the built form seems to be of lesser importance, or rather, the material infrastructure that counts in the making of the city is of a very specific nature.

First of all, the infrastructure and architecture that function best in Kinshasa are almost totally invisible on a material level. Under the trees along most of the city's main roads and boulevards one finds all kinds of places: garages, carpenter's workshops, showrooms for sofas, beds and other furniture, barber shops, cement factories, public scribes, florists, churches and a whole pleiad of other commercial sites offering a variety of services. Yet, none of these take place in built structures. What one needs in order to operate a garage is not a building named "garage," but rather the idea of a garage. The only material element needed to turn an open space into a garage is a used automobile tire on which the garage owner has written the word quado (supposedly after the name of a well-known Belgian garage owner in the colonial period). One cord between two trees suffices to hang up the newspapers of the day, thereby creating a meeting place for the *parlementaires debout*, the people who gather under the trees to comment on the newspapers' content and erect their agora, their parliament, by means of a rhetorical architecture, the built form of the spoken word. These vibrant urban spaces teeming with all kinds of activities generate an infrastructure of paucity, defined by its material absence as much as by its presence. The built form that makes this possible is not made of cement and bricks but consists of the body of the tree under which people gather and meet, and the space between the trees that line the city's main roads. Shortcutting any dependence on unstable infrastructure and technologies, this is the level of infrastructural accommodation, the only level of accumulation also, that really seems to work.

REFERENCE

Douglas, M. (1989) "Distinguished lecture: The Hotel Kwilu—A model of models," *American Anthropologist*, 91(4): 855–865.

FOUR

"'Going South' with the Starchitects: Urbanist Ideology in the Emirati City"

Dubai, the City as Corporation (2011)

Ahmed Kanna

Most companies start with a clean sheet of paper. Ours was a city.

Advertisement for EMAAR (2004 advertisement for EMAAR in the Dubai *Gulf News*)

[. . .]

STARCHITECTURE

Along with China, the contemporary Arabian Gulf is undergoing an urbanization of massive proportions. Possessing a disproportionate share of the world's proven oil reserves, the six countries of the Gulf Cooperation Council were devoting on the order of hundreds of billions to trillions of dollars to construction projects before the financial crisis of 2008, which did indeed slow construction, especially in Dubai, the economy most affected by the crisis (Brown 2008). Like China, architects view the Gulf, whose member states are run by tiny elites disposing of immense wealth and nearly nonexistent labor and environmental regulations, as a liberating place in which to work. No Gulf country has been as aggressive in advancing top-down, large-scale, institutional urbanism (Lefebvre 2003: 79) as the United Arab Emirates (UAE). The rhetoric of government officials, architects, and the media, what Lefebvre has called *urbanists* (Lefebvre 2003: 156–60), employs various discourses of progress and architectural radicalism to justify such projects.

Following places such as Riyadh, Jeddah, and Kuwait, which pioneered the practice of using big name architects in the Gulf, individual emirates of the UAE, especially the wealthiest two, Abu Dhabi and Dubai, have recently tapped so-called starchitects in a large project of urban entrepreneurialism (Harvey 1989; Broudehoux 2007). The list of the famous firms which were "rushing" (Brown 2008) to participate in the UAE's architectural "Xanadu" (Fattah 2007) was a who's who of global north starchitecture: Tadao Ando, Norman Foster, Frank Gehry, Zaha Hadid, Rem Koolhaas/OMA, Jean Nouvel, Skidmore Owings Merrill (already a veteran on the UAE scene), Snøhetta, and many lesser names. Looked at in the wake of the 2008 economic meltdown, in which many brand-name projects, such as Hadid's Dubai opera house and Koolhaas's Dubai Waterfront City, have either been cancelled or put on hold, this rush seems only to have been a brief flash of collaboration between the Emirati and global urban worlds. Serious scholarly observers have dismissed the Emirati use of starchitecture as superficial and merely profit-driven. I agree with this, but regardless of its motives or "depth," the politics of starchitectural representation during this brief, pre-crisis moment remain of significant interest to the study of cultural politics in rapidly developing, globalizing cities. Moreover, such representational politics take on new meaning in an Emirati, and especially Dubai, context in which the main players in urban development are the family states with their history of anti-reformism.

[. . .]

Renzo Piano and Richard Roger's Centre Pompidou arguably initiated the era of starchitects in the mid-1970s, establishing a prototype for urban development in the context of the emerging neoliberal consensus of the times. The Centre Pompidou showed how cities could recruit foreign architects to spearhead projects of urban renewal centered on spectacular spaces of consumption (McNeill 2009; Zukin 2008). According to Sharon Zukin, clients also saw this patronage of foreign talent as a way to display their sophistication and flare for the innovative (Zukin 2008). Since then, the starchitect has become a species of architect whose mere name is so enveloped in the mystique of genius that urban elites the world over now consider the commissioning of a building by a brand-name architect to be a municipal priority, regardless of the aesthetic or design quality of the project, its expense, or the demands it makes on local resources.

[. . .]

THE PRODUCTION OF SPACE AS A CLEAN SLATE

"When we started there was nothing but sand, birds, and sky. Today, lakes have replaced the sand. Tall towers reach out to the stars. Golfers sink birdies on world-class golf courses. . . . We were the first to see the dream of Dubai as a modern, world-class city. And we were the first to see it through. We're changing Dubai's future."

When (Frank) Gehry and other urbanists mention the "clean slate," they are referring to the wider social world in which architect and client operate. It is a space evacuated of history (as human struggle) and culture (as richly textured everyday practice). Such space in turn becomes the arena for the unlimited creativity of the architect-genius. Henri Lefebvre argues that the elitist collaboration between state, market, and architect is a major obstacle to the breakthrough to the urban revolution, in which space is liberated from "the imperialism of know-how" (Lefebvre 2003: 59). Class urbanism creates inert space, space as nothing more than a container for objects, be they animate or inanimate, and for fetishized "needs." The city, in this view, becomes "a simple spatial effect of a creative act that occurred elsewhere, in the Mind, or the Intellect" (Lefebvre 2003: 28). Most disturbingly, "ever since its origins, the State expressed itself through the void: empty space, broad avenues, plazas of gigantic proportions open to spectacular processions" (Lefebvre 2003: 109). Lefebvre's insights from over forty years ago are surprisingly perspicacious even today, in the context of urban entrepreneurialism in the global south.

The case of Dubai is instructive. Like Los Angeles, it is developing into an extreme version of the polycentric city. Since the early 1970s, when it had a definite center around a natural inlet, the *Khor Dubai* (Dubai Creek), the city has exploded into its present form. Dubai's post-1990s sprawl into its surrounding hinterland, the area often called New Dubai, has been characterized by disconnected enclaves with their own amenities, security details, and infrastructures. The further from the traditional town center one goes, the more this is the case. Iconic architecture is not, moreover, always intended to announce Dubai as a world city but to signify the family state's or the landlord's power, vision, and so on. Another aspect of Dubai urbanization is that land is not at a premium. Instead of, for example, clearing out entire neighborhoods for urban renewal, the city simply expands outward into the undeveloped terrain on what is, for the given moment, the existing city's hinterland, creating new poles for urbanization. This situation gives the impression that urbanism is simply about the unconstrained willing into being of new buildings and enclaves. In reality, a specific kind of city is created, in which space becomes increasingly fragmented, subject to privatized instrumentalities of control, and mobility without the use of superhighways becomes impossible. Therefore, Dubai sprawl is not a simple product of technical requirements, nor is its landscape a canvas for politically innocent aesthetic experimentation. Sprawl enables and consolidates the political power of landlords and the class power of wealthier city residents. This reality is not reflected in starchitectural and urbanist discourse.

[. . .]

THE DISCOURSE OF SUSTAINABILITY: POWER IN A SHADE OF GREEN

[. . .]

One good example of the ways in which reifying spatial and cultural representations promote elite interests is the discourse of sustainability in the UAE.

Over the past five to ten years, the notion of sustainability has emerged as a priority for UAE urbanists. This is ostensibly a salutary development. The UAE is one of the world's largest per capita consumers of fossil fuels and emitters of greenhouse gasses. "If the world lived like each individual in the UAE, we would need 5.5 planets" (AME Info 2005). Abu Dhabi, the federation's largest emirate, is the world's top emitter of greenhouse gasses, per capita (Vidal 2008). Moreover, UAE urbanists have recently referred not only to issues of ecological impact, but to economic viability and social equitability as goals which, they argue, their projects, not least, starchitectural brand-name projects, will help them achieve.

Urbanists in the UAE are beginning to make noises about each kind of sustainability—ecological, economic, and social. A 2007 workshop at Harvard's Center for Middle Eastern Studies, which I attended, was devoted to the theme of sustainability in the Gulf (CMES 2007). The workshop's organizer, a prominent Boston architect with a great deal of experience in the Gulf, declared in his opening remarks that in the context of rapid development, it is a moral, professional, and aesthetic quest to formulate a program of sustainable architecture (CMES 2007). Among those in attendance were leading American and British specialists in sustainable architecture and two representatives (non-Emirati Arab expatriates) of a large Abu Dhabi-based real estate developer, Sorouh. What struck me about the discussion over the workshop's two days was the abstract, technical framing of problems of sustainability, and the conflation of orientalist or other ethnocentric stereotypes with locally situated problems of environmental impact that the UAE seemed to present. While expressing concern (doubtless sincere) for the sustainability of UAE architecture and urban development, many of the assembled experts almost automatically resorted to stereotypes about culture, politics, and economy to frame their urbanist programs.

[. . .]

[Below] I would like to focus on how UAE urbanism, in the guise of expertise on sustainability, erases local specificity and implicitly legitimizes the elite production of space. Like the Harris plan modification of 1971, with its uncritical uses of "Arab democracy" and conflation of the Al Maktoum foundation myth with the history of Dubai, the participants at the Harvard workshop generally conflated elite narratives of space and locality with local space and culture per se.

In the ideological and discursive work of sustainability, the development of large leisure and retail enclaves does multiple duties, allegedly contributing to economic, ecological, and socio-cultural adaptation. Architects and real estate developers working in the UAE often make the argument that such enclaves, and most notably their "themed" or iconic architecture, are a rational, long-term adaptation to the economic demands of globalization. As one architecture professor at the American University of Sharjah argues in a paper on Dubai as a "global city," urban entrepreneurialism is what globalization is "all about" (Mustafa 2004). At the 2007 Harvard workshop, the two representatives of the Abu Dhabi real estate developer Sorouh made a similar case for their enclaves. These were necessary for Abu Dhabi to join the global economy; developing such enclaves was a matter of urgency because the emirate was already behind. Abu Dhabi suffered 30 years of stigma "that they never developed anything." The political leadership is committed to moving forward, they said. Primary among Sorouh's desiderata are sustainable buildings and planning, they claimed. However, "with all this pressure to deliver, something has to give, and yes, sometimes we have to compromise." Interestingly, as an aside, one of the presenters added, "We only follow what (Western consultants) tell us." The implication was clearly that any deviation from the developers' sustainable ideal was the result of outsiders' extra-ecological (i.e., monetary) influence. Thus, the implication, often repeated by local UAE urbanists, is that the ruler, the landlord, or the real estate developer possesses agency when the narrative is positive (Arab democracy, liberal policies, the UAE's visionary urbanists) and that they are passive when projects manifest flaws.

In spite of narrow-minded Western consultants, UAE developers are creating sustainable urbanism, according to the Sorouh public relations men. They gave the example of Nakheel, the Dubai developer that is owned by the ruler Muhammad Al Maktoum (and for which one of the Sorouh men had previously worked). Nakheel (famous for their brash iconicism, exemplified by projects such as the World, an archipelago of artificial islands mimicking the world map) "caters to a wide spectrum of society," by which the Sorouh man meant, simply, many nationalities. His implication was that this catering to diverse nationalities is somehow equivalent to "sustainability." Other Nakheel projects, such as the Ibn Battuta mall, a giant amusement park and shopping mall whose theme is based on the famous Arab traveler's journey from North Africa to China, blends cultural and heritage elements, thereby allegedly contributing to "cultural sustainability." In doing so, the Sorouh men argued, it also provides a valuable educational service. (This is a common theme of Nakheel's public relations.)

"Sustainability" has recently become a point of articulation between UAE urbanists and starchitects. Political, along with the aforementioned ecological, cultural, and economic, notions of sustainability are prominent here. As a member of the management consortium for Gehry's Guggenheim Abu Dhabi put it (echoing the Sorouh public relations men that Abu Dhabi is "moving forward"), the emirate's rulers are "very conscious here that [the arts and culture enclave at Saadiyat Island] can change the cultural climate in the region. . . . To be able to add high culture at the high end of international culture, this [is] a tremendous change" (Fattah 2007). Meanwhile, press materials for Norman Foster's Masdar ("the Source") project in Abu Dhabi, "the world's first sustainable city," list "equity," "fair trade," and "fair wages for all workers who are employed to build the city" among the priorities of the developers (Vidal 2008).

If the social reality of labor practices in the UAE is considered, the point about equity and fair wages is particularly difficult to take seriously. The 2006 Human Rights Watch (HRW) report on the UAE construction sector is a scathing critique of labor exploitation in the UAE construction sector, which is composed exclusively of foreign workers (Human Rights Watch 2006). . . . [T]he abuses detailed in the report are systematic and structural, directly connected to the Dubai state's ruling bargain with local corporations and tacit protection agreements with foreign private interests. Not coincidentally, starchitects depend on these laborers to make their projects realizable within accelerated timeframes and at immense scale. These workers constitute the real social basis for the so-called industry laboratory celebrated by starchitects such as Schumacher. Among other things, the report, entitled "Building Towers, Cheating Workers," details widespread corruption among labor recruiters and employers, and practices such as expropriation of workers' passports and nonpayment of wages for months or even years, as well as the inertia of UAE state authorities in intervening on behalf of workers. The report also suggests that risk in the UAE labor market is borne almost entirely by the workers. Employers find various ways of evading serious government oversight (labor recruiters even more so). Worker revolts and suicides were also reported by HRW (as they are, to some extent, in the UAE press). Approximately one in five of the nearly 2.75 million migrant workers in the UAE (2005 estimate) is employed in the construction sector (Human Rights Watch 2006: 7). These workers are largely illiterate and from impoverished rural backgrounds. They are housed in dilapidated labor camps far from the city limits. The camp I visited in 2006–2007, called Sonapoor, the Hindi word for "City of Gold," actually a small town consisting of the camps for numerous construction companies, is at least five kilometers into Dubai emirate's desert interior. School buses transport the workers back and forth between work sites and the camps. For daily necessities, the workers shop at overpriced camp shops selling goods marked up even from the prices charged at upscale Dubai shopping malls. Although there is some anecdotal evidence (supplied mostly by the local press and therefore of limited reliability) that conditions in Dubai camps have improved (owing, among other things, to efforts such as HRW's), conditions in other emirates, according to a credible local source I interviewed, remained as of early 2007 abysmal (intermittent electricity, lack of infrastructure, and non-payment of wages, to cite a few examples).

[. . .]

[T]he permanent UAE representative to the United Nations argued against HRW's recommendations for reforming the construction sector. In a letter to HRW, he claimed that laws applicable in the Western countries cannot be applied to workers in the UAE (Human Rights Watch 2006: 70). For the ambassador, there is apparently no contradiction between drawing on the prestige and expertise of Western starchitects while rejecting Western human rights laws. The same can be said of UAE urbanists.

A good argument can be made that with ventures such as Masdar, famous architects such as Foster are using the symbolic power of their names in genuinely progressive ways, ensuring that projects bearing their imprimatur contribute positively not only to a state's reputation but also to the social relations prevailing within that state. One should not gainsay the counterargument to the one I am making here, that if all architectural projects followed Masdar's lead, UAE urbanism may be far the better. It remains unclear, however, how far-reaching Masdar's sustainable innovations will be simply an isolated enclave of relatively good conditions (or encouragement to other projects to do the same) as well as how local authorities will seek to contain the project's progressive labor policies. However, even the most charitable reading of Masdar, which discounts the opportunism usually characterizing the sudden discovery of sustainability and justice by state–elite formations in the context of the recent global image economy of "green" symbolism, does not have a good response to the question. Are enclaves, with all that they imply in terms of elite political control, economic inequality, and the top-down rule of expertise, the best possible solution to contemporary urban problems that starchitecture can propose? Moreover, witnessing the repackaging by urbanists, such as Sorouh and Nakheel, of large enclave spaces of bourgeois gratification as "green projects," one cannot help but recall the following insightful observation by Donald McNeill (made in relation to similar efforts to "green" airports).

> As an envelope, yes, green airports are a possibility. But only if one ignores what is happening out on the runways and link roads and car parks that surround these light green structures. Indeed, turning these environmental disasters into things of beauty has a parallel with some of the debates over the use of architectural monuments to prop up unpopular political regimes. (McNeill 2009: 144)

In the UAE, the emergence of green ideology and the propping up of the political regime are not only analogous. They are often part of the same process.

[. . .]

Although I am clearly skeptical of the phenomenon of starchitecture, this chapter should not be read as an indictment of individual starchitects, still less of architecture *tout court*. Nor do I believe that architects alone created starchitecture. The emergence of the architect as a marketable commodity (something with which, to be sure, individual architects are complicit) is a complex phenomenon, connected with the rise of urban entrepreneurialism in the era of neoliberalism.

[. . .]

REFERENCES

AME Info. (2005) "Facing up to our ecological footprint," 14 Feb. Online. Available HTTP: http://www.ameinfo.com/53883.html (accessed 27 May 2008).

Broudehoux, A. (2007) "Spectacular Beijing: The conspicuous construction of an Olympic metropolis," *Journal of Urban Affairs* 29, 4: 383–399.

Brown, M. (2008) "Hadid leading architectural rush to the Emirates," *International Herald Tribune*, 3 April.

CMES (2007) "Reconceiving the built environment in the Gulf," Workshop, Center for Middle East Studies, Harvard University, 28–29 April.

Fattah, H. (2007) "Celebrity architects reveal a daring cultural Xanadu for the Arab world," *New York Times*, Feb. 1. Online. Available HTTP: http://www.nytimes.com/2007/02/01/arts/design/01isla.html?pagewanted=all (accessed 28 May 2008).

Harvey, D. (1989) "From managerialism to entrepreneurialism: The transformation in urban governance in late capitalism," *Geografiska Annaler. Series B, Human Geography* 71, 1: 3–17.

Human Rights Watch (2006) "Building towers, cheating workers: exploitation of migrant construction workers in the United Arab Emirates," *Report* 18, 8 E. New York: Human Rights Watch.

Kanna, A. (2011) *Dubai, The City as Corporation*, Minneapolis: University of Minnesota Press.

Lefebvre, H. (2003) *The Urban Revolution*. R. Bononno (trans.), Minneapolis: University of Minnesota Press.

McNeill, D. (2009) *The Global Architect: Firms, Fame and Urban Form*, New York: Routledge.

Mustafa, A. (2004) "Make no little plans," Paper presented at seventh annual Sharjah Urban Planning Symposium, 4–6 April.

Vidal, J. (2008) "Desert state channels oil wealth into world's first sustainable city," *The Guardian*, 21 Jan. Online. Available HTTP: http://www.guardian.co.uk/environment/2008/jan/21/climatechange.energy (accessed 27 May 2008).

Zukin, S. (2008) "Destination culture: how globalization makes all cities look the same," *Trinity College Center for Urban and Global Studies Inaugural Working Papers Series*. Online. HTTP: http://www.trincoll.edu/NR/rdonlyres/8FE6BF06-7F6E-4B7A-823C-CA42D688E054/0/ZUKINATLAST.pdf (accessed 24 January 2010).

SECTION 8
Cities at Risk

Editors' Introduction (with Andrew Rumbach)

In this section we consider cities as sites of *risk*. Risk, broadly defined, is the potential harm a person might suffer from natural and man-made hazards: earthquakes, floods, war, violence, economic collapse, and so on. One of the most durable characteristics of risk is that it is highly uneven, with some people and groups much more likely to suffer than others. The concept of risk helps us to plan for, respond to, cope with, and adapt to a range of everyday and extreme events, with the ultimate goal of building safe, peaceful, and resilient cities.

KEY ISSUES

Cities and "natural" disasters

Cities have always been at risk from natural hazards. In AD 79, the eruption of Mt. Vesuvius so completely buried the cities of Pompeii and Herculaneum in ash that they remained lost for 1,500 years. An earthquake in the Persian region of Qumis in AD 856 killed more than 200,000 people and forced the abandonment of the capital city (Ambraseys and Melville 1982). The Great Lisbon Earthquake of 1755 destroyed one of Europe's greatest colonial capitals and besides causing a crisis of faith for many Enlightenment thinkers also spurred major advances in seismology, earthquake engineering, and city planning (Shrady 2008). More recently, cities like Chengdu (2008), Port-au-Prince (2010), and Bangkok (2011) have suffered terrible disasters with widespread human, economic, and environmental loss. Urban history is tragically replete with such cases, and disasters have played an important role in shaping urban development.

One reason that cities are so often afflicted by disaster is because they tend to be built in hazard-prone places. *Hazards* are events that have the potential to harm people or the things we value (Cutter et al. 2009). Hazards may be natural, like an earthquake, tropical cyclone, or drought, man-made, or something in between. Mark Pelling (2003) prefers *environmental hazard,* recognizing that most urban disasters have both human and natural dimensions. There are many benefits to locating cities in hazardous geographies, of course: rivers, coasts and ports are essential for accessing global trade and communications networks and for projecting military power; floodplains and volcanoes have rich soils and are prime agricultural land; mountains provide ready access to minerals and precious metals; and so on. Urban development concentrates people and capital in these places, exposing them to a wide range of hazards. Some, like earthquakes or cyclones, come suddenly. Others, like droughts, sea-level rise, or pollution, are caused by human actions and arrive slowly or affect populations over longer periods. Some hazards are "everyday," like the environmental problems described by Hardoy and Satterthwaite (1991; see also excerpt in this volume). Others occur infrequently. The Haiti earthquake in 2010, for instance, was the first large earthquake to strike Port-au-Prince in over 150 years.

Risk assesses the potential losses a person, community, or place might suffer from those hazards (Turnbull, Sterrett, and Hilleboe 2012). Two trends point toward significantly higher urban risk in the twenty-first century particularly in the global South. First, urbanization and urban populations are increasing, and much of the growth is happening in low-lying coastal zones and other hazardous areas (e.g. Brecht et al. 2012). Second, global climate change will likely increase the frequency, severity, and/or extent of many natural hazards that affect cities, including tropical storms and storm-surges, drought, heat waves, and landslides (IPCC 2012). Reducing risk from hazards and climate change has emerged as a central challenge for elected officials, city and regional planners, and urban development organizations.

What drives risk?

The roots of urban risk are complex and varied, and our understanding of disasters has changed significantly in the past several decades. Until relatively recently, and in many different places and religious contexts, disasters were seen as "Acts of God," divine intervention to punish the wicked or reward the faithful. The origins of the word disaster itself, from the Italian *disastro* meaning "ill-starred," implies that disasters are caused by unfavorable alignments of stars and are supernatural in origin (OED 1988). Over time, divine explanations gave way to more naturalistic and scientific ones, with the primary focus on the origins and characteristics of natural hazards. Until the mid-twentieth century, disaster research and policy remained largely focused on natural hazards and the delivery of emergency assistance in times of crisis. Most of our planned responses, like levees, storm drainage, and seismic codes were designed to physically shield people and communities from natural events.

The rising costs of disasters spurred a growing number of social scientists to question the field's physically deterministic views. In their seminal article, "Taking the Naturalness out of Natural Disasters," O'Keefe, Westgate, and Wisner (1976) asked a simple question: if the number of climatological and geological hazards had remained relatively constant, what could explain the escalating costs of disasters in developing countries? The answer, they argued, was the "growing vulnerability of the population." Disaster results when a hazard comes into contact with a vulnerable population, hence broadening the study of disasters to include both hazards and vulnerability would require some "radical rethinking on the nature of 'natural' disasters" (ibid.: 566–567).

This more radical approach to disasters, now commonly described as the social vulnerability turn, argues that natural disasters are not natural at all; they are rooted in social, economic, and political systems and structures that systematically marginalize and/or disempower certain individuals and groups. *Vulnerability*, the characteristics that make a person or group of people susceptible to the damaging effects of a hazard, reflects their ability to access different types of resources: natural, human, social, physical, economic and/or political (Wisner, Gaillard, and Kelman 2012). While every disaster is unique and risk is context specific, social characteristics like class, gender, age, religion, sexuality, race and immigration status often influence the potential loss a person or group faces when exposed to a hazard (ibid.: 8–9). Similar to environmental justice and sustainability arguments, the social vulnerability paradigm finds that certain groups bear unequal burdens of environment. Disasters are moments in time that "reveal" broader truths about the city, society, and the economy.

The growing body of evidence collected across a wide range of social and natural sciences points to a few areas of consensus about risk and vulnerability. First, and most important, risk is highly uneven, and social inequality tends to get reflected in disaster outcomes. While it is true that some hazards are so large or intense that they overwhelm the capacity of an entire city, and thus impact the entire population, such events are exceedingly rare, and even when they do occur, the recovery rates of survivors vary greatly based on wealth and access to various types of capital (e.g. Olshansky 2009; Davis, this volume). In the vast majority of cases, and similar to other environmental issues, disasters disproportionately affect the poor and marginalized.

Second, urban risk, like environmental risk more generally, is partially driven by a lack of access to basic urban services like drainage, sanitation, safe drinking water, and waste disposal (Hardoy and Satterthwaite 1991). Lack of basic services exposes people to a wider range of hazards, reduces their ability to cope with hazards, and negatively affects their ability to cope, recover and adapt to future events. Something as common as monsoon rains, easily managed by those with access to services, can cause significant harm to those living in communities without adequate support. In fact, these "everyday" disasters, largely invisible to outsiders, may have a larger cumulative impact than large events (Bull-Kamanga et al. 2003).

Third, risk and vulnerability have strong spatial dimensions. The root causes of vulnerability, like limited access to resources and marginalization in political processes, result in fragile livelihoods and unsafe conditions at the time and place that hazards occur. These conditions of vulnerability are as varied as inadequate housing and infrastructure, low-income levels, insecurity of tenure, and lack of local institutions (Wisner, Gaillard, and Kelman 2012). In cities, many of these conditions of vulnerability are present in low-income and informal settlements, and geographies of vulnerability tend to correlate with larger patterns of spatial inequality in the built environment. Hazard exposure is also very often tied to the political economy of urban land markets. Even in very hazard-prone cities, some pieces of land are safer than others, and unsafe land is often the cheapest or least developed. With few affordable options available in the formal housing market, the poor are forced to settle on these marginal pieces of land, in places where hazards strike most often or are most severely felt. During the 2005 floods in Mumbai, for example, some of the hardest hit communities were located in low-lying areas near the Mithi River, many of which were informal settlements that lack basic infrastructure or access to emergency services (UN-Habitat 2007). Over time conditions tend to improve, as basic investments in infrastructure and services mitigate hazards. The irony of such improvements is that they increase the value or desirability of land and property in those areas, driving lower income people toward ever more marginal and hazardous places.

Cities and climate change

Climate change will be a major driver of urban risk in the coming decades. Climate describes the typical behavior of the atmosphere (Donaghy 2007). The climate is naturally in a constant state of change, but human activity since the Industrial Revolution has dramatically increased the concentration of greenhouse gases (GHGs) in the atmosphere and, as a result, raised global mean temperatures. Scientists have linked human activity to climate change since at least the late 1960s, and the scientific community is alarmed at the pace and scale of warming and its impacts. This means that the temperature of the air and oceans will increase significantly in the coming years, leading to sea-level rise, more frequent and severe storms, droughts and heat waves, changing weather patterns, salinization of land and ground-water supplies, and numerous other hazards that will threaten global cities (IPCC 2012). The impacts of climate change will be complex and many; some will be direct, like elevated storm surges or more frequent tropical cyclones. Others will be indirect, like increases in migration due to climatic hazards. Despite mounting evidence, little progress has been made internationally on climate change mitigation, which mostly aims at curbing greenhouse gas emissions and reducing atmospheric concentrations of CO_2. Examples of mitigation strategies include reducing vehicular emissions, shifting from fossil fuels to renewables like wind, solar, or geothermal, or capturing atmospheric carbon through reforestation.

Climate change will likely impact every person in some way in the future, but residents of cities in the global South are particularly at risk for at least two reasons. First, many of the largest cities built in low-lying and coastal areas are located in Asia, Africa, and the Pacific (Brecht et al. 2012), and within cities, large populations live in areas that are environmentally hazardous. In Jakarta, for instance, millions of urban residents live along river banks, canals, and in newly built communities on the urban periphery that are extremely prone to flooding. In cities across the Himalayan region of Pakistan, India, Nepal and Bhutan, climate change is affecting the pattern and intensity of monsoon rains, increasing risk to deadly landslides. Second, developing cities have fewer economic resources to use for adaptation, and often lack the institutional capacity to effectively plan for climate change impacts. Similar to natural hazards, the impacts of climate change within cities will be highly uneven, with the largest share of the environmental burden carried by the poor and the marginalized.

The risk of violence

Risk is a broad concept, and encompasses a wide range of other potential sources of harm beyond environmental hazards and climate change. Many of the same socio-spatial characteristics that shape disaster and climate risk also shape risk to urban violence. It results in the systematically unequal experience of violence by different groups of people and differential coping mechanisms that individuals and communities adopt. Considered structurally, people who suffer violence often lack access to the basic institutions and resources meant to ensure safety and security: to honest and well-trained police, to a fair and transparent legal system, or to basic urban infrastructure like street lighting or access to safe spaces for toilets. Violence and the threat of violence is also perpetrated to systematically discipline the behavior of certain groups of people from women at the scale of households and neighborhoods to particular ethnic and religious communities, and low-income communities at the scale of the city and the nation.

Urban violence, insecurity, and conflict are intensely contested concepts and the subject of increasing scrutiny in the field of urban studies. It includes understanding violence as the unlawful exercise of force through multiple lenses at different scales. In the broadest sense (which also captures our focus in this section on cities at risk) institutionalized forces that impose and maintain difference and disparities in societies produce intense conflicts and contestations. This includes urban policies of planning and displacement aimed at cleaning up cities, installing infrastructure and protecting those who hold power (Agbola 1997; Caldeira 2000; Baviskar, this volume), acts of war and political violence (Graham, this volume) and the protracted lower-threshold violence brought on by poverty and ill-health through marginalization and the ill-effects of development (Farmer 2003; Moser and McIlwaine 2004; Winton 2004). Structural violence also provokes and shapes direct violence, inflicting physical or psychological harm to another, which is also the form of violence that is best understood commonly though not always in analytical terms. Self-harming behaviors including substance abuse, street assault, gender-based violence, organized crime and other forms of collective violence in the city all come under the category of direct violence (World Bank 2011). Our focus here is on structural violence at the scale of the city where risk accumulates and is shaped by persistent inequalities, the unequal provision of services and the differential imposition of policy and planning provisions. As Graham (this volume) notes acts of violence targeted at urban infrastructure that "permeates the everyday life of every modern urbanite" shifts the life of cities into a "massive struggle against darkness, cold, immobility, hunger, the fear of crime and violence, and . . . a catastrophic degeneration in public health levels."

Toward resilient cities?

With so much of this introduction centered on disasters, violence and loss, it is worth ending on an encouraging note. While cities are often sites for risk, they are also potentially the sites for *resilience*. Resilience is the capacity of people or groups to "anticipate, absorb, and recover from hazards and/or the effects of climate change and other shocks and stresses" (Turnbull, Sterrett, and Hilleboe 2012: 160). The same density of people and institutions that make cities vulnerable also make them resilient. Complex infrastructure is sometimes an "Achilles heel" for urbanites (Graham, this volume) but more commonly lessens the impacts of hazards, at least for those who have access. Rapid advances in communications technology have made information about hazards more readily available, especially in cities where communications infrastructure is more widespread and robust. Freely available geospatial services like Google Earth have begun to break the state's monopoly on environmental and spatial data and have improved community-based disaster planning. Some particularly heinous acts of violence have been widely publicized via social media and met with mass mobilization and protest, with positive impacts on urban change (Khamis and Vaughn, this volume).

ARTICLE SELECTIONS

In this section, we explore the accumulated risks cities face through four selections, each focusing on a different generator of risk (natural hazards, war, planning and policy, and climate change). Each selection also offers different perspectives on vulnerability and resilience. The first is an excerpt from an article by Diane Davis, a prominant political sociologist and Professor of Urbanism and Development at Harvard University who is the author of *Urban Leviathan* (1994) and *Discipline and Development* (2004). In this section she captures the complex ways in which city space and governance institutions shaped the 1985 Mexico City earthquake and its aftermath. Davis argues that understanding urban disasters requires us to "recognize and embrace" the idea that cities are more than the sum of their physical parts; we must also understand urban institutions, economies, cultures, histories, and symbolic meanings. Davis also reminds us that resilience, in and of itself, is not necessarily desirable. In the case of Mexico City, multiple "resiliences" emerged after the earthquake, including a resilience of corruption and crony capitalism. Planners, policy makers, and urban scholars who are coalescing around the idea of resilience would be wise to always ask: resilience of what? and for whom?

The second excerpt, focusing on collective violence aimed at an entire nation and emerging from understandings of geo-political risk is from Stephen Graham's chapter in his edited book, *Disrupted Cities, When Infrastructure Fails* (2009). Graham examines how the United States' war strategy in Iraq deliberately aimed to "demodernize" the country by destroying its urban infrastructure, resulting in casualties and long-ranging effects on health, security, and well-being. Graham, a well-known British planning academic with a doctorate in science and technology, studies the relationship between cities, technologies, infrastructure, and urban life. His major books (Graham and Marvin 2001; Graham 2009, 2010) and prolific publications explore these issues in a range of different contexts.

The third excerpt is from Amita Baviskar, a sociologist in Delhi, India, who studies the politics of environment and development (1995, 2007). Baviskar relates the story of Dilip, a visitor to Delhi staying in one of the city's many *jhuggi* (shanty house) settlements. In 1995, Dilip was beaten to death by a group of house owners and police, enraged over his perceived trespass into a public park while he searched for a toilet. Baviskar's analysis links violence to broader processes of social exclusion brought on by middle-class and elite desires to clean up cities, revealing the contradictory forces at play in many global cities. This contradictory logic gets enshrined in such formal urban institutions as the Delhi Master Plan, and in Supreme Court decisions to expel polluting industries and noxious land uses from certain areas. Even as cities need the types of labor and services that poor people provide to thrive, planners, policy makers, and wealthier urbanites simultaneously seek to exclude the poor from formal spaces or discipline their behavior and movements through various forms of legal and extra-legal violence.

The fourth excerpt is from Jon Barnett and W. Neil Adger, environmental geographers who study the real and likely effects of natural hazards and climate change. In this piece Barnett and Adger focus on atoll countries, the most at-risk countries in the world to climate change. Atoll countries and other small-island developing states (SIDS) are disproportionately at risk from rising sea levels and increased storm activities, because of their location, low-lying geographies, and low levels of infrastructure and development. For some atoll countries, rising sea levels threaten the very existence of the nation. As Barnett and Adger note in the selection below: "the point at which sovereign atoll countries become effectively uninhabitable clearly constitutes a dangerous level of climatic change." While not directly about cities, this excerpt allows us to consider the complicated interdisciplinary analysis required to understand climate change risks and the difficult questions adaptation planning raises in terms of justice and human rights.

[**Editors' Note**: Andrew Rumbach, is an assistant professor of planning and design at the University of Colorado, Denver. Rumbach studies the social and spatial dimensions of urban risk in India, the Pacific Islands, and the United States.]

REFERENCES

Agbola, T. (1997) *The Architecture of Fear: Urban Design and Construction Response to Urban Violence in Lagos, Nigeria,* Ibadan, Nigeria: IFRA.

Ambraseys, N. N. and Melville, C. P. (1982) *A History of Persian Earthquakes*, Cambridge: Cambridge University Press.

Barnett, J. and Adger, N. (2003) "Climate dangers and atoll countries," *Climate Change*, 61(3): 321–337.

Baviskar, A. (1995) *In the Belly of the River: Tribal Conflicts over Development in the Narmada Valley,* Delhi: Oxford University Press.

Baviskar, A. (2003) "Between violence and desire: Space, power, and identity in the making of metropolitan Delhi," *International Social Science Journal,* 55(175): 89–98.

Baviskar, A. (ed.) (2007) *Waterscapes: The Cultural Politics of a Natural Resource,* Delhi: Permanent Black.

Baviskar, A. (ed.) (2008) *Contested Grounds: Essays on Nature, Culture and Power,* New Delhi: Oxford University Press.

Brecht, H., Dasgupta, S., Laplante, B., Murray, S., and Wheeler, D. (2012) "Sea-level rise and storm surges: High stakes for a small number of developing countries," *The Journal of Environment & Development*, 21(1): 120–138.

Bull-Kamanga, L., Diagne, K., Lavell, A., et al. (2003) "From everyday hazards to disasters: The accumulation of risk in urban areas," *Environment and Urbanization,* 15(1):193–204.

Caldeira, T. P. R. (2000) *City of Walls: Crime, Segregation, and Citizenship in São Paulo,* Berkeley: University of California Press.

Cutter, S. L., Emrich, C. T., Webb, J. J., and Morath, D. (2009) "Social vulnerability to climate variability hazards: A review of the literature," *Final Report to Oxfam America.*

Davis, D. E. (1994) *Urban Leviathan: Mexico City in the Twentieth Century,* Philadelphia: Temple University Press.

Davis, D. E. (2004) *Discipline and Development: Middle Classes and Prosperity in East Asia and Latin America,* Cambridge, UK: Cambridge University Press.

Davis. D. E. (2005) "Reverberations: Mexico City's 1985 earthquake and the transformation of the capital." In L. J. Vale and T. J. Campanella (Eds.), *The Resilient City, How Modern Cities Recover from Disaster*, Oxford: Oxford University Press.

Donaghy, K. (2007) "Viewpoint: Climate change and planning: responding to the challenge," *Town Planning Review,* 78(4): i–xiii.

Farmer, P. (2003) *Pathologies of Power: Health, Human Rights, and the New War on the Poor,* Berkeley: University of California Press.

Graham, S. (ed.) (2009) *Disrupted Cities: When Infrastructure Fails,* New York: Routledge.

Graham, S. (2010) *Cities Under Siege: The New Military Urbanism,* London: Verso.

Graham, S. and Marvin, S. (2001) *Splintering Urbanism: Networked Infrastructures, Technological Mobilities and the Urban Condition,* London: Routledge.

Hardoy, J. E. and Satterthwaite, D. (1991) "Environmental problems of third world cities: A global issue ignored?" *Public Administration and Development*, 11(4): 341–361.

The Intergovernmental Panel on Climate Change (IPCC) (2012) *Managing the Risks of Extreme Events and Disasters to Advance Climate Change Adaptation,* New York: Cambridge University Press.

Moser, C. O. N. and McIlwaine, C. (2004) *Encounters with Violence in Latin America: Urban Poor Perceptions from Colombia and Guatemala,* London and New York: Routledge.

O'Keefe, P., Westgate, K., and Wisner, B. (1976) "Taking the naturalness out of natural disasters," *Nature,* 260: 566–567.

Olshansky, R. B. (2009) "The challenges of planning for post-disaster recovery." In U. Paleo (Ed.), *Building Safer Communities: Risk Governance, Spatial Planning and Responses to Natural Hazards*, Washington, DC: IOS Press, pp. 175–181.

The Oxford English Dictionary (OED) (1988) "Disaster," New York: Oxford University Press.

Pelling, M. (2003) *The Vulnerability of Cities: Natural Disasters and Social Resilience,* London: Earthscan Publications.

Shrady, N. (2008) *The Last Day: Wrath, Ruin, and Reason in the Great Lisbon Earthquake of 1755,* New York: Viking.

Turnbull, M., Sterrett, C., and Hilleboe, A. (2012) *Toward Resilience: A Guide to Disaster Risk Reduction and Climate Change Adaptation*, Rugby: The Schumacher Centre.

UN-Habitat (2007) *Global Report on Human Settlements: Enhancing Urban Safety and Security,* London: Earthscan.

Winton, A. (2004) "Urban violence: A guide to the literature," *Environment and Urbanization*, 16(2): 165–184.

Wisner, B., Gaillard, J. C., and Kelman, I. (2012) *The Routledge Handbook of Hazards and Disaster Risk Reduction,* London and New York: Routledge.

World Bank (2011) *Violence in the City, Understanding and Supporting Community Responses to Urban Violence*, Washington, DC: World Bank.

"Reverberations: Mexico City's 1985 Earthquake and the Transformation of the Capital"

The Resilient City: How Modern Cities Recover from Disaster (2006)

Diane E. Davis

On September 19, 1985, at 7:14 a.m. an earthquake reaching a magnitude of 8.1 on the Richter scale and lasting almost two full minutes hit the coast of Mexico, rocking its capital city and shaking its buildings and its people. The next day, at 7:38 p.m., Mexico City experienced a second tremor of an almost equal magnitude on the Richter scale, 7.5. What has come to be known as the Mexico City earthquake, then, was in actuality two earthquakes, although those who experienced it lived through a single disaster whose longer-term reverberations were as powerful as the first set of tremors that hit the city on that initial day in September.

[...]

This chapter explores this mixed record by explaining what was reconstructed or recovered in Mexico City after its earthquake, what wasn't, and why, and asking whether—given its mixed record of reconstruction—Mexico City should qualify as a resilient city. The inquiry is organized around three interrelated propositions that guide the discussion of Mexico City's post-earthquake recovery and resilience and situate it in the larger themes of this volume.

I begin with the initial proposition that a city is more than its buildings, and thus resilience must be understood as more than physical reconstruction. I then turn to two additional propositions: reconstruction is not necessarily recovery (or vice versa), and resilience is not always a good thing (and can even be a bad thing). The chapter concludes with more general remarks about how, given the record on recovery and resilience presented here, we might analyse and rethink the longer-term significance of trauma and physical crisis in cities. In order to know whether or not a city is resilient, and in order to make sense of what was or was not recovered or restored after a disaster, it is absolutely critical to both recognize and embrace the idea that a city is much more than its buildings. Put another way, cities are not just built environments: they are composed of people, social and political institutions, economic activities, and infrastructure; they have histories and symbolic meanings; and they generally are internally differentiated in social and spatial terms, so that different parts of the city often host different concentrations of these attributes or activities. Mexico City's earthquake exposed this claim in ways that no text or body of theories could.

The 1985 earthquake had multiple reverberations in all aspects of life and livelihood in Mexico City, ranging from political and economic to social and spatial. And this meant that struggles over and plans for post-earthquake reconstruction and recovery were very much struggles over the city itself: its meaning and the institutions and practices that were to give it life, form, and character in the days, months, and years after the earthquake. When the September 19 and 20 quakes shook Mexico, it was immediately obvious to both the government and Mexico City citizens that one of the most pressing problems to be addressed was the fact that the daily routine of urban life for almost all of the city's residents had been completely disrupted if not irredeemably transformed. This entailed disruption of key commercial services (food, consumer goods, and so on), the provision of urban services throughout the metropolitan area, and the routine patterns of urban governance. For days and sometimes weeks or longer, hundreds of thousands of people had no homes, no work, no transportation, no food, no water, no telephones, no hospitals to visit for treatment of the wounded, no place to bury their dead, and no reliable authorities to whom they could turn for assistance (Poniatowska, 1995; Zinser, Morales and Peña, 1986). This was the case not just because of the high Richter scale magnitude of the earthquake, but because by hitting the center of the city, most of the services and institutions that sustained the city as a whole were disabled if not destroyed. This state of affairs owed to the history of urban development in Mexico, which had preserved the traditional character of the center city, centralized its key institutions and services, and prevented the outward movement of its residential populations and public institutions. Had an earthquake of the exact same enormity pounded any other spatial location in the metropolitan area, the magnitude of the urban disruption would not have been as great. Perhaps then the main task at hand would have been to think about replacing buildings. But because the hardest hit area was the metropolitan center—which was also the center of the entire country—a multiplicity of essential services, economic activities, and institutions were at risk.

Important communications and electricity providers were concentrated in the area decimated by the earthquake, and their disruption affected the administrative capacity to restore services to the entire metropolitan area. Through the damage incurred by one single building in downtown Mexico City, local, national, and international telephone service for the city was completely suspended. As Elena Poniatowska asked incredulously in her compelling ethnography of the earthquake, "How is it possible that 55,000 branches that connect the south with the north of the country and the whole country with the world were all concentrated in one single old building on Victoria Street?" (Poniatowska, 1995: 32) A large majority of the city's electricity substations were also in these central areas, leading to power outages for an extended period of time. Also, almost all of the metropolitan area's transport services were paralysed, especially buses, since they crisscrossed the center of the city. Subway service was also disrupted, albeit temporarily, because it depended on electricity.

The city's main medical institutions were also affected by the earthquake, since many of these public medical services were located in an area that housed the Centro Médico. Along with most of the city's secondary schools (Dynes, Quarantelli and Wenger, 1990), all of the city's principal medical services were concentrated in the hardest-hit areas. Five major hospitals were destroyed and twenty-two were damaged, leading to disruption of 30 percent of the entire city's medical capacities. With the city's major public health services suspended or disrupted, the government lost much of its capacity to respond to the medical crisis, creating yet a new set of social and health problems and concerns for authorities.

Granted, officials and citizens ultimately constructed makeshift locations for treatment of the injured, and the three most damaged medical buildings were by no means the only hospitals in the city. But they were the principal ones. Moreover, these particular institutions had special symbolic presence and meaning for the city's residents, representing one of the most lauded social services provided to the public by Mexico's post-revolutionary state. When the medical center's obstetrics unit was all but destroyed, and a few newborns and orphaned infants were found clinging to their lives in the face of death, dust, and destruction, scores of residents streamed to the familiar halls of these public hospitals to offer their services and to aid the doctors who, historically, had played an important role in both serving the public and challenging the Mexican state in years past.

The fact that so many key services were disrupted or destroyed posed a recovery problem for the authorities, as did citizens' natural instincts to become involved in restoring them, since this created an additional management challenge. Differences of opinion further stalled the process.

[. . .]

Some disagreement about recovery priorities should be expected, of course. But with the Mexico City earthquake, there were a large number of conflicting priorities because of the location of the earthquake and the spatial concentration of its effects in downtown areas. If the area hit by physical destruction had preserved only one key function, or if the earthquake had destroyed only houses or businesses or services, authorities would have been better able to prioritize and develop a coherent or easily implementable plan for reconstruction. In the Mexican case, reconstruction was determined as much by political pressures as by questions of efficiency or a coherent rebuilding plan.

[. . .]

Mexico's governing authorities clearly believed that recovery of the city's economic functions, which depended on reassuring external lenders and investors, was as important (if not more so) than the reconstruction of buildings. But it is also important to emphasize here that—from the citizens' perspective—reconstruction was not the same as recovery. Citizens wanted to recover or restore many other things besides their homes. In particular, what became most evident during the days and weeks after the earthquake was that citizens were eager to recover what they called "dignity," as well as government accountability. These two important aspects of government legitimacy mattered as much as buildings in the everyday dynamics of urban life.

To be sure, the active struggles of citizens to recover both dignity and accountability ultimately helped ensure that the government would eventually reconstruct their houses; the same social movements that gave life to these accountability concerns also empowered the struggle for housing. And in that sense, there is some relationship between physical reconstruction and recovery. But for a series of reasons, including the fact that the earthquake exposed in ways never seen before the existing institutional structures of power, authority, and abuse that destroyed the meaning of the city for many of its residents, attending to corrupt political and institutional practices became recovery priorities in and of themselves. One of the most high-profile discoveries in the aftermath of the earthquake was the exposure of clandestine garment factories peppered through downtown areas. Many were concentrated on one street, San Antonio Abad, where single buildings of seven or eight stories each held as many as fifty-five different garment factories. When the earthquake hit, these buildings collapsed under the weight of the sewing machines and heavy rolls of textiles crammed into old dilapidated structures. [. . .] The earthquake did more than expose the horrendous conditions under which they [the garment workers] were working; citizens became enraged because, when the earthquake hit, the local governing authorities immediately sent in police to protect the garment factory owner as he—with police help—salvaged the machines and sewing equipment, leaving scores of injured women and bodies buried under the rubble without any attempt to extricate them. This highly publicized incident came to represent the ways that residents of the city had been exploited in life, through collusion between downtown business interests and authorities, and were exploited even in death. As one survivor put it, "My family was not killed by the earthquake; what killed them was the fraud and corruption fostered by the government" (Dynes, Quarantelli and Wenger, 1990: 78). This sort of realization not only spurred a delegitimization of government authorities, but also motivated citizens to shun government assistance and attempt to recover their own forms of urban justice.

The earthquake also exposed serious violations in construction standards, motivating residents to shun government help even in this time of crisis. Many of the buildings that collapsed were not old buildings, but those that had been constructed relatively recently, between 1950 and 1970, during the heyday of government-led public construction. A good number of the most decimated buildings, in fact, were the massive public housing projects built for middle-class workers [. . .] This destruction, tragic in itself, highlighted a corrupt government's failure to comply with building standards and its use of low-quality construction materials.

The earthquake also exposed two other noteworthy examples of government exploitation and abuse, which further intensified citizens' efforts to restore dignity and accountability. The first was the discovery of numerous bodies, many of them showing evidence of torture, in the basements of various collapsed police stations and other public buildings in the city. The second was the initial decision by authorities to use some of the earthquake reconstruction aid to repay Mexico's foreign debt, something that was seen as helping the country's financial institutions and elite at the expense of the thousands of (mainly poor) citizens who were left homeless.

[. . .]

At its core, the earthquake conveyed a larger message about the meaning and character of the city itself: Mexico City had been treated for far too long as a place—or conglomeration of spaces and practices—where privileged people got rich on the backs of modest residents, and where people were abused (even to death) by authorities who acted in collusion with the nation's economic elite. This surely oversimplifies the situation but, for many citizens, the earthquake's trauma generated just this kind of interpretation and lasting memories.

[...]

Moreover, until a few years after the earthquake, Mexico City citizens had been denied the rights of mayoral elections and democratic participation in the governance of the capital city. Thus the earthquake also became a catalyst for Mexico City's citizens to do something about this political situation and to actively recover urban accountability, justice, and dignity. The reconstruction of buildings paled in relation to these larger goals.

[...]

When they [the people] stopped believing in the current government, they began to struggle for a new one, forming more expanded social movements, new tenants' rights organizations (especially in Guerrero), and new economic co-ops—thirty-seven in the badly hit downtown barrio of Tepito alone—with the aim of governing and caring for themselves (Poniatowska, 1995: 270).

The proliferation of social movements and self-help organizations also had longer-term impacts, including a slow but steady change in both the practices and the form of Mexico City's government, not to mention the housing composition and spatial structure of the city itself.

[...]

In the face of trauma, there are always multiple resiliences, and different people and different activities may be resilient in ways that do not necessarily contribute to recovery. Not only can conflict emerge in the face of multiple resiliences, but the more specific evidence from the case of Mexico City shows that it was precisely the resilience of some of the most corrupt and unjust people and institutions in the capital that made the post-earthquake recovery and reconstruction efforts so dreadful, at least initially. Indeed, among the most resilient forces were the local police, the army, and the political leadership of the PRI [Editiors' note: the Institutional Revolutionary Party is a political party that has held power in Mexico uninterrupted for 71 years since 1929], which governed the city and the nation. In the days and weeks immediately after the earthquake, these groups actively struggled to reinforce their own positions and practices, using the earthquake and the trauma it produced among citizens to strengthen their power base.

This unabashed resiliency of some of the city's most authoritarian and least laudable forces took several forms. Even after so much death, destruction, and human and physical trauma, business as usual seemed to continue. Police looted cadavers and buildings, and government authorities made no effort to stop them. The army distributed blankets and other rescue supplies among its soldiers rather than passing them to earthquake victims. Government-appointed authorities (in customs and elsewhere) took bribes for the delivery of rescue equipment, sold rescue materials, and skimmed money out of the coffers of rescue aid (Poniatowska, 1995: 83–5). Many citizens reported being forced to pay bribes to police in order to get the bodies of their loved ones or to return to their homes.

Yet it was not just the resilience of corruption that was so troubling in the aftermath of the earthquake. The ruling party also seemed intent on showing that it was business as usual, not just with respect to macroeconomic policy, but in its efforts to wield power and authority over citizens. The PRI's resolve in this regard seemed to strengthen as citizens themselves began organizing in reaction to the failures and corruption in the clean-up. Less than a week after the initial quake, the government officially declared that only governmental authorities—not neighborhood organizations—had the right to assist with rescue and clean-up (Aranda, Granovosky, and Hurtado 1985). Some of this concern surely owed to the confusion and chaos generated by thousands of volunteers who showed up to aid in the efforts (Dynes, Quarantelli and Wenger, 1990: 40). But much of it owed to the government's fear that social mobilization would be used to lay blame at the evident incompetence of authorities and, worse yet, to politically challenge the government for its failures. President de la Madrid, in fact, publicly acknowledged this when he announced that, in addition to establishing emergency programs, the government was going to work actively to stop citizens who "used the legitimate demands of residents for purposes of social agitation" (cited in Zinser, Morales and Peña, 1986: 100, trans. by author).

[...]

If resilience cannot inherently be seen as good or bad, what should we be talking about when we examine cities that experience major disasters? It may make best sense to focus on what gets transformed rather than on what parts return to the pre-disaster status quo. But how then should we understand the character or extent of change in a post-disaster situation? This is where the notion of resilience is still helpful, especially if that transformation comes in response to conflicting or multiple resiliences, which together set parameters for how cities recover from disaster.

The Mexican case can be conceptualized as a competition among citizens and authorities, each of them resilient in some fashion. This enabled the earthquake to be truly transformative of urban life, politics, and society. The massive tremors themselves—and the ways in which the authorities responded to them—empowered urban citizens to mobilize on their own behalf to challenge a corrupt and highly bureaucratized local government. It also exposed the political biases of government authorities, diminishing their legitimacy in the eyes of citizens. Together, these developments hastened a grassroots challenge to the power of the ruling party in Mexico City, bringing about an urban democratic reform and, eventually, the defeat of the old guard politicians and the election of a leftist mayor committed to housing and employment for the city's low-income residents. Both of these changes helped to signal the end to one-party rule in the nation as a whole, owing to the demise of the PRI in the social, political, and economic center of the capital. The reverberations of the earthquake, in short, were deep and long lasting, and they extended far beyond the built environment to the social and political life of the city.

Of course, all change is neither unassailable nor only for the good. The earthquake also exposed fissures in the capacity of downtown residents to retain central areas of the city for small-scale commerce and low-income housing, and led to large-scale investment in tourism and financial activities, which are starting to displace long-standing residents. The earthquake—and the renovated housing that was eventually constructed under the new property rights regime—also produced new conflicts and divisions among downtown residents, including an explosion of violence among street vendors that has reached new heights in recent years. As a result, Mexico City now faces new pressures to gentrify local property markets, and the displacement of residents has become a problem of great proportions, as big investors now find opportunities in downtown properties. The direction of these changes in the built environment is still unclear, and it is not certain that the major reconstruction of hotels, tourist establishments, and financial headquarters, which many investors want, will actually materialize. It all depends on the continued resilience of advocates and downtown opponents. In any event, Mexico's capital might never have arrived at this point if the earthquake had not shaken the city's foundations so deeply.

BIBLIOGRAPHY

Aranda, A. O., Granovosky, L. and Hurtado, R. P. (1985) *19 Négro: Genesis del terrmoto en México*, Mexico City: Editorial Roma.

Dynes, R., Quarantelli, E. and Wenger, D. (1990) "Individual and organizational response to the 1985 earthquake in Mexico City, Mexico," Available HTTP: http://udspace.udel.edu/handle/19716/2259

Poniatowska, E. (1995) *Nothing, Nobody: The Voices of the Mexico City Earthquake*, Philadelphia: Temple University Press.

Presidencia de la República (1986) *Terremotos de Septiembre: Sobretiro de las razones y las obras crónica del sexenio 1982–1988*, Mexico City: Fondo de Cultura Economica.

Zinser, A., Morales, C. and Peña, R. (Eds) (1986) *Aún tiembla: Sociedad Política y Cambio Social: El terremoto del 19 de Septiembre 1985*, Mexico City: Grijalba.

FOUR

"Disruption by Design: Urban Infrastructure and Political Violence"

Disrupted Cities: When Infrastructure Fails (2009)

Stephen Graham

> If you want to destroy someone nowadays, you go after their infrastructure. You don't have to be a nation state to do it, and if they retain any capacity for retaliation then it's probably better if you're not. (Agre, 2001)

URBAN ACHILLES

In a 24/7, always-on, and intensively "networked" urban society, urbanites, especially those in the advanced industrial world, become so reliant on infrastructural and computerized networked systems that the disruptions become much more than a matter of mere inconvenience. Rather, they creep ever closer to the point where, as Bill Joy puts it, "turning off becomes suicide" (Joy, 2000: 239). Processes of economic globalization, which string out sites of production, research, data entry, consumption, capitalization, and waste disposal across the world, merely add to the "tight coupling" of infrastructure. This is because such systems rely ever-more on complex combinations of logistical, information, and infrastructure systems, working in intimate synchrony, in order to function.

But it is important to remember that the absolute dependence of human life on networked infrastructures exists throughout the modern urban world cities, not just in "high-tech" cities. This has been shown in horrifying detail when states deliberately "de-electrify" whole urban societies as a putative means—influenced by the latest "air power theories"—of coercing leaders and forcing entire populations into suddenly abandoning resistance. Of course strategic bombing rarely, if ever, has had such an effect. Instead, as we shall see, the effects of urban de-electrification are both more ghastly and more prosaic: the mass death of the young, the weak, the ill, and the old over protracted periods of time, as water systems and sanitation collapse and water-borne diseases run rampant. No wonder that such a strategy has been called a "war on public health" or one of "bomb now, die later."

Urban everyday life everywhere is stalked by the threat of interruption: the blackout, the gridlock, the severed connection, the technical malfunction, the inhibited flow, the network unavailable sign. During such moments, which tend to be fairly normal in cities of the Global South, but much less so in cities of the Global North, the vast edifices of infrastructure become so much useless junk. The everyday life of cities shifts into a massive struggle against darkness, cold, immobility, hunger, the fear of crime and violence, and, if water-borne diseases are a threat, a catastrophic degeneration in public health levels. The perpetual technical flux of modern cities becomes, in a sense, suspended. Improvisation, repair, and finding alternative means of being warm and safe, drinking clean water, eating, moving about, and disposing of wastes quickly become the overriding imperatives of everyday life. Very quickly the normally hidden background infrastructure of urban everyday life becomes, fleetingly, palpably clear to all.

[. . .]

Of course, anxieties surrounding risks of infrastructural disruption and destruction are not entirely new. Warfare and political violence have long targeted the technological and ecological support systems of cities. World War II bombing planners evolved complex methods of perpetuating "strategic paralysis" through destroying transport systems, water infrastructures, and electricity and communications grids. Car bombs, of course, have been the staple of every insurgency and terrorist campaign for at least the past four decades.

Nevertheless, it is clear that the sophistication with which everyday infrastructures are attacked and exploited to project lethal power is dramatically escalating.

In such a context, this chapter critically explores the relations between urban infrastructure and state and non-state political violence. Discussion concentrates first on the better known efforts of terrorists and non-state insurgents to appropriate or disrupt networked infrastructures as means to massively amplify the power of their political violence. The chapter then goes on to discuss the ways in which the military doctrine of nation-states legitimizes and practices the systematic demodernization of entire urbanized societies that are deemed to be adversaries. The use of U.S. air power and Israeli military doctrine against the Occupied Territories are the two most pertinent recent examples. A brief penultimate section discusses emerging ideas of inflicting paralysis on adversary nations and cities through the electronic distribution of malign computer code. The chapter concludes by reflecting on the implications of the centrality of networked demodernization within contemporary practices and theories of war.

[. . .]

U.S. STRATEGIC BOMBING: MODERNIZATION'S MIRROR

> The mystique of imperial air power depended on relegating the "savage," colonized target population to an "other" time and space, where morale could be decisively shattered by shock tactics and the "moral effect" of aerial bombing. Backward conditions in the colonial laboratory or testing ground guaranteed the technological progress and ongoing modernization of the imperial war machine. (Deer, 2006: 3)

Whilst it has received much less attention than infrastructural terrorism, the targeting of the vital networked systems that sustain modern urban life by states is a great deal more devastating. Such targeting works by "placing a logistical value on targets through their carefully calibrated, strategic position within the infrastructural networks that are the very fibers of modern society." The site specific bombing of infrastructure, thus, is designed so that effects and impacts "ripple outwards through the network, extending the envelope of destruction in space and time" (Gregory, 2006: 92). Such cascading destruction, however, is rarely captured by popular and media commentary, seduced as they are by the contemporary military euphemisms of "precision" targeting and unwanted, accidental, "collateral damage."

The history of the strategic bombing of cities, in particular, can be read at least in part as a history of trying to disrupt their vital systems and infrastructures to bring paralysis to an urbanized adversary. Underpinning the long-standing targeting of civilian infrastructure by aerial bombing—our first and most important example of state infrastructural warfare practices—is a long-standing belief that, in effect, subjecting societies to systematic, aerial bombing is a form of demodernization which is the exact reverse of post-World War II theories of modernization. Just as such theories see "development" as leading societies toward "progress" through successive ages defined by their infrastructure—coal age, electricity age, nuclear age, information age, and so on—so bombing can effectively lead societies, backwards as it were, through this linear chain of economic stages by effectively de-modernizing them.

Thus, just as late twentieth century "development" programs for "developing" nations employed economists and civil engineers, so, bombing programs employ such experts to ensure that destruction brings such imagined reversals into being. One civil engineer advising on U.S. Marine bombing targets during the 2003 invasion of Iraq, for example, noted that "working from satellite photos and other intelligence, supplied pilots with very specific coordinates for the best place to bomb [Iraqi bridges], from a strategically structural point of view" (Wright, 2003).

FOUR

[. . .] The intimate connections between modernization and development theory on the one hand, and on the other hand demodernization or infrastructural bombing theory, is hammered home forcefully when it becomes clear that, sometimes, the very same experts preside over both. Most notorious here looms the figure of perhaps the most influential U.S. economist of the Cold War: Walt Rostow. On the one hand, he was father of the most important development model of the late twentieth century, outlined in his book, The Stages of Economic Growth (Rostow, 1960). In this, Rostow suggested a linear and one-way model through which "traditional" societies managed to achieve the "pre-conditions for economic take-off" and then enjoyed the fruits of modernization through the "drive to maturity" and, finally, an "age of mass consumption."

On the other hand, Rostow also played a key part in the U.S. strategic bombing surveys of Japan and Germany and was the most influential National Security Adviser to both the John F. Kennedy and Lyndon B. Johnson administrations from 1961 to 1968 (Milne, 2007). In the latter role, Rostow's incessant lobbying was crucial in gradually extending and increasing the systematic bombing campaigns against civilian infrastructure in the "Rolling Thunder" campaign against North Vietnam. As well as "bombing . . . countries back through several 'stages of growth," (Gilman, 2003), this was seen as a means of undermining the communist challenge to U.S. power. [. . .]

This wider notion that bombing, as a form of punitive demodernization, can usher in a simple reversal of conventional models of linear economic and technological progress, is now so widespread as to be a cliché. Curtis LeMay, the force behind the systematic fire-bombing of urban Japan in World War II, famously urged that the U.S. Air Force, which he led at the time, should "bomb North Vietnam back into the Stone Age." He later added that the U.S. "air force . . . [has] destroyed every work of man in North Vietnam."

Despite the waning fashionability of the modernization theories to which it provides a dark but much less well known shadow (Gilman, 2003), theories of demodernizing societies through air power remain as popular as ever. [. . .] Influential right-wing globalization journalist Thomas Friedman, for example, used a similar argument as NATO cranked up its bombing campaign against Serbia in 1999. Picking up a variety of historic dates that could be the future destiny of Serbian society, post bombing, Friedman urged that all of the movements and mobilities sustaining urban life in Serbia should be brought to a grinding halt. "It should be lights out in Belgrade," he said. "Every power grid, water pipe, bridge, road and war-related factory has to be targeted . . . We will set your country back by pulverizing you. You want 1950? We can do 1950. You want 1389? We can do that, too!" (Skoric, 1999) Here the precise reversal of time that the adversary society is to be bombed "back" through is presumably a matter merely of the correct weapon and target selection.

Finally, as U.S. aircraft pummeled Afghanistan in 2002, Donald Rumsfeld famously quipped, with his usual sensitivity, that the U.S. military were "not running out of targets. Afghanistan is" (Rumsfeld, 2004) [. . .] One Afghan critic immediately responded, to the parallel claim that the U.S. Air Force would bomb Afghanistan "back into the Stone Age" by suggesting, caustically, that "you can't . . . We're already there" (Ansary, 2001).

Of course the politics of seeing the bombing of infrastructure as a form of reversed modernization also plays into the longstanding depiction of countries deemed "less developed" along some putatively linear line of modernization as pathologically backward, intrinsically barbarian, unmodern, even savage. "As long as modernization was conceived as a unitary and unidirectional process of economic expansion . . .," writes Nils Gilman, "backwardness and insurgency would be explicable only in terms of deviance and pathology" (Gilman, 2003).

"THE ENEMY AS A SYSTEM"

The central idea shaping the regular U.S. devastation of urban infrastructure systems by aerial bombing in the past two decades has been the notion of the "enemy as a system." Modifying World War II ideas of targeting the "industrial webs" of Germany and Japan to create "strategic paralysis" in war production, this was devised by a leading U.S. Air Force strategist, John Warden (1995), within what he termed his strategic ring theory. This systematic view of adversary societies provides the central U.S. strategic theorization that justifies, and sustains the rapid extension of that nation's infrastructural warfare capability. The theory has explicitly provided the basis for all major U.S. air operations since the late 1990s. The latest U.S.

Air Force document on targeting doctrine states that the technique of infrastructure targeting "often yields useful target sets" and encourages target planners to bomb "infrastructure targets across a whole region or nation (like electrical power or petroleum, oil, and lubricants [POL] production) . . ., non-infrastructure systems such as financial networks [and] nodes common to more than one system" (Air Force, 2006).

Moreover, special weapons have even been developed to destroy civilian infrastructure more effectively. Prime amongst these are so-called "soft" or "blackout bombs, which [are] label[ed] the 'finger on the switch' of a country" (Barriot and Bismuth, 2008). These rain down thousands of spools of graphite string onto electric transmission and power systems, creating catastrophic short-circuits in the process.

As part of the continuous mythology of humanitarianism that pervades military discussions of "precision strikes," these weapons have been widely lauded in the military press as "non-lethal" weapons which create "minimal risk of collateral damage" (i.e., dead civilians) even though their use means that "whole countries, their soldiers and civilian populations can be cast into an isolated darkness within hours of war" (Carroll, 2002).

As we shall see, though, in the words of Patrick Barriot and Chantal Bismuth, such "high-technology weapons are a long way from proving their innocuousness" (2008). Rather, they work to hide the civilian deaths they create by distancing them in time and space from the point of impact. This is convenient as it tends to remove them from the capricious gaze of the media.

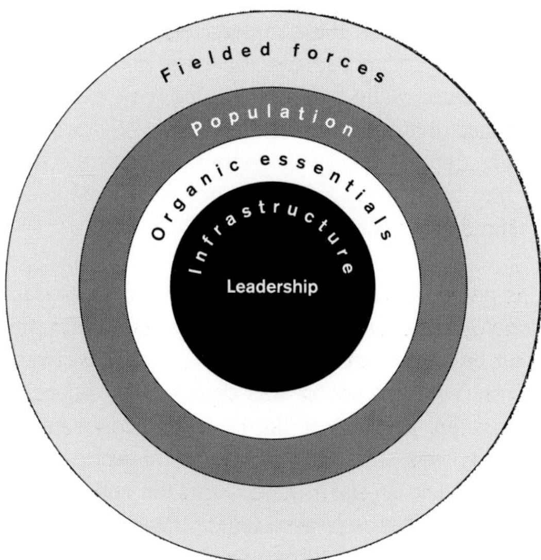

Figure 26 John Warden's 1995 Five-Ring Model of the strategic make-up of contemporary societies—a central basis for US military doctrine and strategy to coerce change through air power.
Source: Felker (1998), p. 12.

"At the strategic level," writes Warden (1995), "we attain our objectives by causing such changes to one or more parts of the enemy's physical system." This "system" is seen to have five parts or "rings": the leadership or "brain" at the center; organic essentials (food, energy, etc.); infrastructure (vital connections like roads, electricity, telecommunications, water, etc.); the civilian population; and finally, and least important, the military fighting force. Rejecting the direct targeting of enemy civilians, Warden, instead, argues that only indirect attacks on civilians are legitimate. These operate through the targeting of societal infrastructures—a means of bringing intolerable pressures to bear on the nation's political leaders, even though this doctrine contravenes a whole host of the key statutes of International Humanitarian Law.

Air power theorists are clearly well aware that the U.S. policy of destroying civilian infrastructure is likely to lead to major public health crises in highly urbanized societies. In a telling example, Kenneth Rizer, another U.S. air power strategist, wrote an article in the official U.S. Air Force Journal, Air and Space Power Chronicles (Rizer, 2001). In it, he seeks to justify the direct destruction of "dual-use" targets (i.e., civilian infrastructures) within U.S. strategy. Rizer argued that, in international law, the legality of attacking dual-use targets "is very much a matter of interpretation" (ibid).

Rizer writes that the U.S. military applied Warden's ideas in the 1991 air war in Iraq with, he claims, "amazing results." "Despite dropping 88,000 tons [of bombs]· in the forty-three-day campaign, only three thousand civilians died directly as a result of the attacks, the lowest number of deaths from a major bombing campaign in the history of warfare" (ibid.). However, he also openly admits that systematic destruction of Iraq's electrical system in 1991 "shut down water purification and sewage treatment plants, resulting in epidemics of gastroenteritis, cholera, and typhoid, leading to perhaps as many as 100,000 civilian deaths and the doubling [of] infant mortality rates" (ibid.).

Clearly, however, such large numbers of indirect civilian deaths are of little concern to U.S. Air Force strategists. For Rizer openly admits that:

> The U.S. Air Force perspective is that when attacking power sources, transportation networks, and telecommunications systems, distinguishing between the military and civilian aspects of these facilities is virtually impossible. [But] since these targets remain critical military nodes within the second and third ring of Warden's model, they are viewed as legitimate military targets ... The Air Force does not consider the long-term, indirect effects of such attacks when it applies proportionality [ideas] to the expected military gain. (Ibid.)

More tellingly still, Rizer goes on to reflect on how U.S. air power is supposed to influence the morale of enemy civilians if they can no longer be carpet-bombed. "How does the Air Force intend to undermine civilian morale without having an intent to injure, kill, or destroy civilian lives?" he asks.

> Perhaps the real answer is that by declaring dual-use targets legitimate military objectives, the Air Force can directly target civilian morale. In sum, so long as the Air Force includes civilian morale as a legitimate military target, it will aggressively maintain a right to attack dual-use targets. (Ibid.)

In 1998 Edward Felker, an air power theorist, like both Warden and Rizer, based at the U.S. Air War College, Air University, further developed Walden's model. This was based on the experience of the 1991 war with Iraq (code-named Desert Storm) and drew directly on Felker's argument that infrastructure, rather than a separate "ring" of the "enemy as a system," in fact pervaded, and connected, all the others to actually "constitute the society as a whole." "If infrastructure links the subsystems of a society," he wrote, "might it be the most important target?" (Felker, 1998).

"BOMB NOW, DIE LATER"

Such sometimes arcane discussions amongst U.S. bombing planners have massive effects. The U.S. 1991 Desert Storm bombing campaign against Iraq, for example, was targeted heavily against "dual-use" urban infrastructure systems, a strategy that Ruth Blakeley has termed "Bomb Now, Die Later" (2003a).

Moreover, because the reconstruction of life-sustaining infrastructures like electricity, water, and sewerage, was made impossible by the sanctions against the regime that were imposed between 1991 and 2003, it is now clear that the 1991 demodernization of Iraqi metropolitan life—in a profoundly urbanized nation—created one of the largest, engineered public health catastrophes of the late twentieth century. Even U.S. Air Force papers admit that the public health disaster created by bombing Iraq's electricity infrastructure killed at least thirty times as many civilians as did the actual fighting (Hinman, 2002:11). Along with military and communication networks, urban infrastructures were amongst the key targets receiving the bulk of the bombing. One U.S. air war planner, Lt. Col. David Deptula, passed a message to Iraqi civilians via the world's media as the planes started going in: "Hey, your lights will come back on as soon as you get rid of Saddam!" (Cited in Rowat, 2003.) Another, Brigadier General Buster Glosson, explained that infrastructure was the main target because the U.S. military wanted to "put every household in an autonomous mode and make them feel they were isolated ... We wanted to play with their psyche." (Cited in Rowat, 2003.) As Colin Rowat suggests, for perhaps 110,000 Iraqis, this "playing" was ultimately to prove fatal (2003). Bolkcom and Pike recall the centrality of targeting dual use infrastructures in the planning of Desert Storm:

> From the beginning of the campaign, Desert Storm decision makers planned to bomb heavily the Iraqi military-related industrial sites and infrastructure, while leaving the most basic economic infrastructures of the country intact. What was not apparent or what was ignored, was that the military and civilian infrastructures were inextricably interwoven. (Bolkcom and Pike, 1993: 23)

The prime target of the air assault was Iraq's electricity generating system. During Desert Storm, the allies flew over two hundred sorties against electrical plants. The destruction was devastatingly effective; almost 88 percent of Iraq's installed generation capacity was sufficiently damaged or destroyed by direct attack, or else isolated from the national grid through strikes on associated transformers and switching facilities, to render it unavailable. The remaining 12 percent was probably unusable other than locally due to damage inflicted on transformers and switching yards (Keaney and Cohen, 1993).

As a result of cascading effects, the destruction of electrical systems meant that Iraq's electrically powered water and sewerage systems collapsed catastrophically. After the war's end, the UN reported that:

> Iraqi rivers are heavily polluted by raw sewage, and water levels are unusually low. All sewage treatment plants have been brought to a virtual standstill by the lack of power supply and the lack of spare parts. Pools of sewage lie in the streets and villages. Health hazards will build in weeks to come. (Cited in Blakeley, 2003a)

Because postwar sanctions prevented Iraq from importing the "dual-use" technologies necessary to repair its electricity, sewerage, and water infrastructures, public health crises, and infant mortality rates, spiraled out of control between 1991 and the end of the 1990s. By 1999, drinkable water availability in Iraq had fallen to 50 percent of 1990 levels (Blakeley, 2003b: 2). As a direct result, UNICEF estimated that, between 1991 and 1998, statistically, over five hundred thousand excess deaths amongst Iraqi children under five—a six-fold increase in death rates for this group—occurred between 1990 and 1994 (UNICEF, 1999). (Such figures mean that, "in most parts of the Islamic world, the sanctions campaign is considered genocidal" (Smith, 2002: 365).) The majority of deaths, from preventable, waterborne diseases, were aided by the weakness brought about by widespread malnutrition. The World Health Organization reported in 1996 that:

> the extensive destruction of electricity generating plants, water purification and sewage treatment plants during the six-week 1991 war, and the subsequent delayed or incomplete repair of these facilities, leading to a lack of personal hygiene, have been responsible for an explosive rise in the incidence of enteric infections, such as cholera and typhoid. (Cited in Blakeley, 2003a)

[...]

STATE CYBERWARFARE

According to William Church, ex-Director of the now-defunct Center for Infrastructural Warfare Studies, the next frontier of state infrastructural warfare will involve the development of capacities to undertake co-ordinated "cyberwarfare" attacks. "The challenge here," he writes, "is to break into the computer systems that control a country's infrastructure, with the result that the civilian infrastructure of a nation would be held hostage" (Church, 2001). Church argues that NATO considered such tactics in 1999 in Kosovo and that the idea of cutting Yugoslavia's Internet connections was raised at NATO planning meetings, but that these options were rejected as problematic. But within the U.S. emerging doctrine of "Integrated Information Operations" and infrastructural warfare—which encompass everything from destroying electric plants, dropping Electronic Magnetic Pulse (EMP) bombs that destroy all electrical equipment within a wide area, developing globe-spanning surveillance systems like Echelon, to dropping leaflets and disabling Web sites—a dedicated capacity to use software systems to attack an opponent's critical infrastructures is now under rapid development.

Deliberately manipulating computer systems to disable opponents' civilian infrastructure is being labeled Computer Network Attack (CNA) by the U.S. military. It is being widely seen as a powerful new weapon, an element of the United States' wider "Full Spectrum Dominance" strategy (U.S. Department of Defense, 2000). Whilst the precise details of this emerging capability remain classified, some elements are becoming clear.

First, it is apparent that a major research and development program is underway at the Joint Warfare Analysis Center at Dahlgren (Virginia) into the precise computational and software systems that sustain the critical infrastructures of real or potential adversary nations. Major General Bruce Wright, Deputy Director of Information Operations at the Center, revealed in 2002 that "a team at the Center can tell you not just how a power plant or rail system [within an adversary's country] is built, but what exactly is involved in keeping that system up and making that system efficient." (Cited in Church, 2001.)

Second, it is clear that, during the 2003 invasion of Iraq, unspecified offensive computer network attacks were undertaken by U.S. forces (Onley, 2003). Richard Myers, Commander in Chief of U.S. Space Command, the body tasked with Computer Networked Attack, admitted in January 2000 that "the U.S. has already undertaken computer networked attacks on a case-by-case basis" (Stone, 2003). A National Presidential Directive on Computer Network

Attack (number 16), signed by George Bush in July 2003, demonstrated the shift from blue-sky research to bedded-down doctrine in this area.

In 2007 it was also announced that the U.S. Air Force had established a new "Cyber Command," located at 8th Air Force at Barksdale Air Force Base, Louisiana. This was tasked with both "cybernetwork defense" for the U.S. Homeland and "cyberstrike" against adversary societies (another term for computer network attack) (Rosenberg, 2007). The purported aim of the five year program, which effectively attempts to militarize the world's global electronic infrastructures, is to "gain access to, and control over, any and all networked computers, anywhere on Earth" (Astore, 2008). Lani Cass, previously the head of the Air Force's Cyberspace Task Force, and now a special assistant to the Air Force Chief of Staff, reveals that these latest doctrines of cyberattack are seen to be a mere continuation of the history of striking societal infrastructure with air power. "If you're defending in cyber [space]," he writes, "you're already too late. Cyber delivers on the original promise of air power. If you don't dominate in cyber, you cannot dominate in other domains." (Cited in Astore, 2008.)

These efforts to bolster the U.S. cyberwar capabilities were significantly bolstered by a series of massive "denial of service attacks" against Estonia in Spring 2007. In apparent revenge for the shifting of a statue dedicated to the Soviet war dead in Talinn, these crippled the Web sites of Estonia's prime minister as well as Estonian banks. They seemed to emanate, at least in part, from hackers linked to the Russian State. U.S. strategic planners are also closely monitoring the growing ability of China's armed forces to launch sophisticated and continuous cyberwarfare attacks as part of Chinese doctrine of "unrestricted" or "asymmetric" warfare. Such risks led NATO leaders to say, in 2008, that they took the risks of cyberwarfare attacks as seriously as a missile strike (Johnson, 2008).

The concern of U.S. military planners is that the proliferation of cyberwarfare might open up advanced and high-tech economies, which rely on the most interdependent, dense, and computerized infrastructure systems, to attack from a wide range of state and non-state organizations operating at a diversity of scales. If state or terrorist cyberattacks were to become common, Steven Metz, of the U.S. Strategic Studies Institute argued, "the traditional advantage large and rich states hold in armed conflict might erode. Cyberattacks require much-less-expensive equipment than traditional ones. The necessary skills can be directly extrapolated from the civilian world . . . If it becomes possible to wage war using a handful of computers with internet connections, a vast array of organizations may choose to join the fray" (Metz, 2000). Metz even suggests that such transformations could lead non-state groups to attain power equivalent to nation-states, commercial organizations to wage cyberattacks on each other, and cyber "gang wars" to be played out on servers and network backbones around the world rather than in "ghetto alleys" (ibid.).

In an interconnected world, however, where infrastructures systems tightly connect both with each other and with other geographical areas, it is strikingly clear that the effects of such cyberwarfare attacks are likely to be profoundly unpredictable. Whilst attacking Iraq in 2003, for example, it is now clear that U.S. Air Force computer network attack staff considered the complete disablement of Iraqi financial systems using computer network attack techniques. But they apparently rejected the idea because the Iraqi banking network was so closely linked to the financial systems network in France. This meant, for example, that an attack on Iraq might easily have led to the collapse of Europe's ATM machines. "We don't have many friends in Paris right now," quipped one U.S. intelligence officer on the emergence of this decision. "There is no need to make more trouble if [then French President] Chirac won't be able to get any Euros out of his ATM!" (Cited in Smith, 2003.)

[. . .]

BIBLIOGRAPHY

Agre, P. (2001) "Imagining the Next War: Infrastructural Warfare and the Conditions of Democracy," Radical Urban Theory, September 14, 2001: 1. http://www.rut.com/911/Phil-Agre.html (accessed February 12, 2004).

Air Force, Secretary of the Air Force Doctrine Document 2–1. (2006), June 8: 22–33.

Ansary, T. (2001) "An Afghan-American Speaks," Salon.com, September 14: http://archive.salon.coml news-feature/2001/09114/Afghanistan/.

Ansary, T. (2002) "A War Won't End Terrorism," San Francisco Chronicle, October 19: http://www.commondreams.org/views02/1019-02.htm.

Astore, W. (2008) "Attention Geeks and Hackers, Uncle Sam's Cyber Force Wants You!" Tom Dispatch, June 5: http://www.tomdispatch.com/post/174940/william_astore_militarizing_your_cyberspace.

Barriot, P., and Bismuth, C. (2008) "Ambiguous Concepts and Porous Borders," in P. Barriot and C. Bismuth, Treating Victims of Weapons of Mass Destruction: Medical, Legal and Strategic Aspects, New York: Wiley: 1–10.

Blakeley, R. (2003a) "Bomb Now, Die Later," Bristol University, Department of Politics, 2003: http://www.geocities.comfruth_blakeley/bombnowdielater.htm (accessed February 2004).

Blakeley, R. (2003b) "Targeting Water Treatment Facilities," Campaign Against Sanctions in Iraq, Discussion List, January 24: 2003: http://www.casi.org.uk/discuss/2003/msg00256.html.

Bolkcom, C. and Pike, J. (1993) "Attack Aircraft Proliferation: Issues for Concern (Federation of American Scientists)": http://www.fas.org/spp/aircraft (accessed February 15, 2004).

Carroll, B. (2002) "Seeing Cyberspace: The Electrical Infrastructure as Architecture," in Sarai Reader: Cities and Everyday Life, Sarai: New Delhi: http://www.sarai.net/publicationslreaders/02-the-cities-of-everyday-life/04cyber_electric.pdf (accessed January 2009).

Church, W. (2001) "Information Warfare," International Review of the Red Cross: 205–16.

Deer, P. (2006) "Introduction: The Ends of War and the Limits of War Culture," Social Text: 25–36.

Felker, E. (1998) Airpower, Chaos and Infrastructure: Lords of the Rings, Maxwell Air Force Base, Alabama: U.S. Air War College Air University, Maxwell paper 14.

Gilman, N. (2003) Mandarins of the Future: Modernization Theory in Cold War America, Baltimore: Johns Hopkins.

Gregory, D. (2006) "'In Another Time-Zone, the Bombs Fall Unsafely . . .': Targets, Civilians and Late Modern War," Arab World Geographer 9(2): 88–111.

Hinman, E. (2002) "The Politics of Coercion," Toward a Theory of Coercive Airpower for Post-Cold War Conflict (CADRE Paper No. 14), Maxwell Air Force Base, Alabama: Air University Press, 36112-6615, August 2002.

Johnson, B. (2008) "NATO Says Cyberwarfare Poses as Great a Threat as a Missile Attack," The Guardian, March 6, 2008.

Joy, W. (2000) "Why the Future Doesn't Need Us," Wired (April): 238–60.

Keaney, T, and Cohen, E. (1993) Gulf War Air Power Surveys (GWAPS), Vol. 2, Part 2, Washington, DC: Johns Hopkins University and the U.S. Air Force, 1993: http://www.au.af.millau/awcgate/awc-hist.htm+gulf (accessed February 15, 2004).

Metz, S. (2000) "The Next Twist of the RMA," Parameters (Autumn): 40–53: http://www.carlisle.army.mil/usawc/Parameters/00autumn/metz.htm.

Milne, D. (2007) "'Our Equivalent of Guerrilla Warfare': Walt Rostow and the Bombing of North Vietnam, 1961–1968," The Journal of Military History: 169–203.

Onley, D. (2003) "U.S. Aims to Make War On Iraq's Networks," Missouri Freedom of Information Center, 2003: http://foi.missouri.edu/terrorbkgd/usaimsmake.html (accessed February 24, 2004).

Rizer, K. (2001) "Bombing Dual-Use Targets: Legal, Ethical, and Doctrinal Perspectives," Air and Space Power Chronicles, January 5, 2001: http://www.airpower.maxwell.af.mil/airchronicles/cc/Rizer.html.

Rosenberg, B. (2007) "Cyber Warriors: USAF Cyber Command Grapples with New Frontier Challenges," C4ISR Journal (August 1, 2007).

Rostow, W. (1960) The Stages of Economic Growth: A Non-Communist Manifesto, Cambridge, UK: Cambridge University Press.

Rowat, C. (2003) "Iraq Potential Consequences of War," Campaign against Sanctions in Iraq Discussion List, November 8, 2003: http://www.casi.org.uk/discuss/2002/msg02025.html (accessed February 12, 2004).

Rumsfeld, D. (2004) News transcript, U.S. Department of Defense, March 22, 2004: http://www.defenselink.mil/transcripts/transcript.aspx?transcriptid=2361.

Skoric, I. (1999) "On Not Killing Civilians," May 6, 1999: http://www.amsterdam.nettime.org (accessed February16, 2004).

Smith, C. (2003) Wrestles with New Weapons," NewsMax.Com, March 13, 2003: http://www.newsmax.com/archives/articles/2003/3/134712.shtml.

Smith, T. (2002) "The New Law of War: Legitimizing Hi-Tech and Infrastructural Violence," International Studies Quarterly 46 (2002): 355–74.

FOUR

Stone, P. (2003) "Space Command Plans For Computer Network Attack Mission," U.S. Department of Defense: Defense Link, January 14, 2003: http://www.defenselink.mil (accessed February 22, 2004).

United Nations Children's Fund (UNICEF). (1999) "Annex II of S/1999/356, Section 18, 1999," at http://www.un.orgiDepts/oip/reports (accessed February 17, 2004).

U.S. Department of Defense. (2000) "Joint Vision 2020," Washington, DC: U.S. Department of Defense.

Warden, J. (1995) "The Enemy as a System," Airpower Journal 9(1): 41–55.

Wright, A. (2003) "Structural Engineers Guide Infrastructure Bombing," Engineering News Record, April 3, 2003.

"Between Violence and Desire: Space, Power, and Identity in the Making of Metropolitan Delhi"

***International Social Science Journal* (2003)**

Amita Baviskar

INTRODUCTION

Delhi, on the morning of January 30, 1995, was waking up to another winter day. In the well-to-do colony of Ashok Vihar, early risers were setting off on morning walks, some accompanied by their pet dogs. As one of these residents walked into the neighborhood "park," the only open area in the locality, he saw a young man, poorly clad, walking away with an empty bottle in hand. Incensed, he caught the man, called his neighbors and the police. A group of enraged house owners and two police constables descended on the youth and, within minutes, beat him to death.

The young man was 18-year-old Dilip, a visitor to Delhi, who had come to watch the Republic Day parade in the capital. He was staying with his uncle in a jhuggi (shanty house) along the railway tracks bordering Ashok Vihar. His uncle worked as a laborer in an industrial estate nearby which, like all other planned industrial zones in Delhi, had no provision for workers' housing. The jhuggi cluster with more than 10,000 households shared three public toilets, each one with eight latrines, effectively one toilet per 2,083 persons. For most residents, then, any large open space, under cover of dark, became a place to defecate. Their use of the "park" brought them up against the more affluent residents of the area who paid to have a wall constructed between the dirty, unsightly jhuggis and their own homes. The wall was soon breached, as much to allow the traffic of domestic workers who lived in the jhuggis but worked to clean the homes and cars of the rich, wash their clothes, and mind their children, as to offer access to the delinquent defecators.

Dilip's death was thus the culmination of a long-standing battle over a contested space that, to one set of residents, embodied their sense of gracious urban living, a place of trees and grass devoted to leisure and recreation, and that to another set of residents, was the only available space that could be used as a toilet. If he had known this history of simmering conflict, Dilip would probably have been more wary and would have run away when challenged, and perhaps he would still be alive.

This incident made a profound impression on me. During my research in central India, the site of struggles over displacement due to dams and forestry projects as well as the more gradual but no less compelling processes of impoverishment due to insecure land tenure, I had witnessed only too often state violence that tried to crush the aspirations of poor people striving to craft basic subsistence and dignity (Baviskar, 2001). Now I was watching a similar contestation over space unfold in my own back yard. I had previously analysed struggles over the environment in rural India; now my attention was directed towards how, in an urban context, the varied meanings at stake in struggles over the environment were negotiated through different projects and practices. This concern has been strengthened over the last two years by two sets of processes, each an extraordinarily powerful attempt to remake the urban landscape of Delhi. Through a series of judicial orders, the Supreme Court of India has initiated the closure of all polluting and non-conforming industries in the city, throwing out of work an estimated 2 million people employed in and around 98,000 industrial units. At the same time, the Delhi High Court has ordered the removal and relocation of all jhuggi squatter settlements on public lands, an order that will demolish the homes of more than 3 million people. In a city of 12 million people, the enormity of these changes is mind-boggling. Both these processes, which were set in motion by the filing of public interest litigation by environmentalists and consumer rights groups, indicate that bourgeois environmentalism has emerged as an organized force in Delhi, and upper-class concerns around aesthetics, leisure, safety, and health have come significantly to shape the disposition of urban spaces.

This bourgeois environmentalism converges with the disciplining zeal of the state and its interest in creating legible spaces and docile subjects (Scott, 1998). According to Alonso (1994: 382), "modern forms of state surveillance and control of populations as well as of capitalist organization and work discipline have depended on the homogenizing, rationalizing and partitioning of space." Delhi's special status and visibility as national capital has made state anxieties around the management of urban spaces all the more acute: Delhi matters because very important people live and visit there; its image reflects the image of the nation-state. As an embodiment of India's modernist ambitions, the capital has been diligently planned since 1962 when the first Master Plan was produced with the help of American expertise supplied by the Ford Foundation. The Master Plan would order Delhi's landscape in the ideal of Nehruvian socialism, and enlightened state control would engineer functional separation, leaving a sanitized slot for history in the form of protection for monuments deemed archaeologically important (Khilnani, 1997). Huge tracts of agricultural land were acquired from the villages close to the city and vested with the Delhi Development Authority (DDA) which had the monopoly of transforming these spaces into zones appropriate for a modern capital: commercial centers, institutional areas, sports complexes, green areas, housing colonies, and industrial estates. Lending urgency to their ambitions was the presence of around 450,000 Hindu and Sikh refugees who had flooded the city from what had become Pakistan, and who had been settled on the periphery of the city in housing colonies, but whose sewage had contaminated the city's water supply, leading to 700 deaths from jaundice in 1955 (Saajha, 2001: 5). Concerns about the physical and social welfare of concentrated human populations were thus channeled into the desire for a planned city, where they converged with the high nationalist fervor for modernization. Fulfilling this desire seemed to be pre-eminently a responsibility of the state: the legitimacy of a national government that had the prestige of fighting for freedom added fresh power to an older development regime established by colonial capitalism (Ludden, 1992) that gave the state primacy in the mission of Civilization and improvement.

[. . .]

NEGOTIATING CONTRADICTIONS

From the interdependence between squatters and their political patrons, profiteering property brokers and those looking for land, and lower-level bureaucrats who benefit through turning a blind eye to violations, there emerge powerful collaborations that undermine the bourgeois dream of re-making the city. The state's Master Plan is undone through resistance both internal and external. The displacement that the creation of a clean and green Delhi entails has been held in check by the delicate political equations on which state legitimacy hinges. The orderly manipulation of people and places cannot rely on brute force alone, even though there have been several violent encounters in the process of enforcing the Supreme Court directives. Politicians across the party divide in the city recognize that their electoral fortunes depend on the support both of financiers and of the numerically important poor. Negotiating the contradictions between these disparate constituencies, the city administration's responses to judicial orders have been heterogeneous: playing for time, pleading to change the rules, placating the judges with new plans, even as it hastens to assure threatened groups that it would protect their interests. The fractures within political authority, partly a consequence of Delhi being not just a city but the capital of India, help to create ambiguous spaces and irregular practices—jurisdictional twilight zones—where the buck can be passed to a bewildering number of authorities and no action taken.

As expected from a heterogeneous group, the responses from the owners of industrial units in the city have been diverse. For a few large industrialists, those who owned factories in the center of the city, this crisis is an opportunity to convert land to more profitable commercial or office space. Others have moved to a new periphery, the industrial estates in nearby Rajasthan, where they will probably continue to pollute without check. Many owners of small firms assert that the installation of pollution control equipment, or the switch to non-polluting technologies, will render their operations economically unviable. That is, their profits depend on exploiting the environment. It is quite likely that some producers and small-scale production will simply go out of business, making way for more capital-intensive technologies.

The ability to weather displacement varies with the material and symbolic capital at one's command. The Supreme Court issued directions about compensating factory owners as well as their employees. However, workers' entitlements are conditional upon their being recognized as employees, their eligibility dependent on officially being on the rolls. Yet the same logic of keeping costs down that makes factory-owners resist the enforcement of pollution laws operates to keep workers off the rolls. The intricacies of contracting and sub-contracting labor, designed to keep labor costs low and capitalists in control, prevent most workers from being recognized as displaced and liable for compensation from specific firms. Workers dependent on daily wages, with no job security and who are the most vulnerable of the city's poor are rendered completely destitute by this process of restructuring the urban economy. The insecure, constantly changing conditions of work that prevented their political organization also make invisible the violence done to these workers. "Free" in Marx's doubly ironic sense to sell their labor wherever they please, without owning any capital, much of Delhi's working class experiences displacement as a constant fact of life.

The trade unions that represent the minority of officially recognized industrial workers have been protesting against the closure of industrial units and the displacement of workers in the courts and through mass demonstrations. Their arguments represent environmental concerns as antithetical to workers' interests. A common accusation is that: *shahar ko sundar banane ke liye ameer log mazdoor ke pet par laat maar rahe hain* (to make the city beautiful, the rich are kicking workers in their belly). But this is only a partial account of the complex politics leading to displacement in which bourgeois environmentalism and Master Plans converge with other processes of capitalist restructuring and real-estate development. Nor is environmentalism an agenda that is antagonistic to working-class interests. Those most vulnerable to environmentally hazardous living and working conditions are most often the working class. The economic compulsion of working in hazardous conditions and the political powerlessness of being unorganized, combined with the state's failure to implement labor and environmental regulations, structure the conflict in terms of a perceived opposition between jobs and the environment. Delhi is a city where the majority scrabble to find a precarious foothold in the race for space and work, their housing concerns focused on getting access to sanitation, water, and electricity in squalid settlements. For them, the sheer uncertainty of employment makes unimaginable the asking of questions about conditions of work, wages, security, and environmental hazard. Workers' organizations have generally been ineffective in pointing out that a safe and clean working and living environment is equally a priority for workers. As Ravindran (2000: 116) observes: "Four decades of urban planning in Delhi, which progressively marginalized both the urban environment and the poor, is now faking an encounter between the two."

[. . .]

CONCLUSION: REFORM OR TRANSFORM DELHI?

In Delhi, the poor have responded to such disciplining attempts by adopting varied strategies of enterprise, compromise, and resistance. They have exercised their franchise as citizens (the "vote banks" that the bourgeoisie holds in contempt), used kinship networks, entered into unequal bargains with politicians and employers, mobilized collectively through neighborhood associations, and most recently, attempted to create a coalition of slum-dwellers' organizations, trade unions, and NGOs. This coalition, called Saajha Manch (Joint Forum), has over the last three years created a powerful critique of Delhi's Master Plan, pointing to the absence of participatory processes in its formulation and highlighting the sharp inequalities in the consumption of urban resources. These multiple practices, simultaneously social and spatial, attempt to democratize urban development even as they challenge dominant modes of framing the environment-development question.

This paper has shown that planned urban development, like other modes of state-making, attempts to transform the relations between populations and spaces, in the process displacing and impoverishing large sections of the citizenry. In the case of Delhi, state-making is not only about reproducing the state nationally and internationally and securing resources for capitalist restructuring, but it also includes interventions aimed at improving the environmental quality of life for Delhi's bourgeoisie. For the bourgeoisie as well as for poor migrants, processes of place-making are marked by both violence and desire (Malkki, 1992: 24), as displacement collides with dreams of a better life. These subjects' strategies to craft work and home, the central axes of social being and identity, are grounded in the negotiation of multiple and shifting fields of power (Moore, 1998). Rather than seeing place-making as a project of rule, I have attempted to direct attention toward the accomplishment of rule (Li, 1999), the contradictions and compromises that radically transform this project. Such an analysis seeks to identify and understand the complexities in the exercise of agency by subaltern subjects, as they attempt to intervene in the unequal processes of creating spaces and identities that are intrinsic to the project of urban development.

REFERENCES

Alonso, A. M. (1994) "The politics of space, time and substance: state formation, nationalism and ethnicity," *Annual Review of Anthropology* 23, 379–405.

Baviskar, A. (2001) "Written on the body, written on the land: violence and environmental struggles in central India," in N. Peluso, and M. Watts, (eds), *Violent Environments*, Ithaca, NY: Cornell University Press, 354–379.

Khilnani, S. (1997) *The Idea of India*, New York: Farrar Straus Giroux.

Li, T. M. (1999) "Compromising power: development, culture, and rule in Indonesia," *Cultural Anthropology* 14(3), 295–322.

Ludden, D. (1992) "India's development regime," in N. B. Dirks, (ed.), *Colonialism and Culture*, Ann Arbor, MI: University of Michigan Press.

Malkki, L. (1992) "National geographic: the rooting of peoples and the territorialisation of national identity among scholars and refugees," *Cultural Anthropology* 7(1), 24–44.

Moore, D. (1998) "Subaltern struggles and the politics of place: remapping resistance in Zimbabwe's eastern highlands," *Cultural Anthropology* 13(3), 344–381.

Ravindran, K. T. (2000) "A state of siege," in *Frontline*, December 22, 116–118.

Saajha, M. (2001) *Manual for People's Planning: A View from Below of Problems of Urban Existence in Delhi*, unpublished report, Delhi, March.

Scott, J. C. (1998) *Seeing Like a State: How Certain Schemes to Improve the Human Condition Have Failed*, New Haven, CT: Yale University Press.

"Climate Dangers and Atoll Countries"

Climate Change (2003)

Jon Barnett and W. Neil Adger

[. . .]

The Small Island States chapter of the Third Assessment Report of the Inter-governmental Panel on Climate Change (IPCC) implicitly concludes that climate change-induced sea-level rise, sea-surface warming, and increased frequency and intensity of extreme weather events puts at risk the long-term ability of humans to inhabit low-lying atolls [. . .]. For climate change research the challenge is to identify the critical thresholds of change beyond which atoll social-ecological systems in particular may be seriously compromised, and we focus here on some important social thresholds. [. . .] We argue that if this occurs, the point at which sovereign atoll countries become effectively uninhabitable clearly constitutes a dangerous level of climatic change.

CLIMATE IMPACTS ON ATOLL COUNTRIES

Atolls are rings of coral reefs that enclose a lagoon. Around the rim of the reef there are islets called *motu* with a mean height above sea-level of approximately two meters (Nunn, 1994). Worldwide there are five countries comprised entirely of low-lying atolls: Kiribati (population 78,000), the Maldives (population 269,000), the Marshall Islands (population 58,000), Tokelau (population 2000), and Tuvalu (population 9000) (Secretariat of the Pacific Community, 2000; UNCTAD, 1999). With the exception of Tokelau (a dependent territory of New Zealand), these are all sovereign states. Kiribati, the Maldives and Tuvalu are official "Least Developed Countries" (LDC) in the United Nations system.

Atolls have common environmental problems that render them particularly vulnerable to climate change. They generally have very high population densities, meaning that large numbers of people are potentially exposed to single events (909 people/km^2 in the Maldives) (UNCTAD, 2002). Water reserves on atolls are restricted to a narrow subterranean freshwater lens easily contaminated by salt water and human and industrial wastes (UNEP, 1999). These freshwater lenses become depleted in times of low rainfall. Atoll islands typically face coastal erosion as a result of exploitation of beaches for building materials, while construction of sea walls and infrastructure and waste dumping on reefs and mangroves undermines the ecological functions on which these island systems depend (Moberg and Folke, 1999). Coastal developments and pollution have also led to depletion of artisanal fisheries (UNEP, 1999).

Overall, their small size, isolation, generally low levels of income, and relatively low levels of physical infrastructure make atoll countries apparently vulnerable to global economic forces as well as to climatic changes (Brautigam and Woolcock, 2001; Commonwealth Secretariat, 1999). The atoll countries are relatively more vulnerable in terms of economic structure and, on average, more food insecure than other small island developing states. [. . .] These parameters of underlying economic vulnerability can, however, be questioned for subsistence economies which are often more resilient, at least to weather

BOX 1 SUMMARY OF POTENTIAL IMPACTS OF CLIMATE CHANGE ON ATOLL COUNTRIES (AFTER NURSE AND SEM, 2001)

Potential loss of land area due to rising sea-levels.
Shifts in species competition and composition.
Coral reefs, mangroves and seagrass adversely affected, with negative effect on reef fish populations.
Increased salinization of soils in coastal margins.
Increasingly variable rainfall, with more intense drought events.
Increase in cyclone intensity with larger storm waves and more intense flooding events.
Adverse effects on staple crops due to changes in soil moisture, salinity and rainfall.
Decline in food security due to adverse effects on crops and declining reef fish populations.
Coastal erosion and changing climatic conditions may adversely affect tourism.
Adverse economic impacts through infrastructure damage from increased intensity of extreme events, coastal protection measures, and decline in tourism income.
Decline in human health through vector-borne diseases and enhanced food insecurity.

extremes and climatic change, than suggested by their economic structure (Smith and Wishnie, 2000). Nevertheless, when combined with sensitivity to climatic changes, this underlying economic vulnerability creates the necessity for large-scale adaptation for the atoll countries.

The "commitment" to climate change caused by greenhouse gases already present in the atmosphere means that the small island states in the Pacific, Indian Ocean and Caribbean regions are projected to experience a certain degree of environmental change. A mean annual warming of 2°C or higher by the 2050s and 3°C for the 2080s is projected, and modest declines in annual precipitation in the Pacific Ocean region are also expected along with heavier rainfall intensity (Lal et al., 2002). Given emissions of greenhouse gases up to 1995, a 5–12 cm rise in sea-level is inevitable (Jones, 1999). However, even if all countries met their Kyoto Protocol commitments, and if all emissions of greenhouse gases ceased after 2020, a sea-level rise of 14–32 cm is very likely (Jones, 1999). Taking into account emissions beyond 2020, sea-level is projected to rise by between 9 and 88 cm by 2100 (Houghton et al., 2001). Atoll countries are the most physically vulnerable to sea-level rise of all small island states because of their high ratio of coastline to land area, relatively high population densities, and low level of available resources for adaptive measures (as measured by national income per capita).

It is not sea-level rise *per se*, but rather projected increases in sea-surface temperature (SST) that pose the greatest long-term risk to atoll morphology. Coral reefs are crucial to the formation and maintenance of atoll motu, and they are highly sensitive to sudden changes in SST. The impact of climate change on SST is arguably already evident, with episodes of mass coral reef mortality through coral bleaching experienced around the world (Reaser et al., 2000). Some evidence suggests that tropical SSTs have been rising over a 50-year period, and in 1998 SSTs reached the highest on record during the change from a major El Niño to a major La Niña event (Reaser et al., 2000). In the same year coral reefs around the world suffered the most severe bleaching on record. Large scale bleaching episodes, such as the 1998 El Niño event, are usually attributable to high sea-surface temperatures and low light conditions. Periodic natural disturbance is increasingly believed to be an important element promoting the diversity and resilience of coral reef ecosystems (Nyström et al., 2000). However, when severe and frequent disturbances above the periodic natural rate occur, coral reef resilience is reduced and the ability of reefs to grow in step with sea-level rise is severely impeded. Rising concentrations of CO_2 in the oceans may also retard the ability of reefs to grow in step with sea-level rise (Kleypas et al., 1999).

So, without coral bleaching, reefs would possibly be able to grow apace with rising sea-levels (Brown et al., 2000). But they are not expected to be able to sustain themselves with the combined impact of projected sea-level rise, with projected increases in bleaching episodes, and with the additional stressors such as increased land-based sources of pollution and increased atmospheric concentration of CO_2 (Westmacott et al., 2000). Due to coastal erosion and a reduction in resilience of corals, the net impact on atoll societies are likely to be increases in flooding events, freshwater aquifers becoming increasingly contaminated with saline water from storm surges and seepage, and decreasing productivity from agriculture and artisanal fishing.

Sea-level rise is a mid- to long-term problem for atoll countries. But the more immediate problems are likely to arise from enhanced climatic variability and extreme weather events (McCarthy et al., 2001). As with coral reef health, the negative impacts from these changes will be exacerbated due to existing stresses caused by unsustainable development. The impacts of enhanced climatic variability and extreme weather events may in themselves constitute a "dangerous" level of climate change to atoll social-ecological systems before and regardless of sea-level rise.

Water resources are likely to be increasingly stressed in the future through a shift to more intense rainfall events and possibly more intense droughts. In 2080 flood risk is expected to be in the order of 200 times greater than at present for Pacific atoll countries (Nicholls et al., 1999). Whilst it is still highly uncertain, some modeling studies suggest that cyclones may be more intense if not more frequent (Cubasch and Meehl, 2001). Climatic variability is linked to El Niño Southern Oscillation (ENSO) events, and whilst also very uncertain, some models suggest that these too may be becoming more frequent and intense as a function of climate change (Cubasch and Meehl, 2001).

Climate variability and the impacts of extreme weather events mean that artisanal fisheries are likely to decline as episodes of coral bleaching increase and the location of deep water fish may become increasingly unpredictable as there is reasonable confidence that changed migration patterns are linked to ENSO events (McLean and Tsyban, 2001). Agricultural output is also susceptible to damage through increased heat stress on plants, changes in precipitation and soil moisture, salt water incursion from rising sea-levels, and increased damage from extreme weather events. In conjunction with the likely increased spread of vector borne diseases and more frequent and severe extreme events, this greater food insecurity is likely to degrade human health in the Pacific region. Climate change will have substantial impacts on the economies of atoll countries. The World Bank estimates that by 2050 Tarawa atoll in Kiribati could face an annual damages bill equivalent to 13–27% of current Kiribati GDP (World Bank, 2000).

So, if they occur, the adverse effects of climate change on atolls are most likely to take the form of multiple and sequential stresses to social-ecological systems arising from extreme events such as cyclones and storm surges, and droughts (Barnett and Busse, 2002). These may have more serious impacts on local and collective well-being when superimposed on climate-induced declines in food security (through a combination of decreased local production of fish and land-based crops, and decreasing ability to pay for food imports through economic contraction), deteriorating human health, reduced availability of fresh water, and coastal erosion. (See also O'Shea, 2001.)

This combination of changes in mean conditions and extreme events may mean that, ultimately, atoll environments may be unable to sustain human habitation, a possibility with even a moderate amount of climate change over the next century. Should low-probability yet high-impact changes occur (such as melting of the Antarctic ice sheet), then this would almost certainly result in atolls becoming uninhabitable (IPCC, 2001). Thus, former IPCC Chair Robert Watson has said that low-lying small island states face "the possible loss of whole cultures" through the impacts of climate change (Watson, 2000). This summary of impacts of climate change on atolls raises questions about the capacity of atoll-societies to adapt to potentially dangerous climate change, and the justice issues that attend these disproportionate risks.

[...]

JUSTICE IN GLOBAL CLIMATE POLICY

In climate policy research, debates over fairness predominantly concern the equitable distribution of costs of emissions reduction measures rather than consideration of equity in the burden of impacts (Azar, 2000; Hamaide and Boland, 2000; Ringius et al., 2002). Analyses of burdens of impacts and emission reduction measures are defined largely in outcome terms often based on the assumption that there can be an optimization of competing values. However, the risk of the loss of atoll countries is incommensurable with economic optimization in that there is no common measure by which to compare the value of atoll sovereignty and cultures with the monetary costs of climate change (Barnett, 2001). To compare values to a common standard for the purposes of seeking an optimal solution implies all items of value are ultimately substitutable and losses can be compensated (O'Neill, 1993). We argue that atoll countries and cultures can never be satisfactorily compensated for the loss of their physical bases. Thus a conception of fairness in this case is in effect based on a deontological rule rather than a consequence or outcome. The difficulties of sufficient compensation for loss of land and the cultural and economic impacts that ensue is evident in ongoing tensions over land rights and self-determination in post-colonial countries such as Australia and New Zealand, countries which are themselves likely hosts for migrants from the Pacific atoll countries given their already large populations of people of Pacific Island origin. It is also evident in the case of Nauru (an uplifted atoll-country), which after ninety years of phosphate mining has been left with 80% of its land area severely degraded and unusable. While partially compensated for these losses, the compensation in no way constitutes an adequate recompense for the environmental and cultural implications of such widespread destruction: indeed it is unlikely that there is some level of "adequate" compensation (McDaniel and Gowdy, 2000).

This incommensurability of loss of rights with monetary values renders decision-making about the costs and benefits of mitigation strategies problematic. Cost-benefit analysis applies more for micro-level decisions and requires standard metrics of value which are unavailable as the comparability between values cannot be sustained in this instance. So, to reduce the problem of lost atoll countries into any such decision-making framework would be to fit philosophically incomparable values into inappropriately technical procedures (O'Neill, 1993). In economic terms, the lack of appropriate compensation means that any decision taken on the basis of winners and losers implies that "some individuals have the right to cause [uncompensated] damage to others" (Azar, 2000). If this is not an acceptable position and the loss of national territory cannot be compensated for, then some new path to the future and some new decision-making rules must be found.

This is not to say the decision-analysis of the problem must be non-rational, or that choice is impossible. It is the latent recognition by some countries of the potential severity of impacts on atoll countries that has given the small island states considerable leverage in negotiations over the UNFCCC [United Nations Framework Convention on Climate Change] (Shibuya, 1997). However, decisions on global climate policy must be made according to a formal conception of justice rather than merely equitable distribution of costs.

"Justice" has no universally shared meaning; its operationalization proceeds only with consensus among parties in specific contexts. At the global scale, difficulties concerning the meaning and practice of justice are at their most complex. The constituencies include at least all states, if not all people, and it requires some standardization across diverse issues. The meaning, let alone practice, of international justice therefore remains elusive (Brown, 1997). The international system, built on the sovereign rights of states, frequently struggles to reconcile sovereign rights with human rights, and at present the rights of states are largely upheld, but at the frequent expense of the rights of people (Linklater, 1999). The case of climate risks to atoll countries is rare in that the rights of atoll countries are largely synonymous with the rights of their constituents. Norms of national justice and social justice are therefore both applicable and mutually reinforcing.

[. . .]

Theories of justice are implicit in such international laws and norms [. . .] In this case, the explicit international rule would be that global decision-making on climate change by nation states would adhere to the maximin principle. [Editors' note: Maximin here means maximizing liberty, minimizing inequality.] Fair actions under this formulation of distributive justice should minimize the impacts of climate change to the most vulnerable state (Adger, 2001; Rawls, 1971). For sovereign states the possibility of extinction would constitute the greatest threat, and this is possible for perhaps five of the 181 states which are Parties to the UNFCCC. Even without this maximin principle, a radically different and just outcome would be arrived at if states negotiated under a "veil of ignorance" whereby the 181 states in the UNFCCC decided on "dangerous climate change" without knowing *a priori* which of the five out of 181 states would cease to exist as a result of continuing anthropogenic interference with the climate system.

[...]

The constitutive principle of international relations—sovereignty—is itself problematized by the possibility of climate change extinguishing atoll countries. Sovereignty is the right of political entities to be free from outside interference. It is an intersubjective construction, meaningful only because it is mutually recognized that states should not act in ways that interfere with other states. It is a founding principle of the Charter of the United Nations (Article 2.1), and is the core value underlying national security practices. For all states to do less than everything possible to prevent the loss of a sovereign entity is to undermine this most essential and powerful norm of international law and politics.

REFERENCES

Adger, W. N. (2001), "Scales of governance and environmental justice for adaptation and mitigation of climate change," *Journal of International Development* 13, 7, 921–931.

Azar, C. (2000), "Economics and distribution in the greenhouse," *Climate Change* 47, 233–238.

Barnett, J. (2001), *The Issue of 'Adverse Effects and the Impacts of Response Measures' in the UNFCCC*, Tyndall Centre for Climate Change Research Working Paper 5, University of East Anglia, Norwich.

Barnett, J. and Busse, M. (2002), "Conclusions," in J. Barnett, and M. Busse (eds.), *Proceedings of the APN Workshop on Ethnographic Perspectives on Resilience to Climate Variability in Pacific. Island Countries*, Macmillan Brown Centre for Pacific Studies, Christchurch: 75–77.

Brautigam, D. and Woolcock, M. (2001), *Small States in a Global Economy*, World Institute for Development Economics Research Discussion Paper 2001/37, Helsinki.

Brown, B., Dunne, R., Goodson, M., and Douglas, A. (2000), "Bleaching patterns in coral reefs," *Nature* 404, 142–143.

Brown, C. (1997), "Theories of international justice," *British Journal of Political Science* 27, 273–297.

Commonwealth Secretariat (1999), *Vulnerability: Small States in the Global Society*, Report of the Commonwealth Consultative Group on the Special Needs of Small States, Commonwealth Secretariat, London.

Cubasch, U. and Meehl, G. (2001), "Projections of future climate change," in J. Houghton, Y. Ding, D. Griggs, M. Noguer, P. van der Linden, X. Da, K. Maskell, and C. Johnson (eds.), *Climate Change 2001: The Scientific Basis*, Cambridge University Press, Cambridge: 525–582.

Hamaide, B. and Boland, J. J. (2000), "Benefits, costs and cooperation in greenhouse gas abatement," *Climate Change* 47, 239–258.

Houghton, J., et al. (eds.) (2001), *Climate Change 2001: The Scientific Basis*, Cambridge University Press, Cambridge.

IPCC (Intergovernmental Panel on Climate Change) (2001), *Climate Change 2001: Synthesis Report, A Contribution of Working Groups I, II, and III of the Intergovernmental Panel on Climate Change*, Cambridge University Press, Cambridge.

Jones, R. (1999), *An Analysis of the Effects of the Kyoto Protocol on Pacific Island Countries: Identification of Latent Sea-Level Rise within the Climate System at 1995 and 2020*, South Pacific Regional Environment Programme and CSIRO, Apia.

Kleypas, J., et al. (1999), "Geochemical consequences of increased carbon dioxide in coral reefs," *Science* 284, 118–120.

Lal, M., Harasawa, H., and Takahashi, K. (2002), "Future climate change and its impacts over small island states," *Climate Res.* 19, 179–192.

Linklater, A. (1999), "The evolving spheres of international justice," *International Affairs* 75, 473–482.

McCarthy, J., et al. (eds.) (2001), *Climate Change 2001: Impacts, Adaptation & Vulnerability*, Cambridge University Press, Cambridge.

McDaniel, C. and Gowdy, J. (2000), *Paradise for Sale: A Parable of Nature*, University of California Press, Berkeley.

McLean, R. and Tsyban, A. (2001), "Coastal zones and marine ecosystems," in J. McCarthy et al. (eds.), *Climate Change 2001: Impacts, Adaptation & Vulnerability*, Cambridge University Press, Cambridge: 343–380.

Moberg, F. and Folke, C. (1999), "Ecological goods and services of coral reef ecosystems," *Ecological Economics* 29, 215–233.

Nicholls, R., Hoozemans, F., and Marchand, M. (1999), "Increasing flood risk and wetland losses due to global sea-level rise: Regional and global analyses," *Global Environ. Change* 9, 69–87.

Nunn, P. (1994), *Oceanic Islands*, Blackwell, Oxford.

Nurse, L. and Sem, G. (2001), "Small Island States," in J. McCarthy et al. (eds.), *Climate Change 2001: Impacts, Adaptation & Vulnerability*, Cambridge University Press, Cambridge: 842–875.

Nyström, M., Folke, C., and Moberg, F. (2000), "Coral reef disturbance and resilience in a human-dominated environment," *Trends in Ecology and Evolution* 15, 413–417.

O'Neill, J. (1993), *Ecology, Policy and Politics*, Routledge, London.

O'Shea, E. L. (ed.) (2001), *Preparing for a Changing Climate: The Potential Consequences of Climate Variability and Change: Pacific Islands*, Report for the US Global Change Research Program, East West Center, Honolulu, Hawai'i.

Rawls, J. (1971), *A Theory of Justice*, Harvard University Press, Cambridge.

Reaser, J. K., Pomerance, R., and Thomas, P. O. (2000), "Coral bleaching and global climate change: Scientific findings and policy recommendations," *Conservation Biology* 14, 1500–1511.

Ringius, L., Torvanger, A., and Underdal, A. (2002), "Burden sharing and fairness principles in international climate policy," *International Environmental Agreements: Politics, Law and Economics* 2, 1–22.

Shibuya, E. (1997), "Roaring mice against the tide: The South Pacific Islands and agenda-building on global warming," *Pacific Affairs* 69, 541–555.

Smith, E. and Wishnie, M. (2000), "Conservation and subsistence in small-scale societies," *Annual Review of Anthropology* 29, 493–524.

SPC (Secretariat of the Pacific Community) (2000), *Oceania Population 2000*, SPC, Noumea.

UNCTAD (United Nations Conference on Trade and Development) (1999), *Statistical Synopsis of the Least Developed Countries*, United Nations, New York and Geneva.

UNCTAD (United Nations Conference on Trade and Development) (2002), *Country Backgrounds*, United Nations Conference on Trade and Development. Available on-line at: http://www.unctad. org/, accessed May 30, 2002.

UNEP (United Nations Environment Programme) (1999), *Pacific Islands Environment Outlook*, UNEP, Apia.

Watson, R. (2000), *Presentation of the Chair of the Intergovernmental Panel on Climate Change at the Sixth Conference of Parties to the United Nations Framework Convention on Climate Change*, November 13, 2000, Intergovernmental Panel on Climate Change, http://www.ipcc.ch/ press/speech.htm, accessed May 30, 2002.

Westmacott, S., Teleki, K., Wells, S., and West, J. M. (2000), *Management of Bleached and Severely Damaged Coral Reefs*, International Union for the Conservation of Nature, Gland, Switzerland.

World Bank (2000), *Cities, Seas, and Storms: Managing Change in Pacific Island Economies*, Washington, DC: World Bank.3

PART V

Planned interventions and contestations

Figure 27 Ministry of Interior barrier wall Cairo/No walls campaign/Various artists, March 2012.
Source: © Bahia Shebab, 2012–2013.

Figure 28 Ministry of Interior barrier wall Cairo/51 Children of Assiut/Bahia Shehab, January 2013
Eight children playing hide and seak.
Translation of speech bubbles (read from right to left):
Child 1: Khalawees (Are you done? Did you hide?)
Child 2: Not yet.
Child 3: Has the revolution succeeded?
Child 4: Not yet.
Child 5: Did we get the rights of the martyrs?
Child 6: Not yet.
Child 7: Has Egypt become heaven on Earth?
Child 8: Not yet.

Source: © Bahia Shehab, 2012–2013.

SECTION 9
Governance

Editors' Introduction

How should societies be governed? What constitutes a fair or effective system of government? Through what institutional arrangements and processes should societies with competing interests be governed? These are age old questions which have assumed crucial importance for the independent nation states that have emerged in much of the global South in the last half a century.

To begin with, it is vital to define three key terms: government, the state, and governance. The government includes politically elected or appointed officials and the bureaucracy, and is largely a time-bound body. In most countries government officials come and go with elections or changes in power. The state, on the other hand, is a term that encompasses government and describes the structures and arrangements through which a society is governed. State structures do not disappear when a newly elected leadership comes into office. The state includes government departments from police and education to the military and diplomacy, as well as national assets like public sector industries and regulatory agencies. By contrast governance is a process, the way in which a society is governed and functions. The institutions of governance include not only the state but also private businesses and civil society organizations like NGOs, churches, unions and membership clubs. Governance should not be confused with governmentality, another important term, coined by French philosopher Michel Foucault, which is often used in debates about the relationship between the state and its ability to govern. Governmentality, refers to norms, values, rationalities, and mentalities that are internalized by members of a society which allow the state to assert social control by people's consent rather than by direct use of force. Moreover, be reminded that "good governance" and "good government" are not synonymous. Some like Rosenau and Czempiel (1992) even call for "governance without government." Good government focuses on a single sector, the government, and government agencies, whereas the notion of good governance stresses the role of multiple sectors and the relationships among them. Good governance, or what some refer to as "networked governance," stresses inclusion of multiple actors beyond the government (for example civil society organizations and the private sector). The concept behind good governance is that inclusiveness will lead to greater transparency and accountability (Tendler 1997).

The issues of the government, the state, and governance have held particular importance for urban development in recent years. They have also been the focus of considerable debate. In this section of the *Reader* we will delve into these debates, focusing on their particular relevance to the cities of the global South.

KEY ISSUES

The role of the state: decentralization and good governance

In the past half century there have been major shifts in thinking about governance. The immediate post-independence years saw the advent of the "developmental state" where national governments spearheaded development projects for industrialization, as well as the expansion of infrastructure and services. In many instances, the state took ownership of industry in addition to controlling the country's financial institutions. Ultimately these developmental states dominated economic, social, and political decisions. In many instances national development plans spelled out these roles and powers explicitly.

In many ways this state-centered path followed the welfare state model that much of the global North had adopted. While this approach may have helped build national unity and provided benefits like education and health services to large numbers of people, the developmental state had numerous shortcomings. In many cases, "state-centered development" concentrated power in the hands of a few political or business leaders. This power frequently became a source of patronage and corruption where leaders used positions of influence to dole out jobs to relatives and friends and at times siphoned off public funds for their personal benefit. In many countries, this led to a bloated and inefficient public sector and a loss of confidence in the integrity of state operations.

The crises of capitalism in the 1980s, along with the weakening of the welfare state model and the shortcomings of the developmental state, led to global re-thinking on the role of government and the state. The idea of breaking down the state to make it more efficient and responsive to the needs of people began to take hold.

Powerful institutional forces and international organizations such as the World Bank and the United Nations also played important roles in promoting decentralization in several parts of the global South. These organizations maintained that decentralization offered potential contributions to improved accountability, inclusion, political stability, as well as increased effectiveness and quality of service provision. They also argued that the decentralized model would enhance the rule of law and effective controls on corruption (World Bank 2001; Burki, Perry, and Dillinger 1999). Many national governments felt some obligation to follow the advice of these powerful international agencies, making decentralization one of the most pervasive trends in the 1980s and 1990s (Cheema and Rondinelli 1983).

The process and outcome of state restructuring is complex. Typically done in the name of promoting "good governance," states began to shed some of their powers and responsibilities and opened the door to more market-oriented entities and ideas. In some instances, local governments or international organizations assumed powers previously borne at the national level. In other cases, these were outsourced to private companies (Miraftab, Beard, and Silver 2008).

Proponents of market-oriented restructuring advanced both political and economic arguments to support their case. The political argument focused on notions of inclusivity, maintaining that the state needed to share its political power with other stakeholders in society. This inclusivity argument, often embodied in a process referred to as administrative restructuring, contended that devolving national state powers to the lower levels of government such as municipalities and counties would bring governance closer to the people and ultimately offer greater accessibility to the disadvantaged. The economic argument advanced a similar logic, alleging that inclusivity also mandated an expansion of the private sector's role. They further buttressed this economic perspective with their beliefs that business was inherently more efficient and effective than government. This led to calls for a "lean" state, which would place cost effectiveness above equity in service delivery to urban inhabitants.

Putting this into practice ultimately meant that the state might outright sell its public sector companies and assets (for example its water company) to a private firm (called privatization), or it might outsource or subcontract an aspect of public sector utilities operation or management–something we often see with urban transportation systems. These privatization initiatives were often accompanied by a shift in philosophy regarding state finance. This shift entailed reducing corporate and property taxes as a way to stimulate business and to increase individual user fees for basic services. User fees were a way to make government services run more like a private business. This market-based model of governance and management, a critical component of neoliberal governance, percolated through to the ways in which bureaucracies and governments began to structure themselves under the label of "New Public Management" (see Osborne and Gaebler 1993).

Boaventura de Sousa Santos (2004) has been a key analyst of this process, emphasizing how political decentralization has often been used to justify or obscure the economic impetus for decentralization. His argument is most clearly demonstrated in the experience of Latin America and post-apartheid South Africa, where private firms have consistently used arguments of inclusivity to promote their increasing role in service delivery and ownership of former state assets.

The role of non-state actors: privatization and partnership

The implication of economic and political decentralization for development and governance of cities is crucial. In this new scenario the roles and responsibilities of the state have changed from ownership, provision, and delivery of basic services and infrastructure to enabling other actors including the private sector and civil society organizations to perform these functions. For provision and delivery of basic services and infrastructure, the state might privatize its public assets or partner with the private sector to reap entrepreneurial profits. A municipality might also enter into partnerships with a non-governmental organization (NGO) to achieve a greater reach and implementation among affected communities. The assumption is that partnership promotes a "win-win situation" benefiting the citizens and the municipality's coffers as well as local businesses (Fiszbein and Lowden 1999; Savas 2000).

Critics, however, have argued that this "win–win" scenario of partnerships overlooks the imbalance in resources and powers among non-state actors (Miraftab 2004). In particular, they stress that non-profit civil society groups, particularly those representing the poor, lack the resources of private companies and corporations to promote their interest. Therefore, this shift in governance might be a win for the private sector actors by expanding the realm of the market but a loss for the poor and marginalized (Petra and Veltmeyer 2001; De Angelis 2005). David Harvey (2005), a prominent geographer, argues privatization and partnership might be new or modern labels for old colonial practices of "accumulation by dispossession." Whereas in its earliest expansion capital took over land and resources in the colonies, dispossessing people there, Harvey argues that in this latest round expansionist capital dispossesses cities, municipalities and the vast majority of citizens from their public goods and assets–be it a public water company or a railroad track. George and Sabelli (1994) add another dimension to this when they interpret "good governance" as yet another rubric under which international development agencies such as the World Bank have encouraged governments in the global South to take on more debt thereby sustaining old dependencies. Mamdani (1996), reflecting on this phenomenon in many African countries, argues that like the chiefdoms of the colonial era the contemporary decentralization policies have recreated decentralized despotism rather than empowering people at the grassroots level.

But not everyone concurs with these conclusions. For example, Misra and Kudva (2008) in their study of electoral quotas for women and other minority groups in Panchayati Raj, or local government in India, conclude that enforcing a quota system that requires inclusion of disadvantaged populations by their caste or gender has had positive effects. In particular they stress that this process brings the state closer to the people and has shifted governance priorities in some local areas. As reflected in the selected readings that follow, restructuring the state for "good governance" is a complex and dynamic process that at times may entail contradictory outcomes. Intentionally or inadvertently, some processes of decentralization for good governance have brought formerly excluded actors into the arena of development decision making and democratized development decisions with a greater downward accountability (Smoke 1994; Agrawal and Ribot 1999).

Non-governmental organizations (NGOs)

The worldwide state restructuring discussed above brought NGOs to the center stage in development processes. Many NGOs moved from their previous roots in humanitarian relief and assistance in time of war and disaster to direct participation in long-term development efforts. Some have celebrated NGOs as a panacea to development processes–as the sector that is independent of the state and the market and best able to reach out to marginalized populations or regions. Some stress the political virtue of NGOs in that they are close allies of grassroots movements and initiatives and praise NGOs as promoters of democratic development from below. Appadurai, in the selection for this section, contends that NGOs have the unique capacity to engage disadvantaged populations and construct a more inclusive development and governance. Others see additional economic virtue in the role of NGOs for being less costly than the profit driven private sector or the wasteful bureaucratic public sector. In short NGOs are touted by many as effective and efficient in delivering of services to disadvantaged communities as well as praised for promoting community self-reliance and participatory processes (Hulme and Edwards 1997; Korten 1990; Clark 1991). Other studies, however, have shown that NGO claims to economic efficiency need to be tempered. Such research portrays NGOs as organizations with limited cost-efficiency or innovation and offer other critiques as well: a lack of breadth of reach, failure to target as deeply as claimed, and tend not to work in areas of highest poverty (Kudva 2005: 236).

The democratic potential and promise of NGOs for inclusive governance continues to be the subject of much debate among practitioners and scholars of development. Critics explain the worldwide mushrooming of NGOs since the 1980s as an intricate part of the neoliberal shift in governance whereby the NGOs "tidy up after the state and the market" (Pieterse 2001: 126; Tvedt 1998; Elyachar 2002) in service sectors that are not profitable to the private firms and are no longer under the ambit of the state. They argue NGOs act as an indirect rule by extending the reach of the state and therefore its control. Another critique holds that NGOs may bureaucratize grassroots movements and depoliticize their struggle–a phenomenon that has been made into a verb, "NGO-ization" (Ferguson 1994; Pithouse 2013). Miraftab amplifies this point with her term, "flirting with the enemy," arguing that many of these organizations have drifted far from their original roots in pedagogy for the oppressed and volunteer activism (Miraftab 1997). Ultimately, many see NGOs as new governmental organizations and most recently as new market ventures. (See selected article by Michael Mascarenhas that follows.)

Two last critiques of NGOs are also important. First, Mohan (2002) bemoans their dependency on private donations. This dependency he claims may influence their agenda as accountability to donors takes precedence over empowering communities. In Mohan's view this leads to relations of patronage and clienteleism with disadvantaged communities rather than solidarity. Second comes the notion of "localist traps." Here Mohan and others stress that NGOs may often be crisis focused, resorting to atomized "firefighting" with local projects while the broader structural inequalities are left unquestioned (Mohan and Stokke 2000; de Wit and Berner 2009).

As this brief survey has shown, the outcome of state restructuring for "good governance" depends on and is influenced by many factors. These might include the history of local and national state formation, the relative strength and levels of organization of various business and civil society groups and social movements, and the ways in which state and non-state actors and political leaders engage with the process of decentralization (Kudva 2005, 2008; Heller 2001). The political regime, as the apparatus that maintains a particular relation with citizens and determines the forms of government that come into power, is an important determining factor in the dynamics and the outcome of state and non-state actors. The question of governance in liberal democratic regimes for example will be somewhat different from that in an authoritarian state. Under authoritarian states NGOs are often not encouraged or used as subcontractors or in completely tokenistic ways allowing little room for expansion of political claim-making. The articles that follow will address a variety of these issues.

ARTICLE SELECTIONS

The selected readings in this section by Ben Kohl and Lisa Farthing, Arjun Appadurai, and Michael Mascarenhas take us to distinct socio-political and historical contexts and offer insights into the varied dynamics among state and non-state actors in development processes. These interpretations stress the contingencies in the relationships that new protagonists of development can facilitate at the local level.

Ben Kohl, a distinguished scholar of urban planning and urban geography, and Lisa Farthing, a prominent investigative journalist of urban contestations, are well known for their work on development and politics in Bolivia (2008, 2012, 2014). In the article selection here, Kohl and Farthing look at state decentralization policies in Bolivia documenting how at the core of Bolivia's decentralization lie the elites' interest in privatizing state enterprise and taking over public assets. However, here they describe the unintended consequences of Bolivia's 1994 Law of Popular Participation (LPP) which attempted to facilitate a greater role for non-state actors by making participation a legal requirement by the state. Marginalized indigenous populations took advantage of the decentralized state and their participation law to gain voice and for the first time ever elect an indigenous president. State decentralization in Bolivia, Kohl and Farthing document, gave rise to greater operation of local government and private entities but also provided greater voice to the well-organized indigenous populations and their movement in Bolivia. Thus, it simultaneously prompted elitist and inclusive governance.

Arjun Appadurai, a leading social-cultural anthropologist, has a large body of work of which *Modernity at Large* (1996) and *Globalization* (2001b) have influenced scholars of cities. In the article selection here, he focuses on an alliance among three civic organizations in Mumbai to address poverty–the Society for the Promotion of Area Resource Centres (SPARC), an NGO, the National Slum Dwellers Federation (SDDF) and Mahila Milan, a cooperative representing women's savings groups. In an article that has greatly inspired, justified, and motivated the rise in international funding for NGOs, Appadurai offers unwavering support to the role of NGOs and their alliance with grassroots organization in addressing problems of the majority poor in cities of the global South. He documents the day-to-day practices and strategies adopted by this alliance for a dignified life–from shelter to toilets to income and saving. He views this as part of a move to achieve inclusion in the city and provide greater access to livelihood resources. Appadurai celebrates the relationship between the NGO and the slum dwellers as one that promotes a politics of self-reliance and patience. He argues that their practices embody a different type of politics that builds lasting and "deep democracies."

Michael Mascarenhas, a Canada-based sociologist who has studied and documented practices of water-related NGOs in Africa, offers a contrasting view. He focuses on the relationships that shape and fuel development projects in the water sector and writes about Water for People, a global NGO with water delivery projects in eleven countries. Like many NGOs, Water for People has turned to "global capital" and "sound financial innovation" to secure its operation and funding. These seemingly independent international NGOs he depicts are deeply enmeshed in the financial interests of capital. From the corporate CEOs who serve on the NGOs' board of directors to the market discourse and ideology that these NGOs promote and adopt, Mascarenhas contends that civil society organizations that serve in the water sector are the new playground and realm of expansion for large companies with a hand in water technology, facilities management, and engineering and consulting services. He calls this "venture philanthropy" whereby "humanitarian efforts become more aligned with investment than opportunities for social change."

FIVE

REFERENCES

Agrawal, A. and Ribot, J. (1999) "Accountability in decentralization: a framework with South Asian and West African cases," *Journal of Developing Areas,* 33: 473–502.

Appadurai, A. (1996) *Modernity at Large: Cultural Dimensions of Globalization,* Minneapolis: University of Minnesota Press.

Appadurai, A. (2001a) "Deep democracy: Urban governmentality and the horizon of politics," *Environment and Urbanization,* 13(2): 23–43.

Appadurai, A. (Ed.) (2001b) *Globalization*, Durham, NC: Duke University Press.

Burki, S. J., Perry, G. E. and Dillinger, W. R. (1999) *Beyond the Center: Decentralizing the State,* Washington, DC: The World Bank. Available online at www1.worldbank.org/publicsector/decentralization/cd/Beyondthecenter.pdf (accessed May 26, 2014).

Cheema, S. G. and Rondinelli, D. A. (Eds.) (1983) *Decentralization and Development: Policy Implementation in Developing Countries,* Beverly Hills: Sage.

Clark, J. (1991) *Democratizing Development: The Role of Voluntary Organizations*, West Hartford, CT: Kumarian Press.

De Angelis, M. (2005) "The political economy of neoliberal governance," *Review*, XXIII(3): 229–257.

De Wit, J. and Berner, E. (2009) "Progressive patronage? Municipalities, NGOs, CBOs and the limits to 'slum dwellers' empowerment," *Development and Change,* 40(5): 927–947.

Elyachar, J. (2002) "Empowerment money: the World Bank, non-governmental organizations, and the value of culture in Egypt," *Public Culture*, 14(3): 493–513.

Farthing, L. C. and Kohl, B. H. (2014) *Evo's Bolivia: Continuity and Change,* Austin: University of Texas Press.

Ferguson, J. (1994) *Anti-Politics Machine: Development, Depoliticization, and Bureaucratic Power in Lesotho*, Minneapolis: University of Minnesota Press.

Fiszbein, A. and Lowden, P. (1999) *Working Together for a Change: Government, Civic, and Business Partnerships for Poverty Reduction in Latin America and the Caribbean,* Washington, DC: The World Bank.

George, S. and Sabelli, F. (1994) "Governance: the last refuge?" in S. George and F. Sabelli (Eds.), *Faith and Credit: The World Bank's Secular Empire*, Boulder: Westview Press, pp. 142–161.

Harvey, D. (2005) *The New Imperialism*, Oxford: Oxford University Press.

Heller, P. (2001) "Moving the state: the politics of democratic decentralization in Kerala, South Africa, and Porto Alegre," *Politics and Society*, 29: 131.

Hulme, D. and Edwards, M. (1997) *NGOs, States and Donors: Too Close for Comfort?* Basingstoke: Palgrave in association with Save the Children.

Kohl, B. and Farthing, L. C. (2006) *Impasse in Bolivia: Neoliberal Hegemony and Popular Resistance*, London: Zed Books.

Kohl, B. and Farthing, L. (2008) "New spaces new contests: appropriating decentralization for political change in Bolivia," in V. Beard, F. Miraftab and C. Silver (Eds.), *Planning and Decentralization: Contested Spaces for Public Action in the Global South*, New York: Routledge, pp. 69–86.

Korten, D. C. (1990) *Getting to the 21st Century: Voluntary Action and the Global Agenda*, Bloomfield, CT: Kumarian Press.

Kudva, N. (2005) "Strong states, strong NGOs," in M. Katzenstein and R. Ray (Eds.), *Social Movements in India: Poverty, Power, and Politics*, Boulder, CO: Rowman and Littlefield/New Delhi: Oxford University Press.

Kudva, N. (2008) "Conceptualizing NGO–State relations in Karnataka: conflict and collaboration amidst organizational diversity," in G. Kadekodi, R. Kanbur and V. Rao (Eds.), *Development in Karnataka: Challenges of Governance, Equity, and Empowerment*, Delhi: Academic Press.

Mamdani, M. (1996) *Citizens and Subjects: Contemporary Africa and the Legacy of Late Colonialism,* Princeton, NJ: Princeton University Press.

Mascarenhas, M. (2014) "Sovereignty: crisis, humanitarianism, and the condition of 21st century sovereignty," in H. E. Kahn (Ed.), *Framing the Global*, Bloomington: Indiana University Press, pp. 296–316.

Miraftab, F. (1997) "Flirting with the enemy: challenges faced by NGOs in development and empowerment," *Habitat International,* 21(4): 361–375.

Miraftab, F. (2004) "Public–private partnerships: the Trojan horse of neoliberal development?" *Journal of Planning Education and Research,* 24(1): 89–101.

Miraftab, F., Beard, V. and Silver, C. (2008) "Introduction: situating contested notions of decentralized planning in the global South," in V. Beard, F. Miraftab and C. Silver (Eds.), *Planning and Decentralization: Contested Spaces for Public Action in the Global South,* New York: Routledge, pp. 1–18.

Misra, K. and Kudva, N. (2008) "En(gendering) effective decentralization: the experience of women in Panchayati Raj in India," in V. Beard, F. Miraftab and C. Silver (Eds.), *Planning and Decentralization: Contested Spaces for Public Action in the Global South,* New York: Routledge, pp. 175–187.

Mohan, G. (2002) "The disappointments of civil society: the politics of NGO intervention in northern Ghana," *Political Geography,* 21: 125–154.

Mohan, G. and Stokke, K. (2000) "Participatory development and empowerment: the dangers of localism," *Third World Quarterly*, 21(2): 247–268.

Osborne, D. and Gaebler, T. (1993) *Reinventing Government: How the Entrepreneurial Spirit is Transforming the Public Sector*, New York: Plume.

Petra, J. and Veltmeyer, H. (2001) *Globalization Unmasked: Imperialism in the 21st Century*, London: Zed Books.

Pieterse, J. N. (2001) *Development Theory: Deconstructions/Reconstructions*, London: Sage Publications.

Pithouse, R. (2013) "NGOs and urban movements: notes from South Africa," *City*, 17(2): 253–257.

Rosenau, J. and Czempiel, E. O. (Eds.) (1992) *Governance without Government: Order and Change in World Politics*, Cambridge: Cambridge University Press.

Santos, B. (2004) "Governance: between myth and reality," paper presented at the 2004 Law & Society Association Annual Meeting, Chicago, May 27–30.

Savas, E. S. (2000) *Privatization and Public–Private Partnerships*, New York: Chatham House.

Smoke, P. (1994) *Local Government Finance in Developing Countries: The Case of Kenya*, Oxford: Oxford University Press.

Tendler, J. (1997) *Good Government in the Tropics*, Baltimore: Johns Hopkins University Press.

Tvedt, T. (1998) *Angels of Mercy or Development Diplomats? NGOs and Foreign Aid*, Trenton, NJ: Africa World Press, Oxford: James Currey.

World Bank (2001) "Making markets work better for poor people," in *World Development Report 2000/2001: Attacking Poverty*, Washington, DC: The World Bank, pp. 61–76.

"New Spaces, New Contests: Appropriating Decentralization for Political Change in Bolivia"

Planning and Decentralization: Contested Spaces for Public Action in the Global South (2008)

Ben Kohl and Linda Farthing

> The United States will use this moment of opportunity to extend the benefits of freedom across the globe. We will actively work to bring the hope of democracy, development, free markets, and free trade to every corner of the world (Bush, 2002).

For forty years, political and administrative decentralization has been integrated as a core component of the global neoliberal ideology described above by US President George W. Bush (Kohl and Farthing, 2006; Peck, 2001; Peet, 2003). As a result, both high- and low-income countries have transferred planning and administrative responsibilities from national to subnational governments (Oyugi, 2000; Samoff, 1979; Wunsch, 2001). These policies are based on the assumption that decentralized governments are not only more efficient and less corrupt than centralized ones, but also more democratic (World Bank, 1997, 2000). In fact, expanding citizen participation in planning is only one of several potential outcomes of decentralization, as experience has demonstrated around the world (Huerta Malbrán *et al.*, 2000; Hutchcroft, 2001; Oxhorn, *et al.*, 2004; Schönwälder, 1997; Wanyande, 2004).

Some critical scholars argue that the focus on formal democracy aims to ensure the political stability that global markets require to operate successfully within national economies. They contend that the focus on formal democratic processes channels citizens' demands to limited local concerns and a tepid and tidy range of choices expressed at the ballot box (Gill, 2002; Kohl, 2002; Robinson, 1996, 2003; Slater, 1989). These scholars also concur, however, that under certain conditions decentralization programs can have the unintended result of opening new political spaces to broaden local control and contest neoliberalism. In Bolivia, decentralization legislation written in 1994 took place in a context of long-established trajectories of political resistance and contributed to the conditions that culminated in the December 2005 election of coca grower Evo Morales, a self- proclaimed socialist and the country's first indigenous president. These events did not take place in isolation but within a context of a growing economic and political crisis after 1999, which was triggered by:

(a) declining government revenues after the privatization of the hydrocarbons company;
(b) the forced eradication of coca; and
(c) declining remittances from the seventeen percent of Bolivians working in crisis-torn Argentina (Kohl and Farthing, 2006).

The power of social movements in Bolivia has been well reported since the 2000 Cochabamba water war (Albro, 2005; Assies and Salman, 2003; Olivera, 2004). Hylton and Thomson (2005) point out that these movements are built on legacies of resistance in the Andes dating from the Spanish Conquest.

The formal political side of the process that laid the groundwork for Morales' election in 2005, however, is not as well understood. In this chapter we address this gap, demonstrating how, in a context of long-standing popular mobilization, political decentralization increased local participation in planning and facilitated the transformation of a contentious social movement (Tarrow, 1998) into a dominant political force.

[. . .]

Two key interrelated legal processes facilitated the political rise of counter-hegemonic movements. First, the 1994 [Law of Public Participation] LPP, along with the related Law of Administrative Decentralization (LAD), created over 250 new, small and largely indigenous and rural municipalities with planning oversight delegated to local organizations. These newly formed municipalities required thousands of council representatives and, as the indigenous and urban poor increasingly assumed office, they acquired some of the formal skills associated with western-style government. This change fundamentally altered the discourse on the rights and roles of citizens in areas long abandoned by the state (Kohl, 2003b). Second, changes in electoral laws in 1996 and 2004 led to greater representation of mostly male indigenous and *campesino* leaders in the national congress, which allowed them to transform traditional peasant and neighborhood organizations into formal political parties. To clarify the changes decentralization has wrought in Bolivia, we briefly consider the nexus between decentralization and broader political participation as reflecting an ideological process with uneven results in practice.

[. . .]

BOLIVIA'S DECENTRALIZATION

Bolivia is about twice the area of France or Texas, is organized into nine departments and has a population of 9 million, the majority of whom are from Quechua and Aymara ethnic groups. Since the mid sixteenth century when the colonial mines of Potosí produced more than half the world's gold and silver (Klein, 1998), Bolivia's natural resource wealth has benefited global, rather than local, economic interests. As with many low-income countries, the successive resource booms driven by silver, quinine, rubber, tin and, most recently, coca (and its derivative cocaine) and hydrocarbons have done little to construct the foundations for continued economic growth (Sachs and Warner, 1999). About half the population still lives as semi-subsistence farmers in rural areas, and these *campesinos* (small scale farmers) provide much of the country's food.

Bolivia has the dubious reputation of having had the largest number of *coups d'état* in the world since winning independence from Spain in 1825 (Morales, 1992). Although frequently controlled by military governments, since the return to civilian rule in 1982, the country has celebrated seven constitutional transfers in administration. (In 2003 and 2005, while presidents were forced to resign in the face of massive popular protests, the transitions followed constitutional procedures.) During the eighteen years of dictatorship prior to 1982, oppositional forces utilized what Beard (2003) calls covert planning to keep a sense of collective agency alive.

Between 1952 and 1985, the government served as the country's prime economic actor. The transition to a market-dominated economy followed the 1986 IMF structural adjustment package that introduced a neoliberal economic model (Sachs, 1987). Structural adjustment led to a sell-off of government firms and, through 2003, a trend towards the privatization of basic services.

In a poorly integrated country where pressures for greater local autonomy have been constant, Bolivia's 1994 LPP reflected longstanding efforts by regional movements, on the one hand, to obtain more resources and decision-making power and, on the other, by NGOs, to shift resources to long-neglected rural communities (CIPCA, 1991; Medina, 1997; Molina Monasterios, 1997). Grindle (2000) contends that for the government in power, the goals were to: extend the reach of the state (a frequent outcome of decentralization programs); to develop a stronger sense of national identity; to control endemic corruption; and to confront regional (mostly urban) elites' demands for political autonomy by focusing resources in rural hinterlands, a strategy that Tulchin and Selee (2004) have discovered has been utilized by national elites elsewhere. It also reflects the notion of planning as a neutral technocratic process that serves as what Sandercock (1998: 24) calls an "ordering tool . . . a kind of spatial police."

Before the LPP, most of the country fell outside any municipal jurisdiction at all. Municipal elections, held every five years since 1987, were relatively unimportant before 1999 as they only took place in the nation's largest cities. Municipal governments only encompassed towns and cities, whose formal boundaries were never registered nationally and often fluctuated in accordance with the interests of the mayor in office. The rural areas, from the perspective of the national government, were largely ignored and, in many areas, the local government was a community organization, whether a "traditional" *ayllu* or *capitania* or a "modern" *campesino* union that operated independent of local municipalities (Albó *et al.*, 1990; Ejdesgaard Jeppesen, 2002: 36). The structure of the national budget also reflected the centralization of formal government: ten percent was targeted for the nine departmental capitals, with other towns and rural areas competing for funds from an additional ten percent channeled through regional development corporations.

In 1994, the LPP and related legislation changed that. The laws combined the funds directed to departmental capitals and regional development corporations and allocated them to municipal governments on a per capita basis. In the process, the laws created over 250 new municipal governments, almost three-quarters of them "rural" municipalities with populations of less than 15,000 people. However, undoubtedly the principal innovation was the mandating of participatory planning and fiscal oversight by neighborhood and indigenous organizations (Kohl, 2003a).

Prior to decentralization, candidates in countrywide elections needed affiliations with a national party, which led to political elites, typically from the largest cities, exerting undue influence on local politics. Through the 1993 national elections, all members of both houses of congress were elected through a proportional-representation system that drew on party lists to field candidates. This style of electoral politics kept indigenous rural and (increasingly indigenous) urban political actors from meaningful political participation. As a result, historically disenfranchised rural communities expressed their voices through the contentious politics associated with social movements.

That began to change owing to a 1996 electoral law that called for one-half of representatives to be chosen from district-level competitions, in a hybrid proportional-representation system following the German model (Domingo, 2001). This required political parties, which traditionally lacked formal rural organizations, to field district-level candidates beginning in 1997.

THE OPENING OF POLITICAL SPACE

The LPP led to a major political accomplishment: for the first time, the government formally recognized traditional organizations, including urban neighborhood organizations, pre-Hispanic indigenous organizations (*ayllus* and *capitanías*) and modern *campesino* unions, and mandated a formal role for them in local planning. The government registered almost 15,000 widely disparate grassroots territorial organizations (GTOs) between 1994 and 1997 and gave them responsibility for creating community development plans, ensuring local oversight of projects, and mobilizing community labor for the construction and maintenance of public works. In rural municipalities, GTOs can have as few as sixty members, whereas in the country's largest cities they number as many as 3,000 (Kohl, 2003a).

[. . .]

The LPP sought to channel Bolivia's traditionally unruly political protest to a local level and contain it within prescribed limits. Medeiros argues that the LPP embodied a "highly regulated construction of a modern participatory citizenry" as part of a hegemonic project that sought to "predefine the limits to what can be achieved" (Medeiros, 2001: 401), and, in the short run, the LPP achieved moderate success in this dimension.

[. . .]

With the introduction of municipal elections, popular movements gained entry to formal political spaces and began to transform social movements into political parties. In the 1995 municipal elections, thirteen political parties fielded candidates, and *campesino* and indigenous representatives were elected to twenty-nine percent (464 of the 1,624) of the seats in 200 of the country's 311 municipalities (MDH-SNPP, 1996). In this first election, however, most successful candidates represented the traditional parties, rather than more populist ones (Albó, 1996: 14).

The large number of political parties made it attractive for the smaller parties that lacked a national organization to open their doors to indigenous and rural people as the best way to field candidates. The success of some of the smaller parties, especially the Assembly for [Indigenous] Peoples' Sovereignty (*Asamblea de Soberania de Pueblos* [ASP]) and the Free Bolivia Movement (Movimiento Bolivia Libre [MBL]), in winning seats in rural municipalities alerted the traditional parties to the need for a stronger rural presence.

Although only four indigenous candidates, all representing the coca growers' movement, reached Congress in the 1997 national elections, their success inspired other *campesino* and indigenous organizations and convinced small political parties to "loan" them their party slates during municipal elections in 1999. Many of the candidates in the 1995 and 1999 municipal elections participated as national candidates for the MAS (*Movimiento al Socialismo*—Movement Toward Socialsim) in 2002 (Healey, 2005). Indigenous councilors won seventy-nine seats in seven of nine departments, although mostly through the traditional parties (Van Cott, 2003: 763).

As Gray Molina (2003: 351) argues, the LPP fundamentally "restructured the rules of the game for political intermediation and policy making in rural areas." The 1995 municipal elections marked an important turning point and signaled the eventual demise of a peculiar form of Andean apartheid that had kept indigenous people from meaningful participation in electoral politics and local planning. Increasingly, *campesino* unions and other indigenous organizations worked with small, progressive political parties and in some cases began to form their own parties. The steady growth in local electoral participation, especially among rural voters, increasingly complemented union politics with party politics (Gray Molina, 2003).

[. . .]

CONCLUSIONS

The processes of change in Bolivia illustrate a particular synergy between decentralization, political restructuring and social movements, reflecting the importance of the social mobilization strand of planning theory in understanding planning processes in countries with strong histories of contentious politics. The LPP, although fundamentally a reform measure incapable of changing the basic material conditions of the majority of Bolivians, did serve as a catalyst to mobilize a marginalized population to adopt a re-energized repertoire of political actions. However, even as the government prescribed spaces of action, it also introduced a new rhetoric of citizenship and participation in planning, as well as creating new expectations of the state.

Political decentralization made municipal planning and electoral politics a proving ground for a growing grassroots democratic opposition to traditional urban political parties. Bolivian opposition movements, centered on the coca producers, consolidated a hold on a small territory before forming broader alliances. Building on initial successes in 1995 and 1999, the opposition movements became increasingly self-assured and effectively combined an indigenous, nationalist and anti-neoliberal discourse to propel the MAS to electoral victory.

Over the short term, decentralization channelled the attention of political groups to the local arena, a strategy similarly used by elites in both Mexico and Kenya seeking to shore up their own legitimacy while simultaneously attempting to prevent greater democratic participation at a national level (Tulchin and Selee, 2004). In Bolivia, decentralization enabled groups that could not successfully compete on a national or departmental scale to occupy new political spaces in the local arena, creating hundreds of laboratories that enabled a largely male leadership to develop. The coca producers in the Chapare gained control of their municipalities and planning early on, but because of their powerful identification with a long history of *campesino* and miner struggles, they always sought national support—and supported broader calls for social justice—as well.

Heller (2002) argues that such historical and political circumstances can determine when popular movements will be able to take advantage of political decentralization. In Bolivia, well-organized and combative social movements, built on a trajectory of resistance that reaches back to the Spanish Conquest, were key elements in enabling the poor to assume a greater role in both local planning and politics. The social movements' ability to nurture astute leaders was also of critical importance. The spiraling economic and political crisis that confronted Bolivia from 1999 on created important political opportunities that these leaders recognized and exploited. All these factors were fundamental in facilitating Bolivians' appropriation of political decentralization for their own ends, rather than those envisioned by national elites and international financial institutions.

REFERENCES

Albó, X. (1996) "Making the leap from local mobilization to national politics," *NACLA Report on the Americas* 29: 15–20.

Albó, X., Libermann, K., Pifarro, F. and Gondinez, A. (1990) *Para comprender las culturas rurales en Bolivia*, La Paz: UNICEF.

Albro, R. (2005) "The water is ours, carajo!: Deep citizenship in Bolivia's water war," in J. Nash (ed.), *Social Movements: An Anthropological Reader*, Oxford and Cambridge: Basil Blackwell: 249–71.

Assies, W. and Salman, T. (2003) "Bolivian democracy: Consolidating or disintegrating?" *Focaal—European Journal of Anthropology* (Netherlands) 42: 141–60.

Beard, V. A. (2003) "Learning radical planning: The power of collective action," *Planning Theory and Practice* 2: 13–35.

Bush, G. (2002) "National security strategy of the United States, September 2002," available online at www.whitehouse.gov/ncs/nss.pdf (accessed July 16, 2006).

CIPCA (1991) *Por una Bolivia diferente*, La Paz: CIPCA.

Domingo, P. (2001) "Party politics, intermediation and representation," in J. Crabtree and L. Whitehead (eds), *Towards Democratic Viability: The Bolivian Experience*, Basingstoke; New York: Palgrave: 141–59.

Ejdesgaard Jeppesen, A. M. (2002) "Reading the Bolivian landscape of exclusion and inclusion: The law of popular participation," in N. Webster and L. E. Pedersen (eds), *In the Name of the Poor*, London: Zed: 30–52.

Gill, S. (2002) "Constitutionalizing inequality and the clash of globalizations," *The International Studies Review* 4 (2): 47–65.

Gray Molina, G. (2003) "The offspring of 1952: Poverty, exclusion and the promise of popular participation," in M. Grindle and P. Domingo (eds), *Proclaiming Revolution: Bolivia in Comparative Perspective*, London: Institute for Latin American Studies: 345–63.

Grindle, M. S. (2000) *Audacious Reforms: Institutional Invention and Democracy in Latin America*, Baltimore, MD: Johns Hopkins University Press.

Healey, S. (2005) "Rural social movements and the prospects for sustainable rural communities: Evidence from Bolivia," *Canadian Journal of Development Studies* 26: 151–74.

Heller, P. (2002) "Moving the state: The politics of democratic decentralization in Kerala, South Africa and Porto Alegre," *Peripherie, Germany*: 337–77.

Huerta Malbrán, M., Presacco Chávez, C. F., Ahumanda Beltrán, C., Velasco Jaramillo, M., Puente Alcaraz, J. and Molina Meza, J. F. (2000) *Decentralización, municipio y participación ciudadana: Chile Colombia y Guatemala*, Bogotá: Central Editorial Javeriano.

Hutchcroft, P. D. (2001) "Centralization and decentralization in administration and politics: Assessing territorial dimensions of authority and power," *Governance* 14: 23–53.

Hylton, F. and Thomson, S. (2005) "The chequered rainbow," *New Left Review* 35: 40–66.

Klein, H. S. (1998) *The American Finances of the Spanish Empire: Royal Income and Expenditures in Colonial Mexico, Peru, and Bolivia, 1680–1809*, Albuquerque, NM: University of New Mexico Press.

Kohl, B. (2002) "Stabilizing neoliberalism in Bolivia: Privatization and participation in Bolivia," *Political Geography* 21, 449–72.

Kohl, B. (2003a) "Restructuring citizenship in Bolivia: *El Plan de Todos*," *International Journal of Urban and Regional Research* 27: 337–51.

Kohl, B. (2003b) "Democratizing decentralization in Bolivia: The law of popular participation," *Journal of Planning Education and Research* 23: 153–64.

Kohl, B. and Farthing, L. (2006) *Impasse in Bolivia: Neoliberal Hegemony and Popular Resistance*, London: Zed.

MDH-SNPP (Ministerio de Desarrollo Humano & Secretaría de Participación Popular) (1996) *La Participación Popular en Cifras: Resulatados y proyecciones para analizar un proceso de cambio*, La Paz: MDH-SNPP.

Medeiros, C. (2001) "Civilizing the popular: The law of popular participation and the design of a new civil society in 1990s Bolivia," *Critique of Anthropology* 21: 401–25.

Medina, J. (1997) *Poderes locales: implementando la Bolivia del próximo milenio, protocolos de gestión de un Subsecretario*, La Paz: FIA/Semilla/CEBIAE.

Molina Monasterios, F. (1997) *Historia de la Participación Popular*, La Paz: MDH-SNPP.

Morales, W. Q. (1992) *Bolivia: Land of Struggle*, Boulder, CO: Westview Press.

Olivera, O. (2004) *Cochabamba! Water War in Bolivia* (translated by T. Lewis), Boston, MA: South End Press.

Oxhorn, P., Tulchin, J. and Selee, A. (2004) *Decentralization, Democratic Governance and Civil Society in Comparative Perspective*, Washington, DC: Woodrow Wilson Center.

Oyugi, W. O. (2000) "Decentralization for good governance and development: The unending debate," *Regional Development Dialogue* 21: iii–xix.

Peck, J. (2001) "Neoliberalizing states: Thin policies/hard outcomes," *Progress in Human Geography* 25: 445–55.

Peet, R. (2003) *Unholy Trinity: The IMF, World Bank and WTO*, New York: Zed Books.

Robinson, W. I. (1996) *Promoting Polyarchy: Globalization, US Intervention, and Hegemony*, Cambridge: Cambridge University Press.

Robinson, W. I. (2003) *Transnational Conflicts: Central America, Social Change, and Globalization*, London: Verso.

Sachs, J. (1987) "The Bolivian hyperinflation and stabilization," *American Economic Review* 77: 279–83.

Sachs, J. and Warner, A. (1999) "The big push, natural resource booms and growth," *Journal of Development Economics* 59: 43–76.

Samoff, J. (1979) "The bureaucracy and the bourgeoisie: Decentralization and class structure in Tanzania," *Society for the Comparative Study of Society and History* 21: 30–62.

Sandercock, L. (1998) *Making the Invisible Visible: A Multicultural Planning History*, Berkeley, CA: University of California Press.

Schönwälder, G. (1997) "New democratic spaces at the grassroots? Popular participation in Latin American local governments," *Development and Change* 28: 753–70.

Slater, D. (1989) "Territorial power and the peripheral state: The issue of decentralization," *Development and Change* 20: 501–31.

Tarrow, S. G. (1998) *Power in Movement: Social Movements and Contentious Politics*, Cambridge and New York: Cambridge University Press.

Tulchin, J. and Selee, A. (2004) "Decentralization and democratic governance," in P. Oxhorn, J. Tulchin and A. Selee. (eds), *Decentralization, Democratic Governance and Civil Society in Comparative Perspective*, Washington, DC: Woodrow Wilson Center: 295–319.

Van Cott, D. L. (2003) "From exclusion to inclusion: Bolivia's 2002 elections," *Journal of Latin American Studies*, 35(4): 751–775.

Wanyande, P. (2004) "Decentralization and local governance: A conceptual and theoretical discourse," *Regional Development Dialogue* 25: 1–13.

World Bank (1997) *The State in a Changing World: World Development Report 1997*, New York: Oxford University Press.

World Bank (2000) *Decentralization: Rethinking Government World Development Report 1999/2000*, New York: Oxford University Press.

Wunsch, J. (2001) "Decentralization, local governance and 'recentralization' in Africa," *Public Administration and Development* 21: 277–88.

"Deep Democracy: Urban Governmentality and the Horizon of Politics"

Environment and Urbanization (2001)

Arjun Appadurai

[. . .]

THE STORY

What follows is a preliminary analysis of an urban activist movement with global links. The setting is the city of Mumbai, in the state of Maharashtra, in western India. The movement consists of three partners and its history as an alliance goes back to 1987. The three partners have different histories. The Society for the Promotion of Area Resource Centres, or SPARC, is an NGO formed by social work professionals in 1984 to work with problems of urban poverty in Mumbai. NSDF, the National Slum Dwellers Federation, is a powerful grassroots organization established in 1974 and is a CBO, or community-based organization, that also has its historical base in Mumbai. Finally, Mahila Milan is an organization of poor women, set up in 1986, with its base in Mumbai and a network throughout India, focused on women's issues in relation to urban poverty and concerned especially with local and self-organized savings schemes among the very poor. All three organizations, which refer to themselves collectively as the Alliance, are united in their concern with gaining secure tenure of land, adequate and durable housing and access to elements of urban infrastructure, notably to electricity, transport, sanitation and allied services. The Alliance also has strong links to Mumbai's pavement dwellers and to its street children, whom it has organized into an organization called Sadak Chaap (Street Imprint), which has its own social and political agenda. Of the six or seven non-state organizations working directly with the urban poor in Mumbai, the Alliance has by far the largest constituency, the highest visibility in the eyes of the state and the most extensive networks in India and elsewhere in the world.

[. . .]

THEORETICAL POINTS OF ENTRY

Three theoretical propositions underlie this presentation of the story of the Alliance in Mumbai.

First, I assume, on the basis of my own previous work (Appadurai, 1996, 2000, 2001) and that of several others from a variety of disciplinary perspectives (Castells, 1996; Held, 1995; Rosenau, 1997; Giddens, 2000), that globalization is producing new geographies of governmentality. Specifically, we are witnessing new forms of globally organized power and expertise within the "skin" or "casing" of existing nation states (Sassen, 2000). One expression of these new geographies can be seen in the relationship of "cities and citizenship" (Appadurai and Holston, 1999), in which wealthier "world cities" increasingly operate like city states in a networked global economy, increasingly independent of regional and national mediation, and where poorer cities—and the poorer populations within them—seek new ways of claiming space and voice. Many large cities, such as Mumbai, display the contradictions between these ideal types and combine high concentrations of wealth (tied to the growth of producer services) and even higher concentrations of poverty and disenfranchisement. Movements among the urban poor, such as the one I document here, mobilize and mediate these contradictions. They represent efforts to reconstitute citizenship in cities. Such efforts take the form, in part, of what I refer to as "deep democracy."

Second, I assume that the nation state system is undergoing a profound and transformative crisis. Avoiding here the sterile terms of the debate about whether or not the nation state is ending (a debate to which I myself earlier contributed), I nevertheless wish to affirm resolutely that the changes in the system are deep, if not graspable as yet, in a simple theory. I suggest that we see the current crisis as one of redundancy rather than, for example, as one of legitimization (Habermas, 1975). By using the term "redundancy," I mean to connect several processes that others have identified with different states and regions and in different dimensions of governance. Thus, in many parts of the world, there has been undoubted growth in a "privatization" of the state in various forms, sometimes produced by the appropriation of the means of violence by non-state groups. In other cases, we can see the growing power in some national economies of multilateral agencies such as the World Bank and the International Monetary Fund, sometimes indexed by the voluntary outsourcing of state functions as part of the neoliberal strategies that have become popular worldwide since 1989. In yet other cases, activist NGOs and citizens' movements have appropriated significant parts of the means of governance.

Third, I assume that we are witnessing a notable transformation in the nature of global governance in the explosive growth of non-government organizations of all scales and varieties in the period since 1945, a growth fueled by the linked development of the United Nations system, the Bretton Woods institutional order and, especially, the global circulation and legitimization of the discourses and politics of "human rights." Together, these developments have provided a powerful impetus to democratic claims by non-state actors throughout the world. There is some reason to worry about whether the current framework of human rights is serving mainly as the legal and normative conscience or the legal/bureaucratic lubricant of a neo-liberal, marketized political order. But there is no doubt that the global spread of the discourse of human rights has provided a huge boost to local democratic formations. In addition, the combination of this global efflorescence of non-governmental politics with the multiple technological revolutions of the last 50 years has provided much energy to what has been called "cross-border activism" through "transnational advocacy networks" (Keck and Sikkink, 1998) These networks provide new horizontal modes for articulating the deep democratic politics of the locality, creating hitherto unpredicted groupings: examples may be "issue based"—focused on the environment, child labor or AIDS—or "identity-based"—feminist, indigenous, gay, diasporic. The Mumbai-based movement discussed here is also a site of such cross-border activism.

Together, these three points of entry allow me to describe the Mumbai Alliance of urban activists as part of an emergent political horizon, global in its scope, that presents a post-Marxist and post-developmentalist vision of how the global and the local can become reciprocal instruments in the deepening of democracy.

SETTING: MUMBAI IN THE 1990s

[. . .]

Mumbai is the largest city in a country, India, whose population has just crossed the one billion mark (one-sixth of the world's population). The city's population is at least 12 million (more, if we include the growing edges of the city and the population of the twin city, New Mumbai, that has been built across Thane Creek). This means a population totaling 1.2 percent of one-sixth of the world's population. Not a minor case, in itself.

Here follow some facts about housing in Mumbai on which there is a general consensus. About 40 percent of the population (about 6 million persons) live in slums or other degraded forms of housing. (The term "slum" is a formally defined settlement category in India and its use here follows that designation.) Another 5–10 percent are pavement dwellers. Yet, according to one recent estimate, slum dwellers occupy only 8 percent of the city's land, which totals about 43,000 hectares. The rest of the city's land is either industrial, middle- and high-income housing or vacant land in the control of the city, the state (regional and federal) or private owners.

The bottom line is 5–6 million poor people living in sub-standard conditions in 8 percent of the land area of a city smaller than the two New York City boroughs of Manhattan and Queens. This huge and constricted population of insecurely or poorly housed people has negligible access to essential services, such as running water, electricity and ration cards for food staples.

Equally important, this population—which we may call "citizens without a city"—are a vital part of the urban workforce. Some of them occupy the respectable low end of white-collar organizations and others the menial low end of industrial and commercial concerns. But many are engaged in temporary, physically dangerous and socially degrading forms of work. This latter group, which may well comprise 1–2 million people in Mumbai, are best described, in the striking phrase of Sandeep Pendse (1995), as Mumbai's "toilers" rather than as its proletariat, working-class or laboring classes—all designations that suggest more stable forms of employment and organization. These toilers, the poorest of the poor in the city of Mumbai, work in menial occupations, almost always on a daily or piecework basis. They are cart pullers, rag pickers, scullions, sex workers, car cleaners, mechanics' assistants, petty vendors, small-time criminals and temporary workers in petty industrial jobs requiring dangerous physical work, such as ditch digging, metal-hammering, truck-loading and the like. They often sleep in (or on) their places of work, insofar as their work is not wholly transient in character. While men form the core of this labor pool, women and children work wherever possible, frequently in ways that exploit their sexual vulnerability. To take just one example, Mumbai's gigantic restaurant and food service economy is almost completely dependent on a vast army of child labor.

Housing is at the heart of the lives of this army of toilers. Their everyday life is dominated by ever present forms of risk. Their temporary shacks may be demolished. Their slumlords may push them out through force or extortion. The torrential monsoons may destroy their fragile shelters and their few personal possessions. Their lack of sanitary facilities increases their need for doctors to whom they have poor access. And their inability to document their claims to housing may snowball into a general invisibility in urban life, making it impossible for them to claim any rights to such things as rationed foods, municipal health and education facilities, police protection and voting rights. In a city where ration cards, electricity bills and rent receipts guarantee other rights to the benefits of citizenship, the inability to secure claims to proper housing and other political handicaps reinforce one another. Housing—and its lack—are the stage for the most public drama of disenfranchisement in Mumbai. In fact, it can be argued that housing is the single most critical site of this city's politics of citizenship.

This is the context within which the activists I am working with are making their interventions, mobilizing the poor and generating new forms of politics. The next [. . .] sections of this essay are about various dimensions of this politics: of its vision, its vocabularies and its practices.

THE POLITICS OF PATIENCE

In this section, I give a sketch of the evolving vision of the Alliance of SPARC, Mahila Milan and the National Slum Dwellers Federation as it functions within the complex politics of space and housing in Mumbai. Here, a number of broad features of the Alliance are important.

First, given the diverse social origins of the three groups that are involved in the Alliance, their politics awards a central place to negotiation and consensus-building. SPARC is led by professionals with an Anglophone background, connections to state and corporate élites in Mumbai and beyond, and strong ties to global funding sources and networking opportunities. However, SPARC was born in 1984 in the specific context of work undertaken by its founders—principally a group of women trained in social work at the Tata Institute for the Social Sciences—among poor women in the neighborhood of Nagpada. This area has a diverse ethnic population and is located between the wealthiest parts of south Mumbai and the increasingly difficult slum areas of central and north Mumbai. Notable among SPARC's constituencies was a group of predominantly Muslim ex-sex trade workers from central Mumbai, who later became the cadre of another partner in the Alliance, Mahila Milan. The link between the two organizations dates from Mahila Milan's founding around 1986, which received support from SPARC.

The link with the NSDF, an older and broader-based slum dwellers' organization, was also made in the late 1980s. The leadership of the three organizations cuts across the lines between Hindus, Muslims and Christians and is explicitly secularist in outlook. In a general way, SPARC contributed technical knowledge and élite connections to state authorities and the private sector. NSDF, through its leader, Arputham Jockin (who himself has a background in the slums), and his activist colleagues, brought a radical brand of grassroots political organization in the form of the "federation" model to be discussed later in this essay. Mahila Milan brought the strength of poor women who had learned the hard way how to deal with the police, municipal authorities, slumlords and real-estate developers on the streets of central Mumbai but had not previously had any real incentive to organize politically.

These three partners still have distinct styles, strategies and functional characteristics. But they are committed to a partnership based on a shared ideology of risk, trust, negotiation and learning among their key participants. They have also agreed upon a radical approach to the politicization of the urban poor that is fundamentally populist and anti-expert in strategy and flavor. The Alliance has evolved a style of pro-poor activism that consciously departs from earlier models of social work, welfare and community organization (an approach akin to that pioneered by Saul Alinsky in the United States). Instead of relying on the model of an outside organizer who teaches local communities how to hold the state to its normative obligations to the poor, the Alliance is committed to methods of organization, mobilization, teaching and learning that build on what the poor themselves know and understand. The first principle of this approach is that no one knows more about how to survive poverty than the poor themselves.

A crucial and controversial feature of this approach is its vision of politics without parties. The strategy of the Alliance is that it will not deliver the poor as a vote bank to any political party or candidate. This is a tricky business in Mumbai, where most grassroots organizations, notably unions, have a long history of direct affiliation with major political parties. Moreover, in Mumbai, the Shiva Sena, with its violent, street-level control of urban politics, does not easily tolerate neutrality. The Alliance deals with these difficulties by working with whoever is in power at the federal and state level, within the municipality of Mumbai or even at the local level of particular wards (municipal sub-units). Thus, the Alliance has earned hostility from other activist groups in Mumbai for its willingness to work with the Shiva Sena, where this is deemed necessary. But it is resolute about making the Shiva Sena work for its ends and not vice-versa. Indeed, because it has consistently maintained an image of non-affiliation with all political parties, the Alliance enjoys the double advantage of appearing non-political while retaining access to the potential political power of the poorer half of Mumbai's population.

Instead of finding safety in affiliation with any single party or coalition in the state government of Maharashtra or the municipal corporation of Mumbai, the Alliance has developed a complex political affiliation with the various levels of state bureaucracy. This group includes civil servants who conduct policy at the highest levels in the state of Maharashtra and who run the major bodies responsible for housing loans, slum rehabilitation, real estate regulation and the like. The members of the Alliance have also developed links with quasi-autonomous arms of the federal government, such as the railways, the Port Authority and the Bombay Electric Supply and Transport Corporation, and to the municipal authorities who control critical elements of the infrastructure, such as the regulations governing illegal structures, the water supply and sanitation. Finally, the Alliance works to maintain a cordial relationship with the Mumbai police— and, at least, a hands-off relationship with the underworld, which is deeply involved in housing finance, slum landlordism and extortion, as well as in the demolition and rebuilding of temporary structures.

From this perspective, the politics of the Alliance is a politics of accommodation, negotiation and long-term pressure rather than of confrontation or threats of political reprisal. This realpolitik makes good sense in a city like Mumbai, where the supply of scarce urban infrastructure—housing and all its associated entitlements—is entangled in an immensely complicated web of slum rehabilitation projects, financing procedures, legislative precedents and administrative codes which are interpreted differently, enforced unevenly and whose actual delivery is almost always attended by an element of corruption.

This pragmatic approach is grounded in a complex political vision about means, ends and styles that is not entirely utilitarian or functional. It is based on a series of ideas about the transformation of the conditions of poverty by the poor in the long run. In this sense, the figure of a political horizon is meant to point to a logic of patience, of cumulative victories and of long-term asset-building that is wired into every aspect of the activities of the Alliance. The Alliance maintains that the mobilization of the knowledge of the poor into methods driven by the poor and for the poor is a slow and risk-laden process; this premise informs the group's strong bias against "projects" and "projectization" that underlies almost all official ideas about urban change. Whether it be the World Bank, most Northern donors, the Indian state or other agencies, most institutional sources of funding are strongly biased in favor of the "project" model, in which short-term logics of investment, accounting, reporting and assessment are regarded as vital. The Alliance has steadfastly advocated the importance of slow learning and cumulative change against the temporal logics of the project. Likewise, other strategies and tactics are also geared to long-term capacity-building, the gradual gaining of knowledge and trust, the sifting of more from less reliable partners and so on. This open and long-term temporal horizon is a difficult commitment to retain in the face of the urgency, and even desperation, that characterizes the needs of Mumbai's urban poor. But it is a crucial normative guarantee against the ever-present risk, in all forms of grassroots activism, that the needs of funders will gradually obliterate the needs of the poor themselves.

Patience as a long-term political strategy is especially hard to maintain in view of two major forces. One is the constant barrage of real threats to life and space that frequently assails the urban poor. The most recent such episode is a massive demolition of shacks near the railroad tracks, which has produced an intense struggle for survival and political mobilization in virtually impossible circumstances in the period since April 2000, a crisis still unresolved at the time of writing. In this sense, the strategies of the Alliance, which favor long-term asset-building, run up against the same "tyranny of emergency," in the words of Jérôme Bindé (2000), that characterizes the everyday lives of the urban poor.

The other force that makes it hard to maintain patience is the built-in tension within the Alliance surrounding different modes and methods of partnership. Not all members of the Alliance view the state, the market or the donor world in the same way. Thus, every new occasion for funding, every new demand for a report, every new celebration of a possible partnership, every meeting with a railway official or an urban bureaucrat can create new sources of debate and anxiety within the Alliance. In the words of one key Alliance leader, negotiating these differences, which are rooted in deep diversities in class, experience and personal style, is like "riding a tiger." It would be a mistake to view the pragmatic way in which all partnerships are approached by the Alliance as a simple politics of utility. It is a politics of patience, constructed against the tyranny of emergency.

[. . .]

CONCLUSION: DEEP DEMOCRACY

One of the many paradoxes of democracy is that it is organized to function within the boundaries of the nation state—through such organs as legislatures, judiciaries and elected governments—to realize one or another image of the common good or general will. Yet its values make sense only when they are conceived and deployed universally, which is to say, global in reach. Thus, the institutions of democracy and its cardinal values rest on an antinomy. In the era of globalization, this contradiction rises to the surface as the porousness of national boundaries becomes apparent and the monopoly of national governments over global governance becomes increasingly embattled.

Efforts to enact or revive democratic principles have generally taken two forms in the period since 1970, which many agree is the beginning of globalization (or of the current era of globalization, for those who wish to write it into the whole of human history). One form is to take advantage of the speed of communications and the sweep of global markets to force national governments to recognize universal democratic principles within their own jurisdictions. Much of the politics of human rights takes this form. The second form, more fluid and quixotic, is the sort that I have described here. It constitutes an effort to institute what we may call "democracy without borders" after the analogy of international class solidarity as conceived by the visionaries of world socialism in its heyday. This effort is what I seek to theorize in terms of deep democracy.

In terms of its semantics, "deep democracy" suggests roots, anchors, intimacy, proximity and locality; and these are important associations. Much of this essay has been taken up with values and strategies that have just this quality. They are about such traditional democratic desiderata as inclusion, participation, transparency and accountability, as articulated within an activist formation. But I want to suggest that the lateral reach of such movements—their efforts to build international networks or coalitions of some durability with their counterparts across national boundaries—is also a part of their "depth."

This lateral or horizontal dimension, which I have touched upon in terms of the activities of Shack/Slum Dwellers International, seeks direct collaborations and exchanges between poor communities themselves, based on the "will to federate." But what gives this cross-national politics its depth is not just its circulatory logic of spreading ideas of savings, housing, citizenship and participation, "without borders" and outside the direct reach of state or market régimes. Depth is also to be located in the fact that, where successful, the spread of this model produces poor communities able to engage in partnerships with more powerful agencies—urban, regional, national and multilateral—that purport to be concerned with poverty and with citizenship. In this second sense, what these horizontal movements produce is a series of stronger community-based partners for institutional agencies charged with realizing inclusive democracy and poverty reduction. This, in turn, increases the capability of these communities to perform more powerfully as instruments of deep democracy in the local context. The cycles of transactions—both vertical (local/national) and horizontal (transnational/global)—are enriched by the process of criticism by members of one federated community, in the context of exchange and learning, of the internal democracy of another. Thus, internal criticism and debate, horizontal exchange and learning, and vertical collaborations and partnerships with more powerful persons and organizations together form a mutually sustaining cycle of processes. This is where depth and laterality become joint circuits along which pro-poor strategies can flow.

This form of deep democracy, the vertical fulcrum of a democracy without borders, cannot be assumed to be automatic, easy or immune to setbacks. Like all serious exercises in democratic practice, it is not automatically reproductive. It has particular conditions of possibility and conditions under which it grows weak or corrupt. The study of these conditions—which include such contingencies as leadership, morale, flexibility and material enablement—requires many more case studies of specific movements and organizations. For those concerned with poverty and citizenship, we can begin by recalling that one crucial condition of possibility for deep democracy is the ability to meet emergency with patience.

REFERENCES

Appadurai, A. (1996) *Modernity at Large: Cultural Dimensions of Globalization*, Minneapolis: University of Minnesota Press.

Appadurai, A. (2000) "Grassroots globalization and the research imagination," *Public Culture* 12: 1–19.

Appadurai, A. (2001) "Spectral housing and urban cleansing: notes on millennial Mumbai," *Public Culture* 12: 627–651.

Appadurai, A. and Holston, J. (1999) "Introduction: cities and citizenship" in J. Holston (ed.), *Cities and Citizenship*, Durham, NC: Duke University Press.

Bindé, J. (2000) "Toward an ethics of the future," *Public Culture* 12: 51–72.

Castells, M. (1996) *The Rise of the Network Society*, Cambridge, MA: Blackwell.

Giddens, A. (2000) *Runaway World: How Globalization is Reshaping Our Lives*, New York: Routledge.

Habermas, J. (1975) *Legitimation Crisis*, translated by Thomas McCarthy, Boston: Beacon.

Held, D. (1995) *Democracy and the Global Order: From the Modern State to Cosmopolitan Governance*, Stanford, CA: Stanford University Press.

Keck, M. and Sikkink, K. (1998) *Activists Beyond Borders: Advocacy Networks in International Politics*, Ithaca, NY: Cornell University Press.

Pendse, S. (1995) "Toil, sweat and the city," in S. Patel and A. Thorner (eds.), *Bombay: Metaphor for Modern India*, Bombay: Oxford University Press.

Rosenau, J. N. (1997) *Along the Domestic–Foreign Frontier: Exploring Governance in a Turbulent World*, Cambridge: Cambridge University Press.

Sassen, S. (2000) "Spatialities and temporalities of the global: elements for a theorization," *Public Culture* 1: 215–232.

"Sovereignty: Crisis, Humanitarianism, and the Condition of 21st Century Sovereignty"

***Framing the Global* (2014)**

Michael Mascarenhas

By most accounts crisis has become the paradigm of modern government. Drought, fires, hurricanes, tsunamis, not to mention fiscal cliffs, financial meltdowns, and regional uprisings dominate both our attention and institutions of government. Crisis continues to be the modus operandi for humanitarian efforts worldwide as most of the world's population still lacks access to basic human rights, such as water, food, and shelter. Increasingly, non-governmental organizations (NGOs) are leading efforts to improve access to water and other human rights. As part of this global humanitarian effort, NGOs have witnessed an unprecedented growth in recent decades. In some places they have replaced government agencies, become strategic business investments, and translated local water and sanitation needs into bureaucratic planning. The relentless expansion of NGOs on the basis of emergency has resulted in a contingent form of sovereignty, where increasingly NGOs are defining which populations to champion or "let live," and by default, which populations to "let die." However, in spite of this profound decision making authority, few scholars, with notable exceptions, have paid attention to the ways in which the policies and practices of non-governmental organizations continue to reassemble the character of modern sovereignty for the majority of the world's population. In this chapter I argue that the policies and practices of NGOs represent a new form of contingent sovereignty among the world's deprived, a form of non-state sovereignty that is transnational in character, emergent in form, and flexible in practice. This new form of sovereignty, while not replacing state-based sovereignty, clearly signals the presence of a growing humanitarian apparatus that exercises sovereign power over transnational territories and populations.

[. . .]

The state of urgency is particularly important in understanding the role of NGOs in development theory and practice. "Emergency," Mark Duffield (2007) argues, "has provided a means of penetrating the world of peoples, ignoring existing laws, conventions or restraints." Drawing on Georgio Agamben's (1998; 2005) work on the state of exception, Duffield (2007) argues that this ability to describe and define a permanent state of exception or emergency amongst the world's poor has allowed NGOs to expand as a non-state or petty sovereign power. However, as Michael Hardt and Antonio Negri (2001: 39) remind us, "we are dealing here with a special kind of sovereignty—a discontinuous form of sovereignty," one that challenges the Westphalian notion that nation states are the only sovereigns and that sovereignty is territorially based. Today, Nancy Fraser (2005; 2007) and other critical thinkers argue, sovereignty is being disaggregated, broken up into several distinct functions and assigned to several distinct agencies, which function at several distinct levels, some global, some regional, some local and subnational. These new transnational assemblages, most often associated with military and security functions and global economic development, are increasingly becoming the dominant institutional structure for humanitarian interventions worldwide.

But as Partha Chatterjee (2004) has carefully illustrated, this form of sovereignty does not reign over citizens, per se, but rather populations of people that are described and organized with particular conditions, such as water insecurity, ill-health, or poverty. The sovereignty of NGOs over these populations lies in their endless decision making authority concerning *how* particular humanitarian conditions are defined, *who* is to be helped, and *how* to go about helping them and, consequently, who can be left behind. These now global humanitarian assemblages, Saskia Sassen (2006) argues, are contributing to different meanings of territory, authority, and rights in the 21st century. Yet in spite of what amounts to sovereign decision-making authority to give life and take it away, we know very little about how water needs are determined and communities selected, how money is raised and spent, how technology is used, how expertise is established, how water programs are supposed to work and why they usually fail. When people set out to change the water crisis conditions in most of the world, the devil is truly in the details.

The central question addressed in this chapter is how do NGOs establish and maintain legitimate sovereign authority over the state of exception, and what sort of sovereignty are they practicing? By insisting that the social and physical life of water development consists of a long labyrinth of associations, I posit that we can open up the possibility of multiple opportunities for politics directed towards social justice ends. One particularly informative sovereign feature of contemporary humanitarian aid has been its ability to function like state agencies. To illustrate this point, we first recall the tragic events of the South Asian Tsunami.

THE TRANSNATIONAL HUMANITARIAN AGENCY

On December 26, 2004 people in the northern hemisphere watched as a massive earthquake triggered a series of devastating tsunamis that inundated the coasts of 14 countries along the rim of the Indian Ocean, killing nearly 230,000 people, injuring tens of thousands more, and displacing more that 10 million men, women, and children. In the days that followed, the South Asian Tsunami became a truly global affair. Bombarded with media reporting and seduced by YouTube videos we watched live as millions of helpless people lost their homes, livelihoods, and in some cases their lives. These horrific images combined with the seemingly arbitrariness of their fate provoked an outpouring of empathy and generosity of global proportions. Governments, humanitarian organizations, community groups, and individuals around the world scrambled to offer aid and technical support. And within six months, official aid and private donations raised over $13 billion for the victims of this natural disaster.

However, in the weeks and months that followed another storyline started to emerge, one that questioned the role of NGOs in this and other humanitarian efforts. Reports started to emerge from the media about donations not reaching victims or donations being diverted by NGOs to other campaigns. We were introduced to the uneasy fact that these civil society organizations were not only involved in matters of state sovereignty by effectively making decisions over who "to make live and to let die" (Foucault, 1997: 241) but were doing so by advancing a particular notion of costs and benefits. In fact, it turns out, the practice of diverting donations to other projects is a usual practice of NGOs. Even more troubling for many donors was the fact that only about half of what is pledged ever ends up reaching the poorest people affected by these disasters. Together these civil society organizations had created a new kind of sovereignty in which they tried to control the population of the world's destitute without having to answer to anyone in particular.

Yet, this form of sovereignty is not without its own organizing logic that grows and stabilizes along particular lines of power relations. One particular feature of contemporary humanitarian aid has been its connection to transnational finance. David Harvey (2010) has suggested that the whole history of capitalism as it has expanded over the globe has been about financial innovation. "Finance capital in rich industrial countries has grown at a rate that is 2.5 times that of the national product. The rate in currencies, bonds, and equities has increased at 5 times the rate of increase in the national product. More and more capital is being invested in the trade of stocks, bonds, and currencies than in manufacturing, presumably because the profits are quicker and greater" (Chatterjee, 2004: 88). It stands to reason, then, that finance has also profoundly transformed twenty-first century humanitarian practice and policy. For example as many progressive and well-intentioned NGOs continue to seek innovative ways to finance their humanitarian efforts, they are also becoming increasingly dependent on the boom and bust risks associated with finance capital. The concern with venture philanthropy, in practice, is that humanitarian efforts become more aligned with investment than opportunities for social change. Moreover, in shifting their humanitarian values to align with donor driven targets and programs, NGOs risk losing substantial funding if specific donor targets are not met or if donor programs become unfashionable. One impact of this relationship between transnational financial capital and humanitarianism is that concerns over efficiency and accountability have become one of "the hottest topics for discussion by NGO practitioners," as they too demand the same performance and accountability criteria of their beneficiaries, as their donors' demand of them (Hilhorst 2003: 125).

FINANCING SOVEREIGNTY AND SOVEREIGN FINANCE

In an effort to secure legitimacy in this highly speculative form of sovereign power many NGOs have taken to enrol financers into various practices of their humanitarian assemblage. I was first introduced to this importance of finance in humanitarian efforts during an interview with a field manager employed by the non-profit, Water for People. Working in 10 countries, and with revenues over 14 million dollars, Water for People is a big player in the water aid world.

[...]

Offices of Water for People are located in nine cities across the globe. Balancing out the centres of "bureaucratic" activity has kept a distributed power-base at the global level as well as a sense of agility to respond to local emergencies, such as the South Asian Tsunami in 2004. This global local presence ensures that Water for People is well positioned to serve (control) local water crises and to transport their local expertise and experiences to other places.

However, the tension of having to be in multiple places at once has also been a concern for many NGOs who struggle to secure their presence amid a permanent state of humanitarian emergency. One strategy to weather the crisis driven demands of humanitarianism has been to reach out to the private sector. For example, corporate support to Water for People has been instrumental in helping them achieve financial independence. Currently, they have eight "global sponsors" who have donated between $100,000 and $999,999 each (Table 6). All except one of the global sponsors (Green Mountain Coffee Roasters, Inc.) are involved in some facet of the water services sector. Water for People is also supported by five "continental sponsors" who have donated between $50,000 and $99,999, over 50 "country sponsors" who have donated between $10,000 and $49,999, and numerous other smaller donors. Total revenue for 2011 was $14,328,110 (Water for People, 2011).

In addition to their financial aid, many corporate donors have chosen to support the efforts of NGOs in other ways. For example, of the eight "global sponsors," three have executives who are also members of Water for People's Board of Directors. Similarly, executives from two of the five "continental sponsors" are also members of Water for People's Board. Many on Water for People's Board of Directors are also board members, CEOs, and presidents of many other large companies with a hand in water technology, facilities management, and engineering and consulting companies. Some corporate donors also supply the volunteer labour force required by NGOs to carry out their programmatic work. For example, in a baseline assessment in Rwanda, in which I was a participant, the project manager was an engineer from CH2M HILL. He was intensely involved in the planning, data gathering, and reporting phases of the project. The three water quality technicians for the baseline assessment were also employees of Nalco, one of the five "continental sponsors," and their two week participation in the project was paid for by their employer.

In addition to corporate donations, foundations have also helped Water for People secure financial independence. In 2010 Water for People received a $5.6 million grant from the Bill & Melinda Gates Foundation to support their Sanitation as a Business program. The goal of this program is to introduce profit incentives into the improvement of sanitation in households and schools in "developing" countries. According to Water for People, this program will transform "unsustainable, subsidy-based sanitation programs" into sustainable, profitable sanitation services" by "merging business principles of market research

Sponsor	Type of business	Additional ties to WfP
AECOM Technology Corporation	A public company providing technical and management support services to a broad range of markets, including transportation, facilities, environmental, energy, water and government	
American Water	The largest public water and wastewater utility company in the U.S.	
American Water Works Association	A non-profit organization that provides water quality and supply information, and other water industry resources to its membership	Founded Water for People in 1991
CDM Smith	A private consulting, engineering, construction, and operations firm in water, environment, transportation, energy, and facilities	Senior Vice President—member of WfP Board of Directors
CH2M HILL	Engineering consulting and service provider	Managing Director—member of WfP Board of Directors; Employees "volunteer" in the regional operations
Green Mountain Coffee Roasters, Inc.	Producer of coffee and coffee brewing devices	
Nalco Company	A public company involved in water treatment and process improvement services	Chief Marketing Officer and VP of Sales, Americas —member of WfP Board of Directors; Employees "volunteer" in the regional operations
Xylem	A public company involved in water and waste water treatment and process improvement services	

Table 6 Water for People global corporate sponsors (donated between $100,000 and $999,999)

Source: Author.

and segmentation" with community involvement and program monitoring (Water for People, 2010). This grant, according to an employee, has put Water for People "on the map." "That just lifts us to a whole new category of non-profit." "People [subsistence farmers in "developing" countries] understand composting, they just don't know [or] understand the business side of it. So, what Water for People is trying to do is branch out of sanitation and make it lucrative for business." Financial security for some NGOs has resulted not only in a shift in development practice but also a transformation in what it means to do "development" in the twenty-first century. An employee of another non-profit informed me, "we in water and sanitation are moving away from specific projects to long-term programs that employ new methods, and modes of analysis and evaluation." "Once we do that," she continued, "then I think there is going to be a bigger hope ... not only to meet development goals, but also [to] go further, and think about development on a larger scale, not just in the community level" (personal interview).

That transnational "thinking" about development has, in part, been facilitated by the strategic use of what has become the material epitome of the global assemblage—the World Wide Web. In effect, the Internet has also been a formidable ally for NGOs, providing a medium to communicate extensive monitoring and evaluation criteria required by their donors and trustees while at the same time providing invaluable networking potential to people from outside of the "development" world. For example, in

2010 Water for People introduced FLOW (Field Level Operations Watch) as part of its commitment to self-sustained long-term water coverage. The FLOW Internet site allows browsers to monitor WfP projects for up to ten years after their completion date. [. . .] FLOW has been described by an employee of WfP as a "game changer" in water development. The irony of a programme like FLOW is that while it uses remote data acquired through Android cell phone technology and Google Earth software to provide information to practitioners and donors, these technologies are entirely foreign to the majority of the people in need of water and sanitation. As a result, the collection tool becomes a one-sided solution, because community participants are unable to understand or engage with these technologies.

[. . .]

CONCLUSION

This new sovereign power seeks to address not only the symptoms but the causes of humanitarian crises. Its strategy involves different types of intervention, including trading of armed forces, advocating for the enactment of international law in domestic legislation, human rights reform, and the strengthening of domestic justice systems (Labbe, 2012). Towards these efforts, NGOs are also becoming integral parts of a transnational decision making apparatus. For example, over 1500 NGOs have consultative status with the United Nations, and this number continues to grow. "Many UN agencies now hold periodic consultations with NGOs on substantive policy and program strategies, and the governing body of at least one agency, UNAIDS, has five seats for NGO representatives" (Opoku-Mensah 2001: 2). Moreover, in 1997 an NGO Working Group was established as part of the UN Security Council. This 30-member consultation group meets regularly with government delegations to collaborate over issues that pertain to local, regional, national and international security. In a similar fashion the transnational networks of NGOs are increasingly gaining access to other multilateral organizations like the World Bank, WTO and the World Economic Forum (Opoku-Mensah 2001).

The relentless expansion of NGOs on the basis of emergency and facilitated by new forms of humanitarian intervention—financial, transnational, and digital, to name only three important features—has resulted in a new and contingent form of sovereignty, where increasingly NGOs are defining and deciding on the state of exception for the world's poor. This new form of sovereignty, I argue, is more global assemblage (Ong and Collier, 2005) than conventional nation-state—a product of multiple determinations that are not reducible to a single logic or space—like providing humanitarian water aid or even challenging global neoliberal reforms. Increasingly, NGOs are embedded in a transnational, if not exactly global, space. These flows of knowledge, expertise, and finance move within the liminal spaces and apparati of control that are constitutive of "the 'non-place' of Empire" (Galloway, 2001: 82; Hardt and Negri, 2001).

REFERENCES

Agamben, G. (1998) *Homo Sacer: Soverign Power and Bare Life,* Stanford, CA: Stanford University Press.

Agamben, G. (2005) *State of Exception,* Chicago: The University of Chicago Press.

Chatterjee, P. (2004) *The Politics of the Governed: Reflections on Popular Politics in Most of the World,* New York: Columbia University Press.

Duffield, M. (2007) *Development, Security and Unending War: Governing the World of Peoples,* Cambridge, UK: Polity Press.

Foucault, M. (1997) *Michel Foucault: "Society Must be Defended," Lectures at the College de France 1975–1976,* Edited by M. Bertani and A. Fontana, Translated by D. Macey, New York: Picador.

Fraser, N. (2005) "Transnationalizing the Public Sphere." Online. Available HTTP: http://www.republicart.net.

Fraser, N. (2007) "Transnationalizing the Public Sphere: On the Legitimacy and Efficacy of Public Opinion in a Post-Westphalian World," *Theory, Culture & Society* 24: 7–30.

Galloway, A. (2001) "Protocol, or, How Control Exists after Decentralization," *Rethinking Marxism* 13: 81–88.

Hardt, M. and Negri, A. (2001) *Empire,* Cambridge: Harvard University Press.

Harvey, D. (2010) *The Enigma of Capital: And the Crises of Capitalism.* Oxford: Oxford University Press.

Hilhorst, D. (2003) *The Real World of NGOs: Discourses, Diversity, and Development,* New York: Zed Books Ltd.

Labbe, J. (2012) "Rethinking Humanitarianism: Adapting to 21st Century Challenges," International Peace Institute, New York.

Ong, A. and Collier, S. (2005) *Global Assemblages: Technology, Politics, and Ethics as Anthropological Problems*, Oxford, UK: Blackwell Publishing.

Opoku-Mensah, P. (2001) "The Rise and Rise of NGOs: Implications for Research." Online Available HTTP: http://www.svt.ntnu.no/iss/issa/0101/010109.shtml.

Sassen, S. (2006) *Territory Authority Rights: From Medieval to Global Assemblages* (Updated Edition), Princeton, NJ: Princeton University Press.

Water for People (2010) "Water for People Receives Funding for Sanitation as a Business Program," Denver, CO: Water for People. Online. Available HTTP: http://www.waterforpeople.org/media-center/press-release/sanitation-as-a-business-funding.html.

Water for People (2011) Financials. Online. Available HTTP: http://www.waterforpeople.org/about/financials/.

SECTION 10
Participation

Editors' Introduction

Today almost no one talks about a planning or development project without intending to incorporate community participation. People find many different ways of expressing this idea: community buy-in, getting the community to "own" the project, ensuring the involvement of "all stakeholders" in the development process. All of these mean roughly the same thing: that in order for a community to develop, the members of that community should take part in the process. Yet community participation is not that simple. This is a concept with a distinct history, a concept that actually means different things to different people. This section introduction will explore some of the complexities and the history of this idea of community participation.

KEY ISSUES

Participation and its rationales

In the past four decades, support for community participation in urban development processes and projects (also referred to as human settlement development) has risen among a vast range of practitioners and scholars. During the 1960s and 1970s participatory development was practiced predominantly by activists associated with the political left who were located outside mainstream development organizations. While the UN endorsed participation in the 1970s, it was not until the 1990s that we saw governments, and multi-/bi-lateral development organizations implementing community participation in their projects (United Nations 1971, 1975). At the close of the twentieth century even large corporations and organizations on the right of the political spectrum joined the call for community participation. This shift requires a more critical examination to acquire a politicized understanding of the historical evolution of the idea of community participation and the considerable purchase it has found in development discourse (Cornwall 2003).

The advocacy for participation in the 1960s and 1970s was by and large motivated by an empowerment rationale. A key thinker in the early days of community participation was Paolo Freire, a Brazilian educator and activist whose famous book, *Pedagogy of the Oppressed* (1970), was a text that was embraced by activists first in Latin America and later in other parts of the world. Freire emphasized that the best way to help the poor was through "conscientization," helping them understand the root causes of their poverty. His writings stressed redemption by participating in one's own salvation and critiqued a charity approach that cast the poor as passive recipients of handouts–be it charity or so-called development projects. At the community level this meant increasing the voice of community members in the decisions that affected their lives. Freire also emphasized the need for people to organize in democratic structures. His ideas were a response to top-down development planning processes for communities which relied solely on the thinking of outside experts to devise projects without consulting those who would be directly impacted.

By the late 1980s global political trends brought about a major shift away from a Freirian approach to community participation. The social movements of the 1960s and 1970s had declined. The free market and neoliberalism were on the rise. Mainstream development practitioners, scholars, and institutions began to re-shape ideas about community participation, arguing that increased participation was in line with cost effectiveness. These free market approaches contributed to the extensive withdrawal of the state from many human settlements and urban development projects across the world by the 1990s. Furthermore, agencies such as the World Bank and the USAID (US Agency for International Development) picked up on the cost effectiveness argument and began not only to promote but require community participation in development projects they funded. These mainstream thinkers also enhanced their argument, focusing not only on the advantages of participation as a way to incorporate communities' unpaid labor, but also as a means of ensuring that community members buy into the project, that they "take ownership" and secure project implementation and maintenance more efficiently and within a greater legitimacy among poor populations.

Over the last three decades, two concepts have been influential in advocacy for community participation in development projects: capabilities and social capital. The former was developed by the Harvard professor and Nobel laureate in Economics, Amartya Sen (1999), who in his most influential work, *Development as Freedom,* argues that the ability to optimize one's capabilities is the core meaning of freedom and hence of development. The latter was developed by the work of yet another Harvard professor, Robert Putnam (1993), who contends that networks of support, trust and reciprocity that exist in a community constitute a form of social capital that members of a community draw on to help each other. Putnam's writings stressed that greater social capital in a community leads to greater economic development. The perspectives informed by the importance of capabilities and social capital have provided considerable theoretical support for the capacity building potential of participatory processes.

But the literature diverges on whether and how participation achieves its promise of empowering and capacity building as opposed to tokenism. These questions have been the focus of an extended debate on the nature and purpose of community participation.

Participatory tyranny and transformation

Sherry Arnstein's (1969) classic article captures this debate by developing a ladder metaphor to describe various forms of participation from mere tokenism to empowerment. While participation at its lowest rung is a means of community manipulation, at its highest rung citizen participation can exercise citizens' control over the process and decisions. Critics have been clear to highlight how participatory processes as mainstreamed and practiced in the development of human settlements often fall short of the ideal, lapsing into a cosmetic and token role. Cornwall and Brock (2005) argue the recent popularity of words like "participation," "empowerment" or "poverty reduction" in the language of mainstream development organizations has an important political function. Beyond mere tokenism they serve as buzzwords which serve to justify old practices of development. In these scenarios community participation does not achieve the participants' expected material gain or a sense of empowerment. It rather replicates similar tyrannies of development practices that often occur through non-participatory development processes (see Cooke and Kothari 2001; Miraftab 2005); or facilitates similar unjust processes which benefit the more powerful and more vocal–often older males or higher caste members (Guijt and Shah 1998; Chambers 1997; Burkey 1996; Bartlett 2010). Youth and children, for example, are often silenced or ignored in these processes as are women or ethnic and linguistic minorities. For instance, while a large share of the communities served or affected by development projects might be younger people, they seldom play a role in participatory processes (Bartlett 2005).

Other scholars argue that we cannot assume community participation will automatically lead to a positive-sum and generative relationship between a community and external actors who take part in a participatory project. This argument also contends that we cannot simply assume the opposite: that community participation necessarily falls victim to the competing interests of external agents and their coopting power. Under certain conditions indeed even skeptics recognize participation can become transformative. (For example, see Cooke (2005) or Hickey and Mohan (2005).) Community participation can move beyond rhetoric and tokenism to become empowering and lead to capacity building processes that ultimately improve citizens' control over development projects. (For example, see Appadurai, this volume or Mitlin (2005).)

An interesting twist to this is the increasing participation of the rising lower middle and middle classes in urban governance. This is particularly true in Asian cities which have seen large numbers of middle class residents organize to take control of their neighborhoods and demand better services from local governments, often at the expense of the poor whose presence and activities go against the vision of the sanitized, beautiful city (see Baviskar, this volume; Mitra, this volume). Participatory strategies employed by these middle class groups vary. Sometimes they involve using their networks of influence and connections with the aim of excluding others. Abu Lughod (1998) uses "good and bad cholesterol" as an analogy to stress the problem with uncritical celebration of participation by organized civil society in the last two decades. Not all participation aims for inclusion. Gated communities, for instance, organize and participate in self-governance to keep others out. Similarly, as goals of pacification vary by hierarchically positioned social groups–e.g. based on class, race, gender, ethnicity, caste–so do the forms and means of citizens' participation diverge. Under authoritarian rule citizens might invent channels outside the legal and formal political structures. In older electoral democracies like India, middle classes might engage in claim-making through legal systems or use their connections to administrators, while the poor and disadvantaged populations often find alliance with political parties and local party functionaries critical to their abilities to participate and have a voice (Harriss 2007). In short, the goals and means of participation vary according to who participates, their goals and the political context.

Two considerations have been central to discussions on conditions that might influence differing outcomes of community participation and participatory processes: (i) the power dynamics and structures within which participation is conceived and practiced and (ii) the institutional design and organizational practices through which participation takes place. We will briefly discuss each in turn.

Commentators that stress power dynamics in the participatory processes argue that the origin of participatory development initiatives is crucial in determining the rules of the game and whether participants are acting on a level playing field (Miraftab 2003; Lankatilleke 1999). Whether a community is well organized and whether it interacts with other actors such as government and NGOs from a position of power makes a difference in the outcome of a participatory process. An empowering participatory process, this perspective stresses, should emerge as a result of the community's definition of its overall objectives and its expectation of the private and the public sectors' contributions to the process. Enforced from above as an institutional requirement, or initiated by an organization outside the community, participation is bureaucratized and ultimately meaningless. But to expect a participatory project as a sole outcome of pure bottom up processes, or expect communities' full self-reliance, other commentators argue, is unrealistic. In practice, these scholars argue, most participatory processes are initiated by a range of agents. For example, NGOs perhaps along with the intellectuals and leaders of civil society organizations are often in an unequal power relationship with the communities concerned (see Dasgupta and Beard 1997). This perspective stresses that the key to the outcome of participatory processes lies in institutional structures and organizational practices (Kudva 2006; Farrington and Bebbington 1993; Burkey 1996; Fung 2006). They maintain that whether community participation leads to empowering and capacity building processes depends on institutional design and organizational practice. Participatory practices "are shaped in concrete organizational sites," Kudva and Driskell argue, and therefore organizations are central to creating and structuring spaces where participation can take place (Kudva and Driskell 2009).

This brings us to the need to look more closely at the formal institutions and organizations involved in community participation.

Institutionalization of participation

Many governmental and non-governmental development organizations have treated institutionalization of participation as a decree from above–for example, an institutional requirement for development projects to include a participatory component. In some instances institutionalized participatory processes make openings that organized civil society groups and movements can then use to further open cracks within the system. A case in point is the Bolivian Law of Popular Participation discussed by Kohl and Farthing in this *Reader*, whereby making citizens' participation a legal requirement ultimately led to unprecedented political gains by indigenous populations.

However, critics take issue with this approach, contending that meaningful participatory processes cannot simply "happen" as the result of a formal or institutional mandate. As Blackburn (1998: 169–171) puts it, institutionalization needs to be understood as "a condition that is built." The formal adoption of community participation by government organizations, national policies, or international agencies does not necessarily increase its impact. Moreover, critics stress that meaningful participatory development cannot simply be achieved by scaling up a local level participatory process–that is inserting participatory practices into larger-scale organization. It also needs to be scaled out. Scaling out refers to expansion in realms of communities' influence and control, for example from project implementation and maintenance to policy design and formulation. Gaventa (1998) argues that a meaningful institutionalization of participatory processes requires both scaling up and scaling out. Ultimately, expanding participatory practices into new realms of decision making has often been overlooked and poses a greater challenge.

This discussion of the range of perspectives on the potential of community participation to promote inclusion and empowerment stresses the complexity of participatory development. The two selected texts below embody the distinct outcomes of community participation in the development process and hence reflect this complexity. Whether community participation becomes a cosmetic and exploitative feature of a development project, as Cornwall (2003) discusses in one selected reading included below, or becomes an empowering process as Baiocchi (2006) discusses in the other reading, all depends on the power relations within which community participation is established and how it is institutionalized.

ARTICLE SELECTIONS

Our two selected texts focus on participatory processes and outcomes at the two ends of what one could call the spectrum of transformative citizen participation.

"The Citizens of Porto Alegre" is authored by Brazilian Gianpaolo Baiocchi, a prominent, US-based political sociologist who discusses a much celebrated participatory process that was institutionalized in 1989 in Porto Alegre, Brazil, one of the strong holds of the Workers' Party (Partido dos Trabalhadores–PT (Baiocchi 2005; Baiocchi *et al.* 2011). Referred to as Participatory Budgeting (PB), in this process ordinary people participate in deciding how discretionary municipal and public funds should be allocated. Despite recent critiques of PB, both within and outside Brazil, this budget allocation process has been celebrated by many as an experiment in institutionalizing citizens' participation that can bring about not only empowerment but also fair distribution of resources. Over the last two decades PB has been adopted by more than 1200 municipalities and wards around the world including other parts of Latin America, Africa, Europe, Canada, and the United States (for example Chicago's Ward 49). The widely attended global conferences on participatory budgeting held in New York in 2012 and in Chicago in 2013, which included aldermen, mayors, academics and activists from around the world, serve as a testimonial to PB's global spread.

"Whose Voices? Whose Choices? Reflections on Gender and Participatory Development" is authored by Andrea Cornwall, a prolific and influential British sociologist and a scholar of community participatory processes. She has published extensively on various aspects of community participation in development projects (Cornwall and Brock 2005; Cornwall 2011). Focusing on gender and development projects Cornwall shares insights on how in some instances participatory projects can merely pay lip service to empowerment (in this case gender empowerment). In the text included here she shows how women's participation in a forest management project was turned into tokenism by using women participants as cheap labor without a meaningful role in the decision-making processes of the project. While in this discussion her point of reference is a forest management project, we draw on the important gendered analysis of participatory tokenism that she offers in this paper. The insights and arguments she offers can aptly apply toward urban projects as well.

REFERENCES

Abu Lughod, J. (1998) "Civil/uncivil society: Confusing form and content," in M. Douglass and J. Friedmann (Eds.), *Cities for Citizens,* Sussex: Wiley and Sons, pp. 227–238.

Arnstein, S. (1969) "A ladder of citizens' participation," *Journal of the American Institute of Planners,* 35(4): 216–224.

Baiocchi, G. (2005) *Militants and Citizens: Local Democracy on a Global Stage in Porto Alegre,* Stanford, CA: Stanford University Press.

Baiocchi, G. (2006) "The citizens of Porto Alegre," *Boston Review*, 1 March.

Baiocchi, G., Heller, P., and Marcelo K. S. (2011) *Bootstrapping Democracy*, Stanford, CA: Stanford University Press.

Bartlett, S. (2005) "Good governance: Making age a part of the equation–an introduction," *Children, Youth and Environments*, 15(2): 1–17.

Bartlett, S. (2010) "Editorial: Responding to urban youth's own perspectives," *Environment and Urbanization,* 22(2): 307–316.

Blackburn, J. (1998) "Conclusion," in J. Blackburn and J. Holland (Eds.), *Who Changes? Institutionalizing Participation in Development*, London: Intermediate Technology, pp. 167–178.

Burkey, S. (1996) *People First: A Guide to Self-reliant, Participatory Rural Development,* London: Zed Books Ltd.

Chambers, R. (1997) *Whose Reality Counts? Putting the First Last,* London: Intermediate Technology Publications.

Cooke, B. (2005) "Rules of thumb for participatory change agents," in S. Hickey and G. Mohan (Eds.), *Participation–From Tyranny to Transformation? Exploring New Approaches to Participation in Development,* London: Zed Books, pp. 42–56.

Cooke, B. and Kothari, U. (2001) "The case for participation as tyranny," in B. Cooke and U. Kothari (Eds.), *Participation: The New Tyranny?* New York: Zed Books, pp. 1–15.

Cornwall, A. (2003) "Whose voices? Whose choices? Reflections on gender and participatory development," *World Development,* 31(8): 1325–1342.

Cornwall, A. (Ed.) (2011) *The Participation Reader*, London: Zed Books.

Cornwall, A. and Brock, K. (2005) "What do buzzwords do for development policy? A critical look at 'participation', 'poverty reduction' and 'empowerment'," *Third World Quarterly,* 26(7): 1043–1060.

Dasgupta, A. and Beard, V. (1997) "Community driven development, collective action and elite capture in Indonesia," *Development and Change,* 38(2): 229–249.

Farrington, J. and Bebbington, A. (1993) *Reluctant Partners? Non-Governmental Organizations, the State and Sustainable Agricultural Development,* New York: Routledge.

Freire, P. (1970) *Pedagogy of the Oppressed,* New York: The Seabury Press.

Fung, A. (2006) *Empowered Participation: Reinventing Urban Democracy*, Princeton, NJ: Princeton University Press.

Gaventa, J. (1998) "The scaling up and institutionalization of PRA: Lessons and challenges," in J. Blackburn and J. Holland (Eds.), *Who Changes? Institutionalizing Participation in Development*, London: Intermediate Technology, pp. 153–166.

Guijt, I. and Shah, M. K. (1998) "Waking up to power, conflict and process," in I. Guijt and M. K. Shah (Eds.), *The Myth of Community: Gender Issues in Participatory Development,* London: Intermediate Technology Publications, pp. 1–23.

Harriss, J. (2007) "Antinomies of empowerment: Observations on civil society, politics and urban governance in India," *Economic and Political Weekly,* 42(26): 2716–2724.

Hickey, S. and Mohan, G. (2005) "Relocating participation within a radical politics of development: Insights from political action and practice," in S. Hickey and G. Mohan (Eds.), *Participation–From Tyranny to Transformation? Exploring New Approaches to Participation in Development,* London: Zed Books, pp. 159–174.

Kudva, N. (2006) "Shaping democracy through organizational practice: The NGOs of the Tribal Joint Action Committee in Karnataka, India," *International Journal of Rural Management,* 2(2): 227–243.

Kudva, N. and Driskell, D. (2009) "Creating space for participation: The role of organizational practice in structuring youth participation," *Community Development,* 40: 367–380.

Lankatilleke, L. (1999) "Resource as prerequisite: Who's participating in whose process?" in *Community and Resource Management*, Nairobi, Kenya: Community Development Programme, United Nations Center for Human Settlements.

Miraftab, F. (2003) "The perils of participatory discourse: Housing policy in post-apartheid South Africa," *Journal of Planning Education and Research,* 22(3): 226–239.

Miraftab, F. (2005) "Making neoliberal governance: The disempowering work of empowerment," *International Planning Studies,* 9(4): 239–259.

Mitlin, D. (2005) "Securing voice and transforming practice in local government: The role of federating in grassroots development," in S. Hickey and G. Mohan (Eds.), *Participation–From Tyranny to Transformation? Exploring New Approaches to Participation in Development,* London: Zed Books, pp. 175–189.

Putnam, R. D. with Leonardi, R. and Nanetti, R. Y. (1993) *Making Democracy Work: Civil Traditions in Modern Italy,* Princeton, NJ: Princeton University Press.

Sen, A. (1999) *Development as Freedom,* New York: Knopf.

United Nations (1971) *Popular Participation in Development: Emerging Trends in Community Development,* New York: United Nations.

United Nations (1975) *Popular Participation in Decision Making for Development,* New York: United Nations.

"The Citizens of Porto Alegre"

Boston Review (2006)

Gianpaolo Baiocchi

Marco is a self-employed handyman in his mid-30s who moved to the city of Porto Alegre from the Brazilian countryside eight years ago. A primary-school-educated son of a farmer, he'd had few opportunities in his small town and had heard about the city's generous social services. He borrowed money for bus fare and landed in Porto Alegre, where he found construction work. But when his wages wouldn't cover rent he headed for one of the squatter settlements on the outskirts of the city. He soon moved in with a *companheira* who sewed clothes and ironed from home. In time his life became more settled, with incremental improvements to the house, small but growing savings, and brisk business owing to his good reputation in the community. Marco's story of migration, squatting, and survival was unremarkable—until he attended a local meeting on how the city government should invest its money in the region.

It is not surprising that Brazilians are, by and large, uninvolved in civic life. Their cities are among the most violent, economically unequal, and problem-ridden in the world. While the elite live in fortified enclaves, one fourth of Brazilian city dwellers live in makeshift slum housing, often without access to any social services and dependent on patrons for survival.

These settlements, which make up as much as a third of some cities' populations, share the mistrust and social disintegration that the political scientist Robert Putnam and his colleagues have documented in areas of "low social capital" in southern Italy; in recent surveys Brazilians have registered some of the world's lowest levels of trust in their democratic institutions.

People like Marco are the most excluded from normal avenues of government decision-making and also the least likely to become involved in formal associations.

So Marco, brought to the meeting by one of his neighbors, was understandably skeptical about what it might accomplish. He was told that a nearby squatter settlement had been able to collectively purchase its land title through similar meetings; but he was sure at first that someone had used a connection to a powerful politician. Yet the meeting, crowded and held at a school gym, appeared genuine. The mayor spoke about the budget, and a dozen of the 2,000 participants got in line for the microphone to question officials about previous projects. Later, the whole group elected delegates for the rest of the year. Though he did not understand most of the technical details at that meeting, Marco became a delegate and started to participate week after week, learning the rules of the process known as participatory budgeting from the parade of city officials who attended. At the end of that first year, he and his fellow delegates elected to invest in paving the streets and adding sewers to the district.

Over the years Marco became increasingly involved, bringing many new faces to meetings, helping to start a neighborhood association, and realizing his dream of legalizing the land title to his settlement. Today he and his neighbors are part of a cooperative that collectively owns the titles to the land. And Marco, who had never before participated in a social movement or association, spends hours in meetings every week and can often be found explaining technical details or the exact role of a certain government agency to newcomers.

Participatory budgeting, popularized in Brazil by the Workers' Party, or PT (Partido dos Trabalhadores), is today practiced in some 200 cities there and dozens of others in Europe, Latin America, and Africa. It has deeply transformed the nature of civic life in Porto Alegre, where one of the first experiments in participatory budgeting was introduced 16 years ago.

By turning over government decisions on investment and spending to local assemblies open to all, participatory budgeting has enabled several thousand Brazilian citizens to make real decisions, to demand accountability, and to monitor the results. The process has become a sort of school of democracy for people like Marco, and an entry into civic life.

* * *

The corruption and cronyism of Brazil's government is legendary. A recent *New York Times* article calls allegations of graft in the national leadership of the Workers' Party the "latest reminder of the unremitting corruption" that has marked the region since colonial times. But among progressives, Brazil is also known for political innovations such as the landless movement and the World Social Forum, a global summit for social-justice activists, as well as participatory budgeting.

In spite of the crisis in the Workers' Party national leadership, experiments in direct democracy by local progressive administrations—including the governments of the state capitals of Belo Horizonte, Porto Alegre, and Belim—continue to inspire onlookers from afar.

These cities work. Some have achieved levels of social-service provision almost unheard of in Brazil, including near-universal clean water and sewers and very high rates of preschool enrollment.

The design of the participatory institutions is clever, enticing the least advantaged to participate by combining an educational component with opportunities to win vital improvements for the community. Unlike the delegated decision-making of representative democracy, direct participation requires active intervention that in effect trains people for citizenship through problem-solving, communication, and strategizing. But the achievement with the greatest lessons for progressives elsewhere is the transformation of the relationship between the government and the governed.

Critics of direct democracy say that it is messy, inefficient, and prone to domination by an articulate few. Defenders note the deficiencies inherent in representative systems and point to instances when direct democracy produces good decisions, such as privileging the use of municipal resources to meet the needs of the poorest. In Brazilian cities, this marks a dramatic break with the patronage-driven politics that has long dominated municipal finance. Having thousands of ordinary citizens voice opinions and observe the process increases transparency, taps into local sources of information, and improves the accountability of elected officials. And by allowing citizens to directly influence the allocation of resources in their communities, participatory budgeting energizes citizen engagement and strengthens civil society.

* * *

Participatory budgeting was haltingly introduced in Porto Alegre in 1990 by an inexperienced and besieged Workers' Party administration, elected just one year earlier and in search of legitimacy. The idea goes back to the 1970s and the social movements that would eventually usher in democracy in the mid-1980s. Radical popular educators and progressive clergy in these movements emphasized the importance of autonomy and participatory democratic procedures; throughout the country citizens formed neighborhood associations and social movements to demand a voice in such local affairs as transportation, health, and housing. The Workers' Party itself was founded in the early 1980s as a party through which movements could speak.

In Porto Alegre, activists from neighborhood associations started demanding direct input into the city budget in 1985. Through a process of trial and error, participatory budgeting evolved into a year-long cycle of meetings that allow participants to decide on projects in their own neighborhoods as well as for the city as a whole. Citizens took over many functions usually reserved for bureaucrats: setting city-wide spending priorities, planning investments, and reviewing payrolls, not to mention setting the rules for the participatory budgeting process itself and monitoring its outcomes. Because since the 1990s Brazilian cities have assumed responsibility for most social-service provision and infrastructure investments, citizens are able to exert significant control over transportation, education, public health, and public works.

The process begins in March of each year with district-level assemblies in each of the city's 16 districts, followed by meetings of delegates elected from each assembly who deliberate about the district's

FIVE

needs and specific projects. By the end of the year, projects and priorities are passed on to the Municipal Council of the Budget, made up of representatives from each district, who then reconcile demands with available resources and propose and approve a municipal budget in conjunction with members of the administration. The municipal legislature then votes on the budget, which is usually approved without modification. As the projects are implemented, street committees monitor their progress. Near the year's end, participants re-draw the rules of the process for the following year based on their experience.

A significant portion of the annual municipal budget (between nine and 21 percent of the total) is decided in this way, funding hundreds of projects with a completion rate of nearly 100 percent. These projects have achieved almost full water and sewer coverage, a threefold increase in the number of children in municipal schools, and significant increases in the number of new housing units provided to needy families.

Porto Alegre's expenditures in certain areas, such as health and housing, are much higher than the national average, yet the municipalities' administrative costs and overhead have declined over the years. And Porto Alegre has managed a redistributive regime that is fiscally responsible and that has remained transparent.

International institutions such as the World Bank have again and again praised it. Preliminary results of a national analysis conducted in the late 1990s show that participatory budgeting tends to lower poverty rates and improve education.

The rate of participation in participatory budgeting in Porto Alegre is also impressive. Once the process started to show results—three or four years after its introduction—the number of participants grew dramatically. By 2004, some 20,000 were attending the first round of meetings, many of them for the first time. A conservative estimate is that ten percent of adults in the city have at one point participated.

Participants, by and large, are like Marco: they lack secondary education, work in manual or service jobs, and earn well below the city average. Women and Afro-Brazilians participate at high rates, as do residents of poor areas. Marco's district consistently turns out very high numbers of participants, despite the fact that they are the least experienced in civic life. In fact, the activation of civil society in neighborhoods such as Marco's has reversed a historical trend and should be considered one of the most significant consequences of participatory budgeting.

Scholars have long agreed that organized collectives are best able to meet collective needs, but they have also noted that those most in need are often the least able to organize. In the case of urban Brazil, government has long frustrated efforts to found and maintain associations organized for the collective good; associations seeking a particular improvement would often come into open conflict with local governments and afterward have a difficult time sustaining themselves. It is for this reason that in Porto Alegre and elsewhere in Brazil, poor people's neighborhood organizations have often been little more than electoral corrals for charismatic politicians who selectively made improvements in exchange for voter loyalty.

In the late 1980s Marco's district of 20,000 had only two active associations, both connected to powerful political figures. Today it has almost 20, all working through the participatory budget. In Porto Alegre as a whole, the number has doubled since 1989, with the biggest increase in the city's poorest areas. And while in the past the main activities of neighborhood associations and civic groups were protests and petitions, today these associations are most likely to organize around deliberative and pragmatic problem-solving rather than protesting or mobilizing for a powerful politician. The strong educational component of the process means that local associations and civic life constantly draw new participants.

The way that decisions are made in participatory budget meetings marks a real break from past models of civic engagement. Participants spend a fair amount of time in deliberative discussions. Though most decisions are made through votes, significant deliberation, in meetings and at the edges of official forums, paves the way for them. This complex process is spread out over a year, and participants regularly resolve conflicts over priorities. A district like Marco's could choose to divide available funds into many small projects, such as paving 100 meters of dirt road in each of the 20 settlements, or spend them all on a major collective priority, such as a thoroughfare or a school. Arriving at a decision sometimes involves tense moments and much negotiation. Active participants such as Marco play key roles in creating solutions and finding ways to balance the needs of neighborhoods and the whole district.

Beyond providing a forum to choose projects and priorities, participatory budget meetings enable other forms of collective action and discussion. Government-sponsored meetings on the technicalities of street-paving projects would seem at first unlikely places for discussions of, say, the setting of poor urban peripheries. Yet in the meetings, participants regularly carve out these spaces for open-ended discussion. People come together in a regular meeting place and address all kinds of needs, fashioning a language of public responsibility and rights that evolves from their work together. Participants bring newspaper clippings to meetings to discuss current affairs; they recruit volunteers for outside projects; they organize protests, some aimed at the administration itself. The participatory budget has been so successful at drawing participants and delivering results that administrators have accepted these other discussions as healthy democratic discourse. For many participants, like Marco, budgeting meetings have become a central part of community life, a place "where the whole community is present" and where "you can discuss a wide range of issues, not just the budget."

It is in these ways that participatory budgeting affects, and ultimately shapes, civil society and civic practice. Much of democratic theory, and much of the policy discussion in the United States today, assumes that democratic influence travels from civil society toward the state. A well-organized and virtuous civil society oversees state institutions and prevents them from falling into corruption. To foster democracy, in this view, is to strengthen the ways that citizens help themselves. But the participatory-budgeting story shows us how reforming the state—by radically increasing its openness to the public mandate—can shape the way civil society functions. A state that is closed to all but the demands of the politically sympathetic creates incentives for cronyism and militancy. A state that responds to direct participation creates incentives for civic organization. In the case of participatory budgeting in Porto Alegre, state reforms have created incentives for the participation of the poor by focusing on basic infrastructure.

Here the state has created not only incentives but the conditions that enable the poor to participate. There is a pedagogical component built into participatory budgeting. By attending meetings and making demands, new participants learn all the skills required for collective action, such as how to run a meeting and how to negotiate compromises. They learn the intricacies of governmental affairs, previously the privileged domain of policy experts. Perhaps most importantly, in a country where the "masses" have traditionally had no voice but to consent to leaders, participatory budgeting asserts the value of their voice.

After the Workers' Party introduced and established the participatory budget in Porto Alegre, the process was exported to dozens of other municipalities throughout the country. By 1992, a dozen Workers' Party municipalities had participatory budgeting; by 1996 the number had increased to 36. And as the Workers' Party continued to increase its electoral strength and regional presence, participatory budgeting spread throughout the country. Over 100 municipalities experimented with it between 1997 and 2000, and at least 200 did so between 2001 and 2004, at least half of which were run by other political parties.

Given its accomplishments, it came as some surprise that the Workers' Party, after 16 years of uninterrupted rule, lost the municipal election in October of 2004 to a competing left-of-center party.

Opposition politicians cheered that Porto Alegre "does not belong to one party." The reasons for the defeat were not straightforward. The opposition candidate ran a well-planned campaign that capitalized on anti-incumbent sentiments, calling on Porto Alegrenses to vote for "democratic alternation" (the tradition of parties alternating in power) and an end to "one-party rule." There were also lingering negative sentiments from the one term of state-level Workers' Party rule, not to mention the dissatisfaction with President Lula's national administration. The opposition ran a campaign of "keeping what is good, improving the rest" that proved particularly effective with middle-class voters who were ideologically opposed to the Workers' Party but who recognized its effective style of governance. Campaign materials promised "change in a safe way, the way we want." The mayoral candidate, José Fogaca, knows that some changes are necessary, but without destroying "the good things the city has achieved in the last few years, such as the participatory budget, and the World Social Forum." Unable to claim that a vote for the opposition was a vote against participatory budgeting, the Workers' Party lost one of its trump cards in its bid for a fifth municipal term. In the first day of campaigning, the opposition candidate met with councilors of the participatory budget, visited their neighborhoods, and discussed ways to preserve and improve the process.

While the vote was polarized, with poorer neighborhoods voting for the Workers' Party, the vote among poor and working-class neighborhoods was not nearly as solid as in previous elections. While Marco voted for the Workers' Party, he knew people who voted for the opposition, confident that the new mayor would devolve more funds to the participatory budget. He himself was skeptical but willing to participate and see before judging.

At the time of this writing, participatory budgeting in Porto Alegre is continuing much as before, with some former staff members even working for the new administration. The previous year's rules, ratified by participants at the end of 2004, remain in force. The number of participants is lower in some districts, but the overall number of participants is comparable to previous years. What happens next to the participatory budget depends in large part on whether the new administration tries to alter the principle of participant-set rules or use its facilitators to manipulate the proceedings.

The Workers' Party leaves behind thousands of participants in dozens of new neighborhood organizations who are connected to their communities and intimately familiar with the ins and outs of government and budgetary affairs. This organized civil society has an immense capacity for monitoring and influencing the government. It would be practically impossible to run a process of participatory budgeting that did not conform to high standards or that tried to manipulate participants.

Experience suggests that it would be difficult to entirely dismantle a successful participatory scheme even after the defeat of its sponsoring party. Parallel cases, as in Recife in the late 1980s, have ended in the eventual re-adoption of the scheme—if not the re-election of the party that promoted it—in response to the organized pressure of former participants. Had participatory budgeting been a different kind of institution, one without the democratic openness that makes it so vibrant, another party might never have been able to claim to be for the participatory budget and against the Workers' Party.

In the end, whatever the future of participatory budgeting in Porto Alegre, its model of effective governance and vibrant civic life offers a striking contrast to newspaper headlines about corruption and civic disaffection. But if these two images seem at odds, it is only because of how we look at democracy.

It is not simply that Brazilian democracy is in shambles and Porto Alegre is an island of civic-mindedness in a sea of apathy. Looking at citizen attitudes and behaviors in isolation from the state institutions and the politics behind them reveals little about democratic possibilities. The story of participatory budgeting is useful beyond the template it offers for other cities in Brazil and throughout the world. It also raises questions about the way we discuss declining civic engagement in the United States. Perhaps we should be asking not only about television-watching habits and other social behaviors but about the nature of our national and local governments. We should be asking how transparent, responsive, and accessible they are to the citizen.

"Whose Voices? Whose Choices? Reflections on Gender and Participatory Development"

***World Development* (2003)**

Andrea Cornwall

[. . .]

Participation has become development orthodoxy. Holding out the promise of inclusion, of creating spaces for the less vocal and powerful to exercise their voices and begin to gain more choices, participatory approaches would appear to offer a lot to those struggling to bring about more equitable development. With the shift in the participation discourse beyond beneficiary participation to wider questions of citizenship, rights and governance (Gaventa, 2002), addressing challenges of equity and inclusion gain even greater importance. Yet claims to "full participation" and "the participation of all stakeholders"—familiar from innumerable project documents and descriptions of participatory processes—all too often boil down to situations in which only the voices and versions of the vocal few are raised and heard. Women, many critics argue, are those most likely to lose out, finding themselves and their interests marginalized or overlooked in apparently "participatory" processes (Guijt and Kaul Shah, 1998; Mayoux, 1995; Mosse, 1995).

[. . .]

Problematizing the way in which "gender" is used is essential for addressing the transformatory goals of participatory development. The practical equivalence between "gender" and "women's issues," and the narrow focus of "gender relations" on particular kinds of male–female relations, obscure the analytic importance of gender as a constitutive element of all social relationships and as signifying a relationship of power (Scott, 1989; Wieringa, 1998). Points of tension between participatory and "gender-aware" approaches to development arise from—and produce—rather different ways of engaging with issues of gendered power. In this article, I explore dimensions of "participation" and "gender" in development, highlighting paradoxes of "gender-aware" and participatory development interventions.

[. . .]

"GENDER-AWARE" PARTICIPATORY DEVELOPMENT: TENSIONS AND OPPORTUNITIES

The ways in which the terms "gender" and "participation" are used in practice cannot be taken for granted; "gender-aware" participatory practice may take different shades depending on where the practitioner or implementing agency situate themselves. Tensions, commonalities and complementarities between approaches to gender and participation complicate any analysis of the gender dimensions of participatory development. Yet it is these very differences and similarities that provoke food for thought and provide entry points for the emergence of new hybrids, new alliances and new tactics for transforming existing practices.

One of the most significant lines of tension runs across—rather than between—approaches to gender and participation. Some participatory approaches, such as Participatory Action Research (PAR) (see Fals-Borda and Rahman, 1991), emphasize the structural dimensions of power, echoing the focus of some versions of GAD [Gender and Development Projects]. These approaches seek to question "naturalized" assumptions, whether discursive or ideological. With the goal of confronting and transforming inequalities, they introduce particular ideas about power and difference, either to create new spaces or transform existing ones. Applying structural models may serve to essentialize gender identities and relations. This can equally produce institutions that "misbehave" (Harrison, 1997), giving voice to elite women who may have little interest in their "sisters" and deepening the gendered exclusion of others—notably, younger, poorer men (Cornwall and White, 2000). They can thus serve to reproduce existing relations of inequality between "women" or "men" (cf. Moore, 1994; Peters, 1995) and strengthen compacts between particular kinds of women and their menfolk (Harrison, 1997), rather than build the basis for more equitable gender relations.

Other schools of thought, such as PRA (Participatory Rural Appraisal), emphasize the importance of tuning into and building on people's own experiences, concepts and categories. Rather than importing concepts from elsewhere, they focus on enabling local people to articulate and analyse their own situations, in their own terms, and focus more on individual agency than on structural analysis. This opens up the potential for a more nuanced and less essentialist approach to issues of power and difference. By seeking to ground analysis and planning in local discourses and institutions, however, PRA-based participatory practices appear to offer the facilitator little scope for challenging aspects of the status quo that feminist practitioners would find objectionable. Local people are presumed to know best, even if they advocate the chastisement of younger women who step out of line or indeed the repression of women considered to be "loose" (Overs, Doezema, and Shivdas, 2002). With their emphasis on consensus, the institutions created as part of participatory development initiatives—whether committees, user groups, community action planning groups and so on—can exacerbate existing forms of exclusion, silencing dissidence and masking dissent (Mosse, 1995; Mouffe, 1992). The voices of the more marginal may barely be raised, let alone heard, in these spaces.

[. . .]

WOMEN'S PARTICIPATION IN PARTICIPATORY DEVELOPMENT PROJECTS

The question of who participates and who benefits raises awkward questions for participatory development. The very projects that appear so transformative can turn out to be supportive of a status quo that is highly inequitable for women. While seeking to avoid the pervasive slippage between "women" and "gender" in development, it is important to emphasize that the marginalization or exclusion of women from participatory projects remains an issue (Mayoux, 1995). Women's involvement is often limited to implementation, where essentialisms about women's caring roles and naive assumptions about "the community" come into play (Guijt and Kaul Shah, 1998; Lind, 1997). The means by which women are excluded, equally, may echo and reinforce hegemonic gender norms, as well as replicate patterns of gendered exclusion that have wider resonance.

[. . .]

Joint Forest Management (JFM) is in many ways a classic example of participatory development. It involves creating or adapting existing community-based institutions in order to devolve (some) opportunities for local people to participate in sectoral governance, as "partners," "stakeholders" and "owners" (Leach, Mearns, and Scoones, 1997; Poffenberger and McGean, 1996). At a time when JFM was being lauded for its prowess with participation, feminist researchers revealed quite a different story (Agarwal, 1997; Sarin, 1998). Their analyses highlighted the shortcomings of JFM, as "gender exclusionary and highly inequitable" (Agarwal, 1997: 1374). Not only were women losing out of benefits from JFM and suffering higher workloads as a result of the difficulties in collecting fuel wood. "To be labeled 'offenders' and forest destroyers into the bargain," Sarin charges, "is making a parody of participatory forest management" (1998: 128).

Women's opportunities to influence decision making in forest protection committees rest not simply on getting women onto these committees, but on how and whether women represent women's interests, whether they raise their voices and, when they do, whether anyone listens. Mohanty (2002) suggests that in Uttaranchal [Ed. Note: A northern Indian state], although there is an emphasis on a certain percentage of women being on the committee, much depends on the good will of its head, who is usually a man, and the forest bureaucrat, also usually a man. "In the lack of any institutional mechanism to ensure this participation, it remains piecemeal, a gesture of benevolence on the part of male members in the committee and the forest bureaucracy" (Mohanty, 2002: 1). Voice, she reminds us, does not automatically translate into influence. Sarin and Agarwal both document the consequences of lack of voice and of influence. Their studies show that, unable to exert influence over the rules for forest protection, women were effectively denied the usufruct rights that they formerly had.

> These rules were formulated by men without either involving the women in framing them or proposing any viable alternatives for how the women could carry out their gendered responsibility of meeting household firewood requirements following forest closure. (Sarin, 1998: 127)

By allocating places on committees to households and assuming equitable intra-household distribution of benefits, JFM institutions largely tended to reproduce existing structures and dynamics of gendered power and exclusion. As such, they served to exemplify "the problem of treating 'communities' as ungendered units and 'community participation' as an unambiguous step toward enhanced equality" (Agarwal, 1997: 1374; Sarin and SARTHI, 1996). While those women who did participate in these new spaces gained new opportunities for leadership and for learning (Mohanty, 2002), those women who were effectively excluded from decision making exercised their agency elsewhere, resisting, rebelling and breaking the rules (Sarin, 1998).

Two sets of issues arise here. First, the very real barriers to women's participation in decision making are worth highlighting. Agarwal (1997) draws attention to familiar constraints: time; official male bias; social constraints about women's capabilities and roles; the absence of a "critical mass" of women; and lack of public speaking experience. She cites a female member of a forest membership group: "I went to three or four meetings . . . No one ever listened to my suggestions. They were uninterested" (Britt, 1993, cited in Agarwal, 1997: 1375). As Mohanty (2002) points out, women end up taking on the burden of implementation instead, patrolling the forests at night and getting even less rest. What solutions does Agarwal suggest? Practical adjustments to meeting times and membership rules would, she argues, be easily enough addressed with gender-aware planning, although this alone would not enable women to exercise decision making.

[. . .]

On issues that do affect women-in-general, such as access to fuel wood, it is important that women qua women are given space to articulate their concerns. Gender-progressive institutions can enable women to challenge their exclusion. Yet here a second issue arises: the extent to which the participation of particular women should be taken as representative of (both in the sense of speaking about and speaking for) women-in-general. Caution may be needed in moving beyond particular concerns that are clearly shared, to identifying female representation with enhancing the position of all women (Phillips, 1991). The essentialisms that lurk behind well-intentioned efforts to increase women's participation as women are dangerous as well as wrong-headed: these can deepen exclusion while providing reassurance that gender inequality has been addressed. Moreover, as Mohanty contends, "the mere presence of women in the decision making committees without a voice can be counter-productive in the sense that it can be used to legitimize a decision which is taken by the male members" (Mohanty, 2002: 1).

Increasing the numbers of women involved may serve instrumental goals, but will not necessarily address more fundamental issues of power. There is no reason to suppose that women, by virtue of their sex, are any more open to sharing power and control than men. Those who represent "women's concerns" may reinforce the exclusionary effects of other dimensions of difference (Mohanty, 1987; Moore, 1994; Moraga and Anzaldua, 1981). Installing women on committees may be necessary to open up space for women's voice, but is not sufficient: it may simply serve as a legitimating device, and may even shore up and perpetuate inequitable "gender relations" between women. Female participants may not identify themselves primarily, or even at all, with other women; their concerns may lie more with their sons and their kin. To assume female solidarity masks women's agency in the pursuit of their own projects that may be based on other lines of connectedness and difference.

[. . .]

Just as the nominal inclusion of women appears to satisfy "gender" goals, so too the use of participatory methods in planning processes may be tokenistic rather than transformative. Participation in planning ranges from more sustained and deliberative processes of engagement to once-off performances: all the way down Arnstein's (1969) ladder of participation from tokenism to delegated control.

[...]

As ... [several examples including projects on Participatory Poverty Alleviation (PPA) and Poverty Assessment show], getting an awareness of gender into the process of generating knowledge for policy is more complex than getting people to use the right tools to gather information. Issues of subjectivity and positionality may have just as much influence on what emerges. Influencing policy, in any case, depends on more than simply feeding information to policy makers (Keeley and Scoones, 1999); getting data on gender issues will not ensure that these issues find their way onto the poverty-alleviation agenda. Goetz's (1994) analysis of the ways in which information about women is taken up in development bureaucracies underlines the point that what policy makers want to know tends to determine how information is used.

Whitehead and Lockwood's (1998) analysis of six World Bank Poverty Assessments (PAs), four with a PPA component, is a powerful example of the limited influence of gender-relevant insights on the shaping of policy recommendations. At every stage, gender issues slipped off the policy agenda. When it came to policy recommendations, gender barely made an appearance.

[...]

These analyses reinforce the point that what is needed is not simply good tools or good analysis, but advocacy, persistence and influence to accompany the process all the way through to the writing stage.

[...]

CONCLUSION: MAKING MORE OF DIFFERENCE

Unless efforts are made to enable marginal voices to be raised and heard, claims to inclusiveness made on behalf of participatory development will appear rather empty. Requiring the representation of women on committees or ensuring women are consulted are necessary but not sufficient. Working with difference requires skills that have been under-emphasized in much recent participatory development work: conflict resolution, assertiveness training (Guijt and Kaul Shah 1998; Mosse 1995; Welbourn 1996). The need for advocacy on gender issues is evident, at every level. Yet there is perhaps a more fundamental obstacle in the quest for equitable development. The ethic of participatory development and of GAD is ultimately about challenging and changing relations of power that objectify and subjugate people. Yet "gender" is framed in both participatory and "gender-aware" development initiatives in ways that continue to provide stumbling blocks to transforming power relations.

[...]

Rather than the "add women and stir" approach to addressing gender, what is needed are strategies and tactics that take account of the power effects of difference, combining advocacy to lever open spaces for voice with processes that enable people to recognize and use their agency. Whether by reconfiguring the rules of interactions in public spaces, enabling once silenced participants to exercise voice, or reaching out beyond the "usual suspects" to democratize decision making, such processes can help transform gender blindness and gender-blinkeredness into the basis for more productive alliances to confront and address power and powerlessness.

REFERENCES

Agarwal, B. (1997) "Re-sounding the alert—gender, resources and community action," *World Development*, 25: 1373–80.

Arnstein, S. (1969) "A ladder of citizen's participation," *Journal of the American Institute of Planners*, 35(7): 216–24.

Cornwall, A. and S. White (eds) (2000) "Men, masculinities and development: politics, policies and practice," *IDS Bulletin*, 31(2).

Fals-Borda, O. and M. A. Rahman (1991) *Action and Knowledge: Breaking the Monopoly with Participatory Action Research*, New York: Apex Press.

Gaventa, J. (2002) "Introduction: exploring citizenship, participation and accountability," *IDS Bulletin*, 33: 1–11.

Goetz, A.-M. (1994) "From feminist knowledge to data for development: the bureaucratic management of information on women and development," *IDS Bulletin*, 25: 27–36.

Guijt, I. and M. Kaul Shah (1998) "Waking up to power, conflict and process," in I. Guijt and M. Kaul Shah (eds), *The Myth of Community: Gender Issues in Participatory Development*, London: Intermediate Technology Publications.

Harrison, E. (1997) "Fish, feminists and the FAO: translating 'gender' through different institutions in the development process," in A. M. Goetz (ed.), *Getting Institutions Right for Women in Development*, New York: Zed Books.

Keeley, J. and I. Scoones (1999) "Understanding environmental policy processes: a 'review'," IDS Working Paper no. 89, Brighton: Institute of Development Studies.

Leach, M., R. Mearns and I. Scoones (eds) (1997) "Community based sustainable development: consensus or conflict," *IDS Bulletin*, 28(4).

Lind, A. (1997) "Gender, development and urban social change: women's community action in global cities," *World Development*, 25: 1205–23.

Mayoux, L. (1995) "Beyond naivete: women, gender inequality and participatory development," *Development and Change*, 26: 235–58.

Mohanty, C. T. (1987) "Under Western eyes: feminist scholarship and colonial discourses," *Feminist Review*, 30: 61–88.

Mohanty, R. (2002) "Women's participation in JFM in Uttaranchal villages," Mimeo.

Moore, H. (1994) *A Passion for Difference*, Cambridge: Polity Press.

Moraga, C. and G. Anzaldua (1981) *This Bridge Called My Back: Writings by Radical Women of Color*, Watertown, MA: Persephone Press.

Mosse, D. (1995) "Authority, gender and knowledge: theoretical reflections on the practice of Participatory Rural Appraisal," KRIBP Working Paper no. 2, Swansea: Centre for Development Studies.

Mouffe, C. (1992) "Feminism, citizenship and radical democratic politics," in J. Butler and J. Scott (eds), *Feminists Theorize the Political*, New York: Routledge.

Overs, C., J. Doezema and M. Shivdas (2002) "Just lip service? Sex worker participation in sexual and reproductive health interventions," in A. Cornwall and A. Welbourn (eds), *Realizing Rights: Transforming Approaches to Sexual and Reproductive Well-being*, London: Zed Books, pp. 21–34.

Peters, P. (1995) "The use and abuse of the concept of 'female-headed households' in research on agrarian transformation and policy," in D. Fahy Bryceson (ed.), *Women Wielding the Hoe: Lessons from Rural Africa for Feminist Theory and Development*, Oxford: Berg, pp. 93–108.

Phillips, A. (1991) *Engendering Democracy*, Cambridge: Polity Press.

Poffenberger, M. and B. McGean (1996) *Village Voices, Forest Choices: Joint Forest Management in India*, Delhi: Oxford University Press.

Sarin, M. (1998) "Community forest management: whose participation?" in I. Guijt and M. Kaul Shah (eds), *The Myth of Community: Gender Issues in Participatory Development*, London: Intermediate Technology Publications, pp. 121–30.

Sarin, M. and SARTHI (1996) "The view from the ground: community perspectives on joint forest management in Gujarat, India," Forest Participation Series no. 4, London: IIED.

Scott, J. (1989) "Gender: a useful category of historical analysis," in E. Weed (ed.), *Coming to Terms*, London: Routledge.

Welbourn, A. (1996) *Stepping Stones: A Training Package on HIV/AIDS, Communication and Relationship Skills*, Oxford: Strategies for Hope.

Whitehead, A. and M. Lockwood (1998) "Gender in the World Bank's Poverty Assessments: six case studies from sub-Saharan Africa," Mimeo, Geneva: UNRISD.

Wieringa, S. (1998) "Rethinking gender planning: a critical discussion of the use of the concept of gender," Working Paper no. 279, The Hague: ISS.

FIVE

SECTION 11
Urban Citizenship

Editors' Introduction

Citizenship is contested terrain. What constitutes citizenship, and how rights of citizenship are defined, achieved, and enjoyed have been the subject of scholarly debate as well as political struggles and even revolutions since the ancient Greek era. In this section of the *Reader* we will focus on citizenship as experienced and constituted in the cities of the global South. The contributions included in this section illustrate the heterogeneity of citizens' actions that shape the city and assert their claims of citizenship. But first we would offer a brief overview of the ideals of citizenship as per their conception in the Western political theory and introduce key debates these ideals and principles evoke. The contemporary critical scholarship stresses the tensions and inconsistencies in claims of citizenship and their varied nature across regions and histories. It also points to cities and localities as key sites through which citizens materialize their abstract rights of citizenship and gain tangible improvement in their living conditions–what we call urban citizenship. The selected readings for this section aim to represent the varied means that citizens use in distinct historical moments and social contexts to assert their rights to the city.

KEY ISSUES

Changing meanings of citizenship and scales of belonging

Since the classical period, the sources of as well as the privileges and duties associated with citizenship have gone through important changes and prompted major political and philosophical debates. For the ancient Greeks and Romans, membership in the city, *polis*, was a source of privileges associated with citizenship, yet women were excluded from possessing such rights. With the French Revolution the source of this protection shifted from city to nationhood. The motto "liberty, equality, fraternity" for all asserted that all men are equal and are protected not based on their allegiance to kings and feudal lords, property ownership, wealth and amount of taxes paid but by membership in a national political community. This liberal democratic model of citizenship expected that by entering a social contract both the state and the citizens assume certain rights and duties: citizens relegate some of their individual freedoms to the state–in exchange the state offers citizens civil and social protection (Rousseau [1762]1968).

In Western liberal democratic societies, the work of T. H. Marshall ([1950]1977) depicts the assumed relationship between these rights and obligations. Marshall divides citizenship rights into three categories–civil, political, and social–and argues that in modern Western societies they develop in sequence and progression. In his view, citizens use their universal civil and political rights and the legal and representative structures set in place during the eighteenth and nineteenth centuries to achieve social rights and fulfillment of basic socio-economic protection. A Marshallian conceptualization of citizenship assumes rights develop cumulatively and in an almost linear progression from one form to another to guarantee universal civil, political, and social protection (see Friedmann 2002; Turner 1990). Because this account of citizenship has been so influential yet controversial, a considerable body of critique of Marshall has arisen. The following are their points of critique:

(a) Civil, political, and social rights are not necessarily linked and sequentially developed. Following their independence many of the colonies formed nation states and granted universal civil and political citizenship to their people. But the membership of their citizens in a national state did not bring them political nor social gains Western models of citizenship promised or assumed (Mamdani 1996). Indeed false promises of this citizenship model are apparent even in the liberal democratic societies of the global North, where class, gender, race and ethnicity have persistently marked hierarchical inclusion/exclusion in claims of citizenship (Purcell 2003).
(b) The promise of inclusion through universal citizenship is deceptive. Less privileged members of a society including women, indigenous groups, or cultural and linguistic minorities, groups of diverse sexual orientation and identities are not always universally included through formal and representative political structures. These critiques highlighted that the rights recognized by the conventional model of citizenship are incomplete, and need to be extended to include cultural rights and identity-based claims of citizenship. (For feminist critique see Lister 1997; Yuval-Davis 1997; Oldfield, Salo and Schlyter 2009; Hassiem 1999. For critique regarding cultural rights and identity-based claims of citizenship see Roberts 1996; Kymlicka 1995; Taylor 1994.)
(c) The formal political arena does not offer an adequate site to renegotiate rights and duties associated with citizenship. Those excluded from the formal project of citizenship use a varied range of political arenas and forms of action to be heard by the state. Indeed they often find informal arenas of community activism as the most effective means of waging their struggle for their rights of citizenship (for example, the 1980s land occupations and neighborhood movements that took place in Chile and Mexico as discussed by Castells below; or recent citizenship claim-making movements that took place through street protests in Egypt discussed by Khamis and Vaughn below).
(d) Membership in a national community is insufficient in determining the rights and duties associated with citizenship. With the global restructuring of capitalism and unprecedented mobility of labor and capital, since the 1980s we see a growing immigrant population that cannot enjoy political inclusion in the societies where they have formal citizenship but they might enjoy fulfillment of certain basic needs in societies to which they have moved without formal political membership. These realities, as argued by Smith and Guarnizo (2009) in the selected reading included below, unsettle "the established view of the close correspondence between nationhood and citizenship" and have transformed people's expectation of the source of citizenship rights and the means of their fulfillment.

In the past few decades these critiques of the Western liberal democratic model of citizenship stressing the fallacies it upholds–namely assuming that the state grants rights and upholds responsibilities toward citizens while citizens uphold their civil and political rights and duties to enjoy social rights based on membership in a national community–have become ever more visible, hard to deny or ignore. In the post-1970s neoliberal era the state and government agencies began to dedicate their services and powers to protect the interests of capital and private companies rather than those of citizens. In this model the responsibility for social and economic welfare is shifted on to citizens. This not only has shifted the terms of the social contract that the old models of citizenship claimed to have in place; but also transformed the possibilities of popular politics and made more visible the limitations of the old analytic categories in respect to citizenship (Chatterjee 2006). These transformations have opened the idea of citizenship to alternative interpretations that dramatically depart from an earlier one. Instead of a mere formal status or a decree granted by the state from above by a set of formal institutions, the emerging interpretations of citizenship embrace a conceptual and practical shift that focuses on construction of citizenship from below.

Urban citizenship from below

Shifting relationships between the state and citizens and the kinds of rights and duties they ensue, imply real policy implications for cities and citizenship. Certain key questions emerge from this changing terrain: what are the rights of citizens in terms of their urban living conditions? How are those rights fulfilled and does the responsibility to make these rights a reality fall on individual citizens, public-sector organizations and institutions or private business? Lastly, who is a citizen–is it only those with legal status or does it include all those who reside in a given city?

A set of more complicated questions arises when we consider disadvantaged urban inhabitants and their living conditions: who has the responsibility to provide shelter, education, or basic urban services for citizens who cannot afford market prices? Do those rights come from above and remain exclusively granted by the state or can those be accessed by the citizens or the grassroots from below? If formal channels of claiming one's rights of citizenship do not respond to disadvantaged citizens, should these citizens try to exercise those rights through practices and processes outside those formal political structures and processes recognized by the state? For example, should they take over land, build their own shelter, bring water and basic infrastructure to their neighborhood outside official channels? Some would call such practices exercising citizenship from below; others would see these as processes that undermine law and order. These shifting meanings of citizenship have had important implications for development policies and planning, and they have a history.

The bulk of urban commentaries in the 1950s through 1970s explained the spread of informal settlements on the peripheries or outskirts of the developing city as an absence of urban citizenship. They saw such settlements as glaring proof of the failure of a social contract between citizens and the state. They argued, and many still do, that informal settlements not only display the political dis-functionality of the state but also the apathy and laziness of their inhabitants. Writers like the anthropologist Oscar Lewis (1961) contended that such settlements provided a breeding ground for a "culture of poverty," through which inhabitants learned to behave as "free-riders" without aspiration to perform their duties of citizenship. The work of Lewis was influential in urban policies during this period. At that time, the major focus of urban policy was the eradication of informal settlements and the relocation of their residents from the "margins" to an environment supposedly more conducive to their development. The assumption was that these populations were politically, culturally, and economically marginal to the fate of the cities.

An important turning point in the evolution of the dominant theoretical and policy frameworks of the time came through a critical urban scholarship that debunked the myth of marginality of poor squatters (Perlman 1980; Castells 1983). Far from apathy and marginality, inhabitants of favelas and squatter settlements, this scholarship revealed, are significant political and socio-economic actors in shaping the cities of the global South and achieving the rights of citizenship the state fails to deliver. In his seminal work, John Friedmann (1989) studies the neighborhood (*barrio*) movements in Latin American cities and the role poor citizens play in economic development of the city. This perspective was further developed by the work of urban planners, urban sociologists, and anthropologists who documented the significant role the grassroots play in developing the city. They achieve the socio-economic protection and the political inclusion the state might promise but does not deliver to populations disadvantaged by race, class, gender, and ethnicity. In other words, these studies showed how citizens and urban dwellers construct the city and citizenship from below.

Seen through this lens commentators argue that the spread of informal settlements on the outskirts of cities are testimonials to a healthy civil society and active practices of citizenship by the urban poor (Holston 2008). Holston argues the vast spread of favelas across urban Brazil is indeed one form of citizens' practices of citizenship. He calls the direct action through which populations excluded from the state's project of citizenship shape and claim the city "insurgent citizenship." This differs from the formal, state-centered kind of citizenship that expected the state to protect the citizens, but is a process in which the citizens' claim to the city and social protection is unmediated (Sandercock 1998; Friedmann 2002; Irazabal 2008). This insurgent citizenship encompasses practices by both the 'minoritized' populations of the Global North and marginalized residents of the Global South, in where citizens refuse "to relegate the defense of their interests to others be they politicians, bureaucrats, or planners but take the matter into their own hands" (Miraftab 2012: 1191). Neither protected by the state nor serviced by the market, these citizens strive to assert their political and social rights outside the formal channels of the state. This is not to remove the state from its obligations but to hold the state accountable through means beyond the state-sanctioned channels of citizen participation (Miraftab and Wills 2005; De Souza 2006). They ultimately invent channels and forms of participation–in other words the "invited and invented spaces of citizenship" (Miraftab and Wills 2009).

But within a neoliberal model of citizenship, it is not only the poor who take satisfaction of their rights into their own hands. Emboldened members of the middle class and the elite also assert their power by mobilizing through the market institutions or civil society organizations. They too shape political societies and seek to practice their rights of citizenship through direct action and civic associations; often at the cost of populations less privileged by class, caste, ethnicity or race (see, for example, the discussion of exclusionary practices using the discourse of "inclusive governance" in the Editors' Introduction to Governance, this volume). Here the state operates by principles of the market, citizens' rights are respected and recognized only as fee paying "responsible" costumers (see Editors' Introduction to Basic Urban Services and Beall et al., this volume). To justify replacing citizenship rights with consumer rights, the rhetoric of "active citizenship" and "choice" looms large. "Active citizenship" is a term former British Prime Minister Margaret Thatcher's administration used to justify massive privatization policies. In this paradigm, the demise of state's responsibilities was framed as the rise of active citizenship whereby citizens exercise their rights by expanding their choices–for example the right to choose their solid waste collector company, their charter school or their subsidized housing. In this neoliberal model of citizenship each of these expanded market "choices" that might liberate the middle and upper classes as "choices-full" market consumers, potentially exclude populations who cannot perform as market citizens.

Clearly, in practice the rights and obligations of citizenship are not fixed. They are the subject of constant negotiations and are "defined by a combination of elite and popular pressures" (Roberts 1996: 39). Meaning and expectations associated with citizenship change as the outcome of these struggles changes. Social movements are the most obvious arenas of struggle to redefine citizenship. But to assert those rights the strategies citizens use vary. Sometimes they involve organizing large-scale demonstrations and confrontations (such as practices documented by Castells in the reading below); other times they use quiet strategies that aim at going under the radar of authorities' oppressive machinery. Bayat (2010), for example, studies the spread of informal settlements of the 1980s in Middle Eastern cities like Tehran and Cairo, and characterizes these urban formations as politically potent but "quiet encroachments" by the poor. Citizens sometimes use their local urban and neighborhood networks (Friedmann 1989; Castells 1983) while at other times draw on their transnational networks and resources (as discussed by Smith and Guarnizo 2009). "Technologies" that undergird the expression of citizenship rights and means of organizing may also vary. Sometimes they are based on block committees as in the cases discussed by Castells in Latin America in the 1970s, or by new social media as described by Khamis and Vaughn within movements that occurred in the Middle East in the 2010s–also known as the Arab Spring. The readings in this section of the *Reader* will capture the varied meanings and practices associated with cities and citizenship.

ARTICLE SELECTIONS

The first reading is a classic piece by Manuel Castells, one of the most influential urban sociologists of our time. His analytic interventions since the early 1970s helped to reframe urban sociology, precipitating a shift from physical determinism to understanding the city as a social process (for a citation of Castells' prominant work see Editors' Introduction to Historical Underpinnings, this volume). In his earlier work he offered an analysis of the city as a historical manifestation of power and production in capitalism. In his later work, *The City and the Grassroots* (Castells 1983), he helped to understand the city as an outcome of struggle between citizens and the state. In that volume from which our reading selected below is drawn, Castells compiled historical and contemporary case studies from around the world to demonstrate how ordinary citizens organize and shape urban spaces around distinct group interests sometimes defined by class relations and other times by cultural, sexual, and racial identities. The city, he demonstrated through these case studies, is the outcome of the struggle grassroots wage vis-à-vis the state for their rights and interests as citizens. In the chapter included below he focuses on grassroots practices in Santiago, Chile; Lima, Peru; and Mexico City in the 1960s and 1970s and the strategies of the urban poor to shelter themselves and their families. He argues in this chapter that it is the practices of these dissident grassroots, who collectively mobilize, liaise with political parties, and engage in confrontational politics, that shape these cities of Latin America.

The second reading is by Michael Peter Smith and Eduardo Guarnizo, prominent scholars and theorists of transnationalism. In particular their work on transnationalism from below (Smith and Guarnizo 1998) and Smith's on *Transnational Urbanism* (2001) have been influential to urban studies. In the excerpt below they challenge the conventional understanding of citizenship as an exclusively nationally based institution and practice. Based on the shifting political economic developments operating at the global scale, they argue we see great transnational mobility of populations, both poor and rich, from the global South and North. These movements have dramatically transformed not only forms of membership in a nation state but also expectations and responsibilities that might be associated with such membership. In the context of the contemporary mass mobility of populations, they argue we need to re-think the meaning and practice of citizenship because of the new and enduring transnational socio-economic and political relationships migrants forge across their communities of origin and destination. They focus on the discourse of "the right to the city"–the concept developed by Lefebvre (1996) to stress people's right to produce, use and inhabit the urban space for a dignified life (for more see Mark Purcell 2003). In practice, in the face of the unsettled terrain of national citizenship, they argue, "the right to the city" movement seems to "create alternative political spaces where variously excluded groups of globally mobile urban inhabitants may empower themselves" (from Smith and Guarnizo, this volume). It provides not only a discourse but also a mechanism for greater inclusion in the city by citizens who claim and achieve their livelihood in the localities, neighborhoods and cities they inhabit. This is not because of but despite their formal status as citizens.

The third reading in this section, commissioned for this volume, is by communication and policy analysts Sahar Khamis and Katherine Vaughn, who write on the more recent means of grassroots organizing through the use of social networking technologies, media, and cyber activism. They report on grassroots action in the 2010–2011 uprising known as the Arab Spring in Egypt. They focus on the strategies behind the demonstrations staged in Tahrir Square, Cairo's main plaza, that took an iconic role in representing the movement. They critique the mainstream idea that the Arab uprising was a "Facebook revolution." Far from it, they argue, the uprising that the worldwide media broadcasted to the world was brewed for a long time through the *kefaya* (enough) movement in which labor unions and dissidents' formal political organizations had played a central role. While Facebook, Twitter, and YouTube did not create the social movement and Arab uprising, they were important in its fast and speedy spread of information and ability to bring millions of protestors to the street and the plaza. This was key in the successful street politics of Egypt in the twenty-first century, a technology not available to previous generations of Cairo's urban dissidents.

The authors in these three readings discuss urban citizens' efforts to make a dignified living in political environments where formal structures do not accommodate their interests. They highlight how the grassroots population makes and claims cities of the global South through practices of citizenship that use distinct mediums and are facilitated differently in distinct historical moments and socio-political contexts.

Figure 29 *Brazilian Protests Illustrated.*
Source: © Carlos Latuff, 2013.

REFERENCES

Bayat, A. (2010) *Life as Politics: How Ordinary People Change the Middle East,* Palo Alto: Stanford University Press.

Castells, M. (1983) *The City and the Grassroots: A Cross-Cultural Theory of Urban Social Movements*, Berkeley: University of California Press.

Chatterjee, P. (2006) *The Politics of the Governed: Reflections on Popular Politics in Most of the World*, New York: Columbia University Press.

De Souza, M. L. (2006) "Together with the state, despite the state, against the state: Social movements as 'critical urban planning' agents," *City,* 10(3): 327–342.

Friedmann, J. (1989) "The Latin American *barrio* movement as a social movement: Contribution to a debate," *International Journal of Urban and Regional Research,* 13(3): 501–510.

Friedmann, J. (2002) *The Prospect of Cities*, Minneapolis: University of Minnesota Press.

Hassiem, S. (1999) "From presence to power: Women's citizenship in a new democracy," *Agenda,* 40: 6–17.

Holston, J. (1995) "Spaces of insurgent citizenship," *Planning Theory,* 13: 35–52.

Holston, J. (2008) *Insurgent Citizenship: Disjunctions of Democracy and Modernity in Brazil*, Princeton, NJ: Princeton University Press.

Irazabal, C. (2008) "Citizenship, democracy and public space in Latin America," in C. Irazabal (ed.) *Ordinary Places/Extraordinary Events: Citizenship, Democracy and Public Space in Latin America*, New York: Routledge, pp. 11–34.

Kymlicka, W. (1995) *Multicultural Citizenship: A Liberal Theory of Minority Rights*, New York: Oxford University Press.

Lefebvre, H. (1996) "The right to the city," in G. Bridge and S. Watson (eds.) *The Blackwell City Reader*, Oxford, UK: Blackwell, pp. 367–374.

Lewis, O. (1961) *Children of Sanchez: Autobiography of a Mexican Family*, New York: Vintage Books.

Lister, R. (1997) "Citizenship: Towards a feminist synthesis," *Feminist Review,* 57(Fall): 28–48.

Mamdani, M. (1996) *Citizens and Subjects: Contemporary Africa and the Legacy of Late Colonialism*, Princeton, NJ: Princeton University Press.

Marshall, T. H. ([1950]1977) *Citizenship and Social Class and Other Essays,* Cambridge: Cambridge University Press.

Miraftab, F. (2012) "Planning and citizenship," in Weber, R. and Crane, R. (eds.) *Oxford Handbook of Urban Planning*, Oxford University Press, pp. 1180–1204.

Miraftab, F. and Wills, S. (2005) "Insurgency and spaces of active citizenship: The story of the Western Cape Anti-Eviction Campaign in South Africa," *Journal of Planning Education and Research,* 25(2): 200–217.

Miraftab, F. and Wills, S. (2009) "Insurgent planning: Situating radical planning in the global South," *Planning Theory,* 8(1): 32–50.

Oldfield, S., Salo, E. and Schlyter, A. (2009) "Body politics and the gendered crafting of citizenship," *Feminist Africa,* 13: 1–13.

Perlman, J. (1980) *Myth of Marginality: Urban Poverty and Politics in Rio de Janeiro,* Berkeley/Los Angeles: University of California Press.

Purcell, M. (2003) "Citizenship and the right to the global city: Reimagining the capitalist world order," *International Journal of Urban and Regional Research,* 27(3): 564–590.

Roberts, B. (1996) "The social context of citizenship in Latin America," *International Journal of Urban and Regional Research,* 20(1): 38–65.

Rousseau, J. J. ([1762]1968) *The Social Contract*, New York and London: Penguin.

Sandercock, L. (1998) "Framing insurgent historiographies for planning," in Leonie Sandercock (ed.) *Making the Invisible Visible: a Multicultural Planning History*, Berkeley/Los Angeles/New York: University of California Press.

Smith, M. P. (2001) *Transnational Urbanism: Locating Globalization,* Malden, Oxford: Blackwell Publishers.

Smith, M. P. and Guarnizo, L. (Eds.) (1998) *Transnationalism from Below*, New Brunswick, NJ and London, UK: Transaction Publishers.

Smith, M. P. and Guarnizo, L. E. (2009) "Global mobility, shifting borders and urban citizenship," *Tijdschrift voor Economische en Sociale Geografie,* 100(5): 610–622.

Smith, M. P. and McQuarrie, M. (Eds.) (2012) *Remaking Urban Citizenship: Organizations, Institutions, and the Right to the City*, London, UK and New Brunswick, NJ: Transaction Publishers.

Taylor, C. (1994) "Politics of recognition," in C. Taylor et al. (eds.) *Multiculturalism: Examining the Politics of Recognition,* Princeton, NJ: Princeton University Press; 6th edn, pp. 25–74.

Turner, B. S. (1990) "Outline of a theory of citizenship," *Sociology,* 24(2): 189–217.

Yuval-Davis, N. (1997) "Women, citizenship and difference," *Feminist Review,* 57(Fall): 4–27.

"Squatters and the State: The Dialectics between Social Integration and Social Change (Case Studies in Lima, Mexico, and Santiago de Chile)"

The City and the Grassroots: A Cross-Cultural Theory of Urban Social Movements (1983)

Manuel Castells

The conditions in Latin American societies force an increasing proportion of the metropolitan population to live in squatter settlements or in slum areas. This situation is not external to the structural dynamics of the Third World, but is connected to the speculative functioning of some sectors of capital as well as to the peculiar patterns of popular consumption in the so-called informal economy.[1] On the basis of their situation in the urban structure, the squatters tend to organize themselves at the community level. Their organization does not imply, by itself, any kind of involvement in a process of social change. On the contrary, as we have pointed out, most of the existing evidence points to a subservient relationship with the dominant economic and political powers.[2] Nevertheless, the fact of a relatively strong local organization is itself a distinctive feature which clearly differentiates the squatters from other urban dwellers who are predominantly organized at the work place or in political parties, when and if they are organized at all.

Furthermore, the state's attitude towards squatter settlements predetermines most of their characteristics. Thus the connection between the squatters and the political process is a very close one. And it is precisely in this way that urbanization, and its impact on community organization, becomes a crucial aspect of political evolution in Latin America. Let us, therefore, explore the relationship between squatters and the state in three major Latin American countries: Peru, Mexico, and Chile.

SQUATTERS AND POPULISM: THE *BARRIADAS* OF LIMA[3]

Lima's spectacular urban growth has mainly been due to the expansion of *barriadas,* periperal[4] substandard settlements, often illegal in their early stage, and generally deprived of basic urban facilities. The population of the *barriadas* came, on the one hand, from the slums of central Lima *(tugurios)* once they had reached bursting point or when they were demolished, and on the other hand, from accelerated rural and regional migration (Weisslitz,1978), the structural causes of which were the same as for all dependent societies (Castells,1971; Safa, 1982). And in Peru, this particular form of urbanization—the *barriadas*—cannot be explained without reference to the action of political forces as well as to the state's policies (Henry, 1977). Given the illegal nature of land invasion by population of the *barriadas,* only institutional permissiveness or the strength of the movement (or a combination of both) can explain such a phenomenon. More specifically, given the way power has been unevenly distributed within Peruvian society until very recent times, the land invasion must be understood to have been the result, in part, of policies that originated from various dominant sectors. Very often landowners and private developers have manipulated the squatters into forcing portions of the land onto the real estate market, by obtaining from the authorities some urban infrastructure for the squatters, thus enhancing the land value and opening the way for profitable housing construction. In a second stage, the squatters are expelled from the land they have occupied and forced to start all over again on the frontier of a city which has expanded as a result of their efforts.

Nevertheless, the main factor underlying the intensity of urban land invasion in Lima has been a political strategy consisting of protection given for the invasion in exchange for poor people's support. Table 7, constructed by David Collier on the basis of his study of 136 of Lima's *barriadas,* clearly shows the political context for the peak moments of urban land invasion between 1900 and 1972 (Collier, 1978: 61). The political strategies and their actual effects on the squatters were really very different from one land invasion to another.

The most spectacular stage in the history of land invasions corresponds to the initiative of General Odria's government in 1948–1956. At a time of political repression against the Communist Party, and particularly against Alianza Popular Revolucionaria Americana (APRA—which was trying to seize power to implement an "anti-imperialist program"), Odria's populism was a direct attempt to mobilize people on his side by offering to distribute land and urban services. The aim was to dispute APRA's political influence by taking advantage of the urban poor's low level of political organization and consciousness, and by mobilizing people around issues outside the work place where the pro-APRA union leaders would be more vulnerable. Nevertheless, APRA's reaction was very rapid: they demanded that Odria keep his promises to the squatters' organizations by accelerating the invasion, and the final outcome was a political crisis and the downfall of Odria's government.

We can understand, then, why Prado's government, supported by APRA, continued to be interested in the *barriadas* in order to eliminate the remaining pro-Odria circles and to widen its popular basis.

President	Number of cases	% of cases	Population	Percentage of population
Before Sanchez Cerro (1900–30)	2	1.5	2,712	0.4
Sanchez Cerro (1930–31, 1931–33)	3	2.2	12,975	1.7
Benavides (1933–39)	8	5.9	18,888	2.5
Prado (1939–45)	8	5.9	6,930	0.9
1945–Information uncertain	5	3.7	24,335	3.2
Bustamante (1945–48)	16	11.8	38,545	5.1
Odria (1948–56)	30	22.1	203,877	26.9
1956–Information uncertain	2	1.5	11,890	1.6
Prado (1956–62)	30	22.1	93,249	12.3
1962–Information uncertain	2	1.5	22,377	2.9
Perez Godoy (1962–63)	2	1.5	1,737	0.2
Lindley (1963)	3	2.2	11,046	1.5
Belaunde (1963–68)	15	11.0	93,407	12.3
Velasco (1968–72 only)	10	7.4	217,050	28.6
Total	**136**	**100.3**	**759,018**	**100.1**

Table 7 Number and population of *barriadas* (squatter settlements) formed in Lima in each presidential term, 1900–1972

Source: David Collier (1978).

Instead of stimulating new invasions, Prado launched a program of housing and service delivery for the popular neighborhoods, trying to integrate these sectors into the government's policy without mobilizing them. In a complementary move, APRA started formally controlling the organizations of squatters (the *Asociaciones de Pobladores*) in order to expand its political machine from the trade unions to the social organizations centered on residential issues.

Belaunde's urban policy was very different. [Ed. Note: Belaunde was president of Peru in 1965–68 and 1980–85.] Although he also looked for some support from the squatters, allowing and stimulating land invasions, he did not limit his activity to the struggle against the APRA, but tried a certain rationalization of the whole process. His Law of *Barriadas* was the first attempt to adapt urbanization to the general interest of Peruvian capitalist development without adopting a particular set of political interests. The activity of his party, *Accion Popular*, was aimed at modernizing the *barriadas* system and facilitating an effective connection with the broader interests of corporate capital. The social control of the squatters was then organized by international agencies, churches, and humanitarian organizations, which were closely linked to the interests of the American government (Rodriguez and Riofrio, 1974). Belaunde's strategy was quite effective in weakening APRA's political influence among the *pobladores*, but it was unable to provide a new form of social control established on solid ground. This situation determined a very important change in the government's strategy after the establishment of a military junta in the revolution of 1968. At the beginning, the military government tried to implement a law-and-order policy, repressing all illegal invasions and putting the *asociaciones de pobladores* under the control of the police. Nevertheless, its attitude towards the *barriadas* changed dramatically on the basis of two major factors: first, the difficulty of counteracting a basic mechanism that determines the housing crises in the big cities of dependent societies, and second, the military government's need to obtain very rapidly some popular support for the modernizing policies once these policies had come under attack from the conservative landlords and business circles.

The turning point appears to have been the *Pamplonazo* in May 1971. An invasion of urban land in the neighborhood of Pamplona was vigorously repressed and provoked an open conflict between the Minister of Interior, General Artola, and Bishop Bambaren, nicknamed the "*Barriadas* Bishop," who was jailed. The crisis between the state and the Catholic Church moved President General Velasco Alvarado to act personally on the issue. He conceded most of the *pobladores'* demands, but moved them to a very arid peripheral zone close to Lima, where he invited them to start a "self-help" community supported by the government. This was the beginning of Villa El Salvador, a new city which in 1979 housed up to 300,000 inhabitants recruited among the Lima dwellers and rural migrants looking for a home in the metropolitan area.

The military government learned a very important lesson from this crisis: not only did it discover the dangers of a purely repressive policy, but it also realized the potential advantages of mobilizing the *pobladores*. Using the Church's experience, the military government created a special agency, the *Oficina Nacional de Pueblos Jovenes* (New Settlements' National Office), charged with legalizing the land occupations and with organizing material and institutional aid to the *barriadas*. At the same time, within the framework of *Sistema Nacional de Movilizacion Social* (SINAMOS), the regime's "social office," a special section was created to organize and lead the *pobladores*. Under the new measures, each residential neighborhood in the *barriadas* had to elect its representatives who would eventually become the partners of the government officials, controlling the distribution of material aid and urban facilities. At the same time, the new institution relied on the existing agencies and voluntary associations (most of them linked to churches and international agencies) to tailor their functions and co-ordinate their activity to the parameters set by state policy. This policy would develop along several paths: economic (popular savings institutions, production and consumption co-operatives); legal (laws recognizing the squatting of urban land); ideological (legitimizing of the *pobladores'* associations, propaganda centers for the government); and political (active involvement in the Peruvian "revolution" through SINAMOS) The *barriadas* became a crucial focus of popular mobilization for the new regime.

As a consequence of the successive encouragements given to squatter mobilization by the state, as well as by political parties, the *barriadas* of Lima grew in extraordinary proportions: their population grew from 100,000 in 1940 to 1,000,000 in 1970 and became an ever larger proportion of the population in most of Lima's districts.

Nevertheless, it would be wrong to conclude that all forms of mobilization were identical save for different ideological stances. In his study, Etienne Henry makes clear some fundamental differences in practice. Odria's and Prado's policies expressed the same relationship to the *pobladores,* patronizing them to reinforce each's political constituencies. In the case of Belaunde, the action to integrate people was subordinated to the effort of rationalizing urban development. The military government's policy between 1971 and 1975 represented a significant change in urban policy. It was not an attempt to build up partisan support for a particular political machine but was, in fact, a very ambitious project to establish a new and permanent relationship between the state and urban popular sectors through the controlled mobilization of the *barriadas* now transformed into *pueblos jovenes* (new settlements). This transformation was much more than a change in name: it expressed the holding of all economic and political functions of the *pobladores'* voluntary associations by the state in exchange for the delivery and management of required urban services. The goal was no longer to obtain a political constituency but to build a "popular movement" mobilized around the values promoted by the revolutionary regime. In this sense, the *barriadas* became closely linked to Peruvian politics and were increasingly reluctant to adapt to the new government's orientation resulting from the growing influence of the conservative wing within the army.

The picture of the Lima squatters' movement appears as one of a manipulated mob, changing from one political ideology to another in exchange for the delivery (or promise) of land, housing, and services. And this was, to a large extent, the case. The *pobladores'* attitude was quite understandable if we remember that all politically progressive alternatives were always defeated and ferociously repressed. So, as Anthony and Elizabeth Leeds (1976) have pointed out, the behavior of the squatters was not cynical or apolitical, but, on the contrary, deeply realistic, and displayed an awareness of the political situation and how their hard-pressed demands could be obtained. Thus it appears that the Peruvian urban movement was, until 1976, dependent upon various populist strategies of controlled mobilization. That is, the movement was, in its various stages, a vehicle for carrying the social integration of the urban popular sectors in the same direction as the political strategies of the different political sectors of the dominant classes.

Now this process, like all controlled mobilizations, expresses a contradiction between the effectiveness of the mobilization and the fulfillment of the goals assigned to the movement. When these goals are delayed as a result of the structural limits to social reform, and when people's organization and consciousness grow, some attempts at autonomous social mobilization occur. A sign of this evolution was, in the case of Lima, the organization of the *Barriada Independencia* in 1972. When the autonomous mobilization expanded, the government tried to stop it by means of violent repression as it did, for example, in March 1974. In spite of repression; the movement continued its opposition, making alliances with the trade unions and with the radical left, as was revealed in the *barriadas'* massive participation in the strikes against the regime in 1976, 1978, and 1979. After the dismantling of SINAMOS by the new military president, Morales Bermudez, the political control of the *barriadas* rapidly collapsed. Ironically, Villa El Salvador became one of the most active centers of opposition to the state's new conservative leadership.

This evolution supports a crucial hypothesis. The replacement of a classic patronizing relationship, ruling class to popular sectors, by controlled populist mobilization expands the hegemony of the ruling class over the popular sectors which are organized under the label of "urban marginals." But the crisis of such a hegemony, if it does happen, has far more serious consequences for the existing social order than the breaking of the traditional patronizing ties of a political machine. In fact, it is this type of crisis that enables the initiation of an autonomous popular movement, the further development of which will depend on its capacity to establish a stable and flexible link with the broader process of class struggle.

Our analysis of the Lima experience, although excessively condensed, provides some significant findings:

1 An urban movement can be an instrument of social integration and subordination to the existing political order instead of an agent of social change. (This is, in fact, the most frequent trend in squatter settlements in Latin America.)

2 The subordination of the movement can be obtained by political parties representing the interests of different factions of the ruling class and/or by the state itself. The results are different in each case. When the movement has close ties with the state, then urban policies become a crucial aspect of change in dependent societies.
3 Since urbanization in developing countries is deeply marked by a growing proportion of squatter settlements (out of the total urban population), it appears that the forms and levels of such urbanization will largely depend upon the relationship established between the state and the popular sectors. This explains why we consider urban politics to be the major explanatory variable of the characteristics of urbanization.
4 When squatter movements break their relationship of dependency *vis-a-vis* the state, they may become potential agents of social change. Yet their fate is ultimately determined by the general process of political conflict.

BETWEEN *CACIQUISMO* AND UTOPIA: THE *COLONOS* OF MEXICO CITY AND THE *POSESIONARIOS* OF MONTERREY

Mexico's accelerated urban growth is a social process full of contradictions (Unikel, 1971). An expression of these contradictions during the seventies were the ever increasing mobilizations by the popular sectors and urban squatters to obtain their demands in the *vecindades* (slums) (Montano, 1976; Pueblo, 1982; Navarro and Moctezuma, 1981). The potential strength of this urban mobilization must be seen in the context of a political system that was perfectly capable of controlling and integrating all signs of social protest.[5]

In traditional squatter settlements on the periphery of big cities, the key element was a very strong community organization under the tight control of leaders who were the intermediaries between the squatters *(colonos)* and the administration officials. In its early stages, this form of community organization may be considered to be dominated by *caciquismo,* that is, by the personal and authoritarian control of a leader, himself recognized and backed by local authorities. Therefore, illegal land invasion, by itself, did not present a challenge to the prevailing social order. Indeed, economically, it represented a way to activate the capitalist, urban land market and, politically, was a major element in the social control of people in search of shelter. What must be emphasized is that *caciquismo* was not an isolated phenomenon, but had a major function to fulfill within both the political system and the state's urban policies. The local leaders were not neighborhood bosses living in a closed world: they were representatives of political power through their relationships with the administration and with the *Partido Revolucionario Institucional* (PRI)—the government's party—from which they obtained their resources and their legitimacy. So Mexican squatters have always been well organized in their communities, and this organization has performed two major functions: on the one hand, it has allowed them to exert pressure for their demands to stay in the land they have occupied and to obtain the delivery of urban services; on the other hand, it has represented a major channel of subordinated political participation by ensuring that their votes and support goes to the PRI. Both aspects have, in fact, been complementary, and the *caciques* (the community bosses) were the agents of this process. They were not, however, the real bosses of the squatters, since they exercised their power on behalf of the PRI. To understand this situation we must remember the historical and popular roots of the PRI, and the need for it to continuously renew its role of organizing the people politically while providing access to work, housing, and services in exchange for loyalty to the PRI's program and leadership.

Thus the new urban movements that developed in Mexico during the 1970s derived from a previous network of voluntary associations existing in the *vecindades* and *colonias* which were, at the same time, channels for expressing demands and vehicles of political integration with PRI. Taking into consideration the ideological hegemony of the PRI and the violent repression exercised against any alternative form of squatter organization, how can we explain the upheaval caused by autonomous urban movements since 1968? And what were their characteristics and possibilities?

Two major factors seem to have favored the development of these movements:

1 President Echeverria's reformism (1970–1976) to some extent recognized the right to protest outside the established channels, while legitimizing aspirations to improve the living conditions in cities.[6]

2 Political radicalism among students, after the 1968 movement, provided militants who tried to use the squatter communities as a ground on which to build a new form of autonomous political organization.

This explains how the evolution of the new urban movements came to be determined by the interaction between the interests of the squatters, the reformist policy of the administration, and the experience of a new radical left, learning how to lead urban struggles.

From the outset of urban mobilization, radicals tried to organize and politicize some squatter settlements, linking their urban demands to the establishment of permanent bases of revolutionary action and propaganda within these settlements. These attempts were often unable either to overcome the squatters' fears of reprisal, or to uproot the PRI's solid political organization. When the radicals did succeed in their attempt to organize a squatter settlement as a revolutionary community, the state resorted to large scale violence having taken prior care to undermine the movement by claiming it had subversive contacts with underground guerillas. The most typical example was the *colonia* Ruben-Jaramillo in the city of Cuernavaca, where radical militants organized more than 25,000 squatters, helped them to improve their living conditions, and raised the level of their political awareness. The *colonia's* radicalism prompted a violent response by the army which occupied and put it under the control of a specialized public agency to deal with squatter settlements.

Nevertheless, other settlements resisted police repression and survived by maintaining a high level of organization and political mobilization. The best-known case is the *Campamento 2 de Octobre* in Ixtacalco in the Mexico City metropolitan area. Four thousand families illegally invaded a piece of highly valued urban land where both private and public developers had considerable interests. Students and professionals backed the movement and some of them went to live with the squatters to help their organization. The squatters kept their autonomy *vis-a-vis* the government, and used their strong bargaining position to call for a general political opposition to the PRI's policies. They became the target of the most conservative sectors of the Mexican establishment. After a long series of provocations by paid gangs, the police attacked the *campamento* in January 1976, starting a fire and injuring many of the squatters. Some days later, several hundred families returned to the settlement, reconstructed their houses, and started to negotiate with the government to obtain the legal rights to remain there. But if repression could not dismantle the *campamento,* it succeeded in isolating it by making it too dangerous an example to be followed by other squatters. When the Ixtacalco squatters tried to organize around themselves a *Federacion de Colonias Proletarias* to unite the efforts of other settlements, they obtained little support given their image of extreme radicalism. In fact, their demands were relatively modest, consisting of the legalization of the settlement and a minimum level of service. But the repression was very severe because the government saw a major danger in the movement's will to autonomy, its capacity to link urban demands and political criticism, and its appeal to other political sectors to build an opposition front, bypassing the political apparatus of the PRI within the communities. At the same time, police action was made easier by the political naïveté of some of the students, who at the beginning of the movement thought of the settlement as a "liberated zone" and spent much of their energy on verbal radicalism. In this sense, Ixtacalco was an extraordinarily advanced example of autonomous urban mobilization, but was also a very isolated experience which went forward by itself without considering the general level of urban struggle elsewhere in Mexico City.

In fact, the most important urban movements in recent years have taken place in northern Mexico, particularly in Chihuahua, Torreon, Madero, and above all in Monterrey, where the movement of the *posesionarios* (squatters) was perhaps one of the most interesting and sizeable in Latin America. Let us examine this experience in some detail.

Monterrey, the third largest Mexican city with a population of 1,600,000, is a dynamic industrial area with an important steel industry. It is dominated by a local bourgeoisie with an old and strong tradition, cohesively organized and closely linked to American capital. The so-called Monterrey Group is a modernizing entrepreneurial class, politically conservative and socially paternalistic. It has always opposed state intervention, often criticized the PRI, and succeeded with its workers through a policy of social benefits and high salaries. In Monterrey, the powerful *Confederacion de Trabajadores Mexicanos* (CTM), the major labor union controlled by the PRI, is relatively unimportant since most workers have joined the *sindicatos blancos* (the white unions) which are manipulated by company management. The city, which is proud of maintaining the highest living standards in Mexico, has experienced a strong urban growth rate since 1940: 5.6 percent annual growth 1940–1950 and 1950–1960, and 3.7 percent in 1960–1970. Urban immigration has been the result of both industrial growth and accelerated rural exodus caused by the rapid capitalist modernization of agriculture in northern Mexico. This urban growth has not, however, been matched by the increase in housing and urban services. The big companies have provided housing for their workers, but for the remaining people (one third of the population) there has been no available housing. The consequence has been, as in other Mexican cities, the invasion of surrounding land and massive construction of their own housing by the squatters. Three hundred thousand *posesionarios* have settled there. The underlying mechanisms were similar to those already described: speculation and illegal development on the one hand, and the role of the PRI's political machine as intermediary with local authorities on the other.

It was against this background that the student militants acted, trying to connect urban demands to political protest in the same mold as the university-based radicalism which began in 1971. The students led new land invasions contrary to the agreement with the administration. To differentiate themselves from the former settlements, the students called the new ones *colonias de lucha* (struggle settlements). In 1971 they founded the first settlement, *Martires de San Cosme*, in the arid zone of Topo Chico. The police immediately surrounded them, but withdrew after a month of violent clashes. Then the squatters built their houses and urban infrastructure and established a very elaborate social and political organization. The same process was renewed in the following years, and the cumulative effect made it extremely difficult to use repression as a means of halting their progress. The participants of each land invasion included not only its beneficiaries but also squatters already settled elsewhere who considered the new invasions as part of their own struggle. The timing of each invasion was extremely important as a means of averting repression: one of the most courageous invasions, in San Angel Bajo, succeeded in occupying good open space close to the municipal park without suffering reprisals because it was carried out on the eve of President Lopez Portillo's arrival in Monterrey during his electoral campaign of 1976. Once the invasion was accomplished, people raised the Mexican flag, running the red flag up a few weeks later. In similarly ingenious ways *Tierra y Libertad, Revolucion Proletaria, Lucio Cabanas, Genaro Vasquez*, and 24 other settlements were born and eventually combined in an alliance, the *Frente Popular Tierra y Libertad*, representing, at the time of our field work in August 1976, about 100,000 squatters (Villarreal, 1979).

A key element in the success of the movement was its ability to take advantage of the internal contradictions of the ruling elite. For example, the Monterrey bourgeoisie openly opposed President Echeverria's reformism, and launched a major attack against the governor, who replied by trying to obtain support from the people. Using the themes of the governor's populist speeches as justification of their actions, the squatters made open repression against them more difficult. Nevertheless, the embittered local oligarchy, which controlled the city police, reacted by organizing continuous provocations. In one of the police actions, on 18 February 1976, six squatters were killed and many others wounded. The movement's protest was impressive, and signs of solidarity came from all over the country. There were street demonstrations in Monterrey for 15 days, some of them attracting over 40,000 people, organized jointly by squatters, students, and workers. For two months, the squatters occupied several public places. Finally, they were personally entertained by President Echeverria in Mexico City. Victims' relatives obtained economic compensation, an official inquiry was opened, the city police chief was ousted, and the government provided strong financial support for the revolutionary squatter settlements.

So, in a critical moment, the movement clearly displayed its strength and political capacity. But it also revealed its limits. To understand this crucial point, we must consider the organizational structure and the political principles of the Monterrey squatter's movement.

The basic idea, shared by all the squatters' leaders, was that struggles for urban demands were meaningful only as far as they allowed people to unite, to be organized, and to become politically aware, because (according to these leaders) such political strength was the only base from which to successfully plead for demands. On the other hand, they wanted to link the squatters' actions to a collective theme aimed, in the long term, to the revolutionary transformation of society. Only if these principles are remembered can some surprising aspects of the movement be understood. For instance, the squatters strongly opposed the legalization by the government of their illegal land occupation. Their reasons were threefold: economic, ideological, and political. Economically, legalization implied high payments for a long time under conditions that many families could not afford. Ideologically, the movement could be transformed into a pressure group *vis-a-vis* the state instead of asserting their natural right to the land. Above all, politically, legalization, by individualizing the problem and dividing the land, would create a specific relationship between each squatter and the administration. Thus, the movement itself could be fragmented, lose its internal solidarity and be pushed towards the integration with the state's machinery. Therefore, to preserve their solidarity, cohesiveness, and strength (which they considered to be their only weapons), the squatters refused the property rights offered by the state, and expelled from the settlements those squatters who accepted the legal property title. A similar attitude was taken towards the delivery of services.

The squatters believed in self-reliance and rejected the state's help in the first stages of the movement. They did not, however, avoid contact with the state, since they were continuously engaged in negotiation, but wanted to preserve popular autonomy in a Mexican context where the political system is quite capable of swallowing up any initiative by a grass-roots organization. So they stole construction materials or obtained them by putting pressure on the administration, but they collectively built the schools, health services, and civic centers, with excellent results (unlike most Mexican squatter settlements). Houses were built by each family but in lots of a collectively decided size, in proportion to the size of the family and following a master plan approved by the settlement's General Assembly. Water, sewerage, and electricity were provided by illegal connections to the city systems. It is interesting to note that several settlements refused electric power in order to avoid television because it was considered a source of "ideological pollution." To overcome transport problems, the squatters seized buses on several occasions, finally forcing the bus company to adapt to the new urban structure. Schools were integrated into the general educational system and paid for by the state, but were controlled and managed by the Parents' Association in collaboration with the children's representatives. A similar organization managed health services. There was also in each settlement an Honor and Justice Committee which passed judgment on conflicts, the most serious of which were handled by the General Assembly. Alcohol and prostitution were strictly forbidden. Settlement leaders organized vigilante groups to protect the squatters. The general organization was based on a structure of block delegates which nominated the settlement committees which, in turn, reported to the General Assembly. There were a variety of voluntary associations, the strongest of which were the Women's Leagues and the Children's Leagues. The ideology of collective solidarity was reinforced. On Red Sundays—in 1976, every Sunday—everybody had to do collective work on shared urban facilities. There was also a high level of political and cultural activity run by "activist brigades."

Nevertheless, in spite of this extraordinary level of organization and consciousness, the *posesionarios'* movement in Monterrey suffered from the shortcomings of its isolation—geographical, social, and political. Geographically, it was the only urban movement of such size and character in the whole country. Socially, the squatters' population consisted almost entirely of unemployed, migrant peasants, having little contact with Monterrey's industrial workers. Politically, the leaders had no national audience and were only important at a local level.

The movement's leaders were well aware of this situation and of the danger of closing themselves into a new kind of communal Utopia. To break this isolation they tried to launch a series of actions to support "fair causes": for instance, each time a worker was unjustifiably fired, the squatters occupied the manager's home yard until the worker was reappointed. Each individual repression was faced by the whole movement, and so it became increasingly politicized. But such political radicalism based only on the squatters' support carried two major risks: first, increasing repression, chiefly from the army; and second, political infighting within the movement.

Two crucial elements emerge from the analysis of this extraordinary experience:

1 The speed and development of an urban movement cannot be separated from the general level of organization and consciousness in the broader process of political conflict.

2 The relationship to the state is not exhausted either by repression or integration. A movement may increase its autonomy by playing on the internal contradictions of the state. Monterrey was able to go further than Ixtacalco mainly because of the type of relationship which the *posesionarios* were able to establish with the state.

Such political, urban movements as Monterrey or Ixtacalco are only able to stabilize if the power relationships between social classes change in favor of the popular classes. But this does not seem to have been true of Mexico, so that the survival of these community organizations ultimately required some alliance with a sector within the state. Thus the experience of Mexico shows, again, the intimate connection between urban movements and the political system. We will now turn to the most important political squatter movement in recent Latin American history—Chile during the *Unidad Popular*—so that we may study this relationship in detail.

URBAN SOCIAL MOVEMENTS AND POLITICAL CHANGE: THE *POBLADORES* OF SANTIAGO DE CHILE, 1965–1973[7]

The historical significance of urban movements in Chile between 1965 and 1973 has been surrounded by a confused mythology. Our respect for the Chilean popular movement requires a careful reconstruction of the facts, as well as a rigorous analysis of the experience.

The squatter movement in Chile was closely linked with class struggle and its political expressions[8] and this explains both its importance and shortcomings. While invasions of urban land had always happened in Chile (Urrutia, 1972), they changed their social implications when they became entrenched with the political strategies of conflicting social classes: urban popular movements reached a peak as a consequence of the failure of the Christian Democratic program for urban reform (Rojas, 1978). The reform, initiated under Eduardo Frei's presidency in 1965, relied on three elements:

1 A program of distribution of urban land (*Operacion Sitio*) combined with public support for the construction of housing by the people.
2 The formation of voluntary associations of *pobladores* and of housewives *(centros de madres)* linked to a series of public agencies, organized around the government's Department of Popular Promotion.
3 The decentralization of local governments after the creation in 1968 of advisory neighborhood councils *(Juntas de Vecinos)*, elected by the residents of each neighborhood (Vanderschueren, 1971).

In fact the program of urban reform failed because of two constraints: the first from the structural limits of the system (the difficulty of redistributing resources without affecting the functioning of private capital)[9] and the second from the pressure of interest groups (mainly the Chilean Chamber of Private Builders, and the Savings and Loan institutions) which used the program as a means of producing profitable housing for middle class families (Cheetham, 1971).

As a consequence of this failure, the Christian Democrats lost control of the *pobladores'* movement and the neighborhood councils became a political battlefield (Rojas, 1978). The movement started then to put pressure on the government in two ways: on the one hand, the residents of the popular neighborhoods started asking for the delivery of promised services, and on the other, thousands of families living with relatives or in shanties gathered to form Committees of the homeless *(Comites Sin Casa)*. These committees, in the late 1960s, took the initiative of squatting on urban land to force the government to provide the housing and urban services promised in the reform program (Alvarado, Cheetham and Rojas, 1973). In the first period of the movement, between 1965 and 1969, the government responded by repressing the invasions, even causing a massacre (Puerto Montt, March 1969) and partially succeeded in stopping the process (Bengoa, 1972). But the presidential elections were scheduled for September 1970, and in the Christian Democratic Party the left had won endorsement for its leader, Tomic, against the wishes of the incumbent President Eduardo Frei. Therefore, an open repression of the *pobladores* might have been politically costly among the urban popular sectors, whose vote had been crucial for the electoral victory in 1964. So, when in 1970 the police were restricted in the use of violence, mass squatting was launched in most cities of the country, taking advantage of the new leniency to establish a new

Site	1966	1967	1968	1969	1970	1971	Sept. 1971–May 1972	1 Jan. 1972–31 May 1972
Santiago	0	13	4	35	103	NA	88	NA
Chile (including Santiago)	NA	NA	8	23	220	560	NA	148*

Table 8 Illegal invasions of urban land, Chile, 1966–71. (Units are acts of land invasion, regardless of the number of squatters involved.)

Source: For Chile: Direccion General de Carabineros (cited by FLACSO); for Santiago: FLACSO Survey on Chilean Squatters, 1972 (as cited in Methodological Appendix , p. 364).

*Source references for these particular figures: Ernesto Pastrana and Monica Threlfall, *Pan, Techo y Poder: El Movimiento de Pobladores en Chile 1970–3* (Buenos Aires: Ediciones SIAP, 1974). (They also relied on the Direccion General de Carabineros.)

NA: Not available

form of settlements called *campamentos* to symbolize their political ideology (see Table 8).

When the newly elected socialist President Salvador Allende took office in November 1970, more than 300,000 people were living in these *campamentos* in Santiago alone. At the end of 1972, by which time the number of urban invasions had stabilized, more than 400,000 people were in the *campamentos* of Santiago, and 100,000 or more in the other cities.[10] The main characteristic of these *campamentos* was that from the beginning they were structured around the *Comites Sin Casa* that led the invasion, each of which were in turn organized by different political parties (Duque and Pastrana, 1973)—so much so that we can say that the Chilean *pobladores'* movement was created by the political parties. Of course, to do so they took into consideration the people's urban needs, and they were instrumental in organizing their demands and supporting them before the government. But we can by no means speak of a "movement" of *pobladores,* unified around a program and an organization; it was not, for instance, like the labor movement, which in Chile was unified and organized in the *Central Unica de Trabajadores* (CUT), in spite of political divisions within the working class.

The majority of the *pobladores* were organized by the *Comando de Pobladores de la Central Unica de Trabajadores* (the urban branch of the trade unions), linked to the Communist Party, and by the *Central Unica del Poblador* (CUP), dependent on the Socialist Party. A very active minority constituted itself as the *Movimiento de Pobladores Revolucionarios* (MPR), a branch of the radical organization *Movimiento de Izquierda Revolucionaria* (MIR). Almost 25 percent of the *campamentos* were still under the control of the Christian Democratic Party, and a few settlements were even organized by the National Party (radical right). This whole situation had two major consequences:

1 Each *campamento* was dependent upon the political leadership which had founded it. Political pluralism within the *campamento* was rare, except between Socialists and Communists. (For instance, the largest *campamento,* Unidad Popular, had a joint leadership of both parties.)
2 The participation of the *campamentos* in the political process very closely followed the political line dominating in each settlement. We should actually speak of the *pobladores'* branch of each party, rather than of a "squatters' movement." While all the parties always spoke of the need for unifying the movement, such unity never existed except in moments of political conflict, such as the distribution of food and supplies during the strike in October 1972 launched by the business sector against the government.

This key feature of the movement explains the findings of the field work study we conducted on 25 *campamentos* in 1971.[11] The social world we discovered did not present any major social or cultural innovation. The only exception was the organization of police and judicial functions,[12] which due to the absence of state legal institutions within the squatter settlements allowed (and forced) the *pob-*

ladores to take a series of measures representing a beginning of popular justice. Yet, concerning the urban issues, the *pobladores'* massive mobilization made it possible for hundreds and thousands to obtain, in a few months, housing and services, against the prevailing logic of capitalist urban development. The urban system was deeply transformed by the *campamentos*. But experiences aimed at generating new social practices were limited by the political institutions where the old order was still the strongest force. A good example of such a situation was the Christian Democrats' congressional veto in 1971, opposing Allende's project to create Neighborhood Courts *(Tribunales Vecinales)* based on existing experiences of grassroots justice.

The dependency of the *campamentos* upon the political parties opened the door to their use by each party for its particular interest, lowering the level of grassroots participation. The most conclusive demonstration on this subject is the careful case study done by Christine Meunier on the Nueva La Habana, one of the most mobilized and organized of all *campamentos,* under the leadership of MIR, where she lived and worked between 1971 and 1973, until the military coup.[13] We crosschecked her information with our own observation and interviews in Nueva La Habana in 1971 and 1972, as well as with the demographic and social research conducted also on the same *campamento* by Duque and Pastrana in 1970 and 1971 (Duque and Pastrana, 1973). All the findings by the three independent research teams converge towards a similar picture, the significance of which explore in some detail impels us to the social universe of a *campamento* in order to analyse as conclusively as possible the complex relationship between squatters and parties in the midst of a revolutionary process.

Nueva La Habana was one of the most active, well organized, and politically mobilized *campamentos.* And it certainly was the most highly publicized, both by the media and by the observers of the Chilean sociopolitical evolution, the main reason being that it was considered the "model" *campamento* under the leadership of MIR. The Ministry of Housing expedited its settlement in November 1970 by relocating 1,600 families (10,000 people), with their consent, from three previous MIR-led land invasions (*campamentos* Ranquil, Elmo Catalan, and Magaly Monserato). MIR accepted the relocation of the three *campamentos* in a new 86 hectare urban unit as a challenge that would demonstrate its capacity to organize, ability to obtain housing and services, and effectiveness in transforming the squatters into a revolutionary force. If there was potential for an urban social movement in the *campamentos* of Santiago, we could expect to see it emerge from the mud and shacks of Nueva La Habana.

The strength of Nueva La Habana came from its tight grassroots organization and militant leadership. All *pobladores* were supposed to participate in the collective tasks of the *campamento,* as well as in the decisions about its management. All residents were included in a territorial organization on the basis of *manzanas* (blocks) that delegated one of their members to a board that elected an executive committee *(jefatura)* of five members. At the same time, the most active *pobladores* were invited to form a functional structure, the "fronts of work," both at the level of each block and in the *campamento* as a whole, to take care of the different services that had to be provided for the residents on the basis of resources made available by the government: health, education, culture, police and self-defense, justice, sports, and so on. As a matter of fact, the capacity of the MIR to agitate and the deliberate purpose of Allende's government to limit confrontations with the revolutionary left, led to the paradox that Nueva La Habana received preferential treatment for housing and social services compared to the average squatter settlement.[14] On the basis of the legitimacy acquired by its very effective delivery of services, particularly in the field of health care, MIR frequently asked the *pobladores* to show support for its policies outside the *campamentos,* and it was usual, in all major political demonstrations, to see buses and trucks from Nueva La Habana loaded with *pobladores* waving the red and black flag of the *Movimiento de Pobladores Revolucionarios.* A few dozens of Nueva La Habana's residents were dedicated MIR militants under the leadership of a charismatic and thoroughly honest *poblador,* Alejandro Villalobos, nicknamed El Mike.[15] For the majority of residents, though they were sincere supporters of left wing politics, involvement in the political struggle depended upon issues like the access to land, housing, and services.

The ideological gap between the political vanguard and the squatters was the cause of continuous tension inside the *campamento* during the three years of its life.[16] This tension was expressed, for instance, in the resistance of the residents to the efforts for a cultural revolution in the children's schools, set up by MIR using old buses as classrooms. When the young teachers tried to change the traditional version of Chilean history or to recast the teaching to follow Marxist themes, many parents threatened to boycott the school, forcing the staff to preserve the "official" teaching program. The reason was not that they were necessarily anti-Marxist. but rather that they did not want their children to become exceptional by virtue of receiving a different education from the rest of the city. The *campamento* with its revolutionary folklore, its popular theater group and its 12 meters high Che Guevara portrait, was clearly seen by most residents as a transitory step towards a more "normal" neighborhood, a neighborhood where one could receive visits from friends and relatives from the outside world, who for a long time had been scared to come and visit the squatters living in areas reported as "dangerous" by the press and proclaimed as "revolutionary" by their leadership.

The careful observations by Meunier about the social use of space and housing by the squatters provide a striking illustration of the individualism of the majority of squatters. Most houses, though tiny (with ground measurements of six by five meters), tried to enclose a piece of land, to mark a front yard as a semi-public space, while refusing space for common yards. The shack itself was divided between the main room, where the man could receive visits, and the kitchen-toilet, the private domain of the woman. Only the more enlightened leadership tried to make some space available for public use, but this practice led to spatial segregation: the shacks of the leaders tended to be concentrated towards the center of the *campamento,* close to the shacks used for public purposes. The discrepancy between the level of involvement and consciousness thus became expressed in the spatial organization of the settlement. Individualism was even more pronounced when the residents were called to decide upon the design of their own houses. While asking for architectural diversity (three types of houses were built to fit the different sizes of families), they emphasized the desire for a standard design, utterly rejecting high-rise buildings. They also asked for the individual connection of each house to the water and electricity supply, restated the convenience of individual yards, and specified that the conventional domestic equipment, including television sets and individual electric appliances, would have to have enough room in the new houses. The real dream of most *pobladores* was that Nueva La Habana would one day cease to be a *campamento* and become an average working class *poblacion.*

[. . .]

Yet it should not be deduced that cultural conservatism and political opportunism were the reasons for this attitude. In fact the residents of Nueva La Habana were ready to mobilize in defense of their houses and in defense of their political beliefs each time it was required. They invaded land against police repression during the hard months of 1970. They worked hard to dig sewerage trenches, connect electricity, provide water, build shacks, set up public services, administer their "city," and help each other when required. When, in October 1972, the economic boycott from external and internal capitalist forces halted the distribution of basic foods, the entire *campamento* mobilized to obtain supplies from the factories and the fields, and to distribute a basket to each family for weeks, without asking any payment from those who could not afford it. They also established a new popular morality, banning prostitution, alcohol and alcoholism in the *campamento,* protecting battered women, and taking care of each other's children when the parents were working or involved in political activity. In sum, Nueva La Habana did not refuse its share of mobilization or cultivate a hypocritical attitude towards socialist ideals in exchange for urban patronage. But what was clear to every observer was that such a struggle was a means, and not a goal, for the great majority of the *pobladores,* that Nueva La Habana was an introspective community, dreaming of a peaceful, quiet, well-equipped neighborhood, while MIR's leadership, conscious of the sharpening of the political conflict, desperately wanted to raise the level of militancy so that the entire *campamento* would become a revolutionary force. Their efforts in this direction proved unsuccessful.

On the basis of 20 focused interviews with residents, Meunier hypothesized the existence of three types of consciousness in the *campamento:*

1 The *individual,* focused upon the satisfaction of urban demands through the participation in the squat.
2 The *collective,* whose goals were limited to the success of the *campamento* as a community

through the collective effort of all residents, closely allied to the government's initiatives.

3 The *political,* emphasizing the use of the *campamento* as a launching platform for the revolutionary struggle.

Although her sample is too limited to be conclusive, similar observations can be drawn from the survey by Duque and Pastrana, as well as from our own study. It would seem that the political level was only reached by MIR's cadres, that the mainstream of residents had some kind of collective consciousness while a strong-minded minority maintained an individualistic attitude, though sympathetic to left wing politics. It is crucial for our analysis to try to understand some of the reasons behind each level of consciousness, since Meunier's study concludes with the connection between levels of consciousness and the behavior of social mobilization. While the "collective consciousness" appears to have been randomly distributed among a variety of social characteristics, the two other types tended to be connected with a few significant variables: the "political consciousness" seemed more likely to happen among men than among women, among individuals with lower income, and among unemployed workers (although the fact of being unemployed in 1971's Chile might be a *consequence* of being a revolutionary worker). The "individualists" seem to have been associated with higher income, better than average housing conditions before coming to the *campamento,* and women.

On the basis of these observations, two complementary themes can be noted:

1 The mainstream of the working class in Nueva La Habana probably followed the same pattern found elsewhere, collectively defending their living conditions but leaving the task of general political leadership to the government. A minority group of higher income families joined the invasion to solve their housing problem without further commitment. The radical vanguard of MIR was composed of unskilled workers whose political leanings could surface more easily through the *pobladores'* movement, given the tight political control by communists and socialists in the labor movement (the CUT). This argument, specific to Nueva La Habana, confirms one of our basic general theses on the *pobladores* movement in Chile. A support for this interpretation can be found in the comparison of the occupational structure between Nueva La Habana and the *campamento* Bernardo O'Higgins, the model squatter settlement organized

	Fidel Castro	26 de Julio	Nueva La Habana	Bernardo O'Higgins
Low income	9	29	18	NA
High level of education (primary completed)	15	20	18	24
Self-employed workers	11	13	15	17
Manufacturing workers	39	35	48	53
Service workers	21	27	24	36
Workers in modern industrial companies	16	25	25	28
Workers in large companies (over 50 employees)	36.6	46.8	42.9	37.4
Unemployed	37.6	22	33.2	19.5
High level of urban experience	56.7	65.9	64.6	66.5
Urbanized	30	43	35	44

Table 9 Social composition of four *campamentos*, Santiago de Chile, 1971. (Percentage of residents over the total of each *campamento* who have the listed characteristics.)

Source: Joaquin Duque and Ernesto Pastrana, Survey of Four *Campamentos* (Santiago de Chile: Facultad Latinoamerica de Ciencies Sociales, 1971).

NA: Not available. But other sources indicate that the income levels in Bernardo O'Higgins seems to be noticeably higher than in the other three *campamentos*.

by the Communist Party. Table 9, constructed by us on the basis of the census that Duque and Pastrana took on four *campamentos*, shows that the Nueva La Habana's residents had a lower proportion of well-educated people, a lower proportion of workers in the "dynamic sector" (modern industry), and a much higher proportion of unemployed (33.2 percent against 19.5 percent in Bernardo O'Higgins), and a much lower proportion of unionized workers.

The apparent contradiction that Nueva La Habana came higher in the proportion of workers it had from large factories is a simple statistical artifact: a substantial number of those from Bernardo O'Higgins were skilled and well-paid bus drivers who could not be counted as working in factories. In sum, Bernardo O'Higgins and the Communist Party seem to have relied on the support of the organized working class while MIR and Nueva La Habana seem to have been more successful among the workers of the informal urban economy. We will develop this argument at a more general level, once the profile of Nueva La Habana is complete.

2 Another major factor in Nueva La Habana was that women appear to have been the most reluctant group to follow MIR's revolutionary ideology and the ones who emphasized the satisfaction of basic needs before general political commitment. In fact, this is the main reason advanced by Meunier for explaining the gap in consciousness and mobilization between the vanguard and the majority of the squatters. Meunier lists many examples about the absence of any real transformation in women's roles and lives. She describes how women cooked and ate in the kitchen while serving their husbands and friends in the main room. She describes the sexual domination some cadres (felt) and the difficulty that women faced because their participation in the running of the *campamento* gave rise to suspicions of infidelity. She goes on to describe the difficulties women had in taking advantage of contraception services provided by MIR because the men felt their virility to be threatened. As a result, in 1972, a majority of the women in Nueva La Habana were pregnant. Furthermore, unable to fully participate in the political mobilization, women saw the absence of the men from the house, their unemployment and their political commitment as threats to family life. Separations were common among political leaders and their women as a result of these tensions. This, in turn, widened the gap between single men, who became full-time political activists, and the majority of families, dominated by women's fears and pragmatic feelings. The situation was paradoxical if we consider that MIR, given its strong student basis, was perhaps the one Chilean party that tried the hardest to liberate women and to integrate them fully into politics. But in Nueva La Habana the form that this liberation took deepened the divide between MIR's militant women and the majority of residents. MIR organized a women's militia that took care of a variety of tasks, particularly to do with health but also in matters of self-defense. But this initiative was not supported by a change of attitude of men towards 'their women', who were still unable to participate in the collective activities. So it further isolated the few political women and exposed them to the criticism and distrust of the housewives. Such a dramatic contrast can be illustrated by two events:

a The general blame put by almost the entire *campamento* on a woman whose unguarded child drowned while she was working at the health center.
b The rejection by women of MIR's proposal to close down the mothers' centers (an inspiration of the Christian Democrats) where women met to learn domestic skills, and to replace them by women's centers which would emphasize women's militant role. Most women felt that such a change would politicize their free space, depriving them of their capacity to autonomously decide how to use these centers. Thus, the mothers' centers continued to function in the heart of revolutionary squatters' settlements.

Although they were unable to challenge the machismo prevalent in the settlements, most resident women rejected MIR's heavy-handed politicization of women's issues and became the prime mover for the use of urban mobilization strictly for the improvement of their conditions.

So Nueva La Habana lived entirely under the shadow of MIR's initiatives. New housing was built, urban infrastructures were provided, health and education services were delivered, cultural activities were organized, goods were supplied, prices were controlled, moral reform was attempted, and some form of democratic self-management was implemented, although under the unchallenged leadership of the *miristas* (MIR militants). Yet the social role of the *campamento* shifted according to the political tasks and priorities established by MIR at the national level. In the first year, MIR supported urban demands as a means of consolidating its position in the squatter movement so as to reinforce its militant power. In May 1971, in the congress of Nueva La Habana residents, the leadership announced that top priority should be given to the penetration of the organized working class by MIR, to counteract reformist forces in the labor movement. So most of the cadres of the *campamento* were sent to other political duties, leaving the *pobladores* in a support role for the main struggle being fought in the work places. The working class ideology of MIR came into open contradiction with its militant presence among the squatters, and there followed in 1971–72 a period of disorientation in the *campamento,* leading to demobilization and in-fighting.

The general mobilization in October 1972 against the conservative offensive in Chile led to a new role for the *campamento,* first, in the battle over distribution and, later on, in the support of the construction of *cordones industriales* (industrial committees) and *comandos comunales* (urban unions) as centers of revolutionary, popular power. Militants from Nueva La Habana tried on 3 April 1973 to occupy the National Agency of Commercial Distribution (CENADI) as a gesture in favor of this strategy: they built barricades in Santiago's main avenue, Vicuna Mackena, and clashed fiercely with the police for a whole day. Nueva La Habana subsequently kept its subordinate role as a branch of a political party, adopting a variety of tactics, corresponding to the different directions taken by MIR's political activity. So if the 'model *campamento*' was an expression of the militant squatters' capacity to build their city and to try new communal ways of life, it was also, above anything else, an organizational weapon of a revolutionary party.

In our own research with the CIDU team we came to a similar conclusion: for all *campamentos,* whatever their political orientation, the practice of the squatters was entirely determined by the politics of the settlement, and the political direction of the settlement was, in turn, the work of the dominant party in each *campamento.*

The same finding was obtained by studies on mobilization in the *conventillos* (central-city slums) (Pingeot 1976) and in the neighborhood-based demand organizations (Alvarado, Cheetham and Rojas 1973). Perhaps the only exception consisted of the committees organized to control prices and delivery of food, the *Juntas de Abastecimientos y Precios,* where a high level of popular autonomy was observed, cutting across the political membership. Even so, the practice of these committees was very different: they either aimed to reinforce the socialist government or to structure a dual power, depending upon the political tendency of their leadership.[17]

The political stand of the *pobladores* was a decisive element in enabling the formerly passive popular urban sectors to join the crucial political battle, initiated by the working class movement for the construction of a new society. Thus we could speak, in this case, of an urban social movement because the popular masses were mobilized around urban issues and made a considerable political contribution to the impetus for social change.

Nevertheless, the partisan and segmented politicization of the *pobladores* made it impossible for the left to expand its influence beyond the borders of the groups it could directly control. The different sympathies held in each *campamento* hardened into political opposition among different groups, a situation quite unlike that of the trade unions, where Christian Democratic workers often backed initiatives from the left such as the protest against the boycott of Chile by the international financial institutions. Furthermore, the separating of political forces in the squatter settlements led each group to seek support in the administration, splitting the whole system into different constellations of state officials, party cadres, and *pobladores,* each with its own political flag.

The social effects of this development became evident in the changing relationships between the *pobladores* and Allende's government. During the first year (November 1970 to October 1971) the difficulties of putting the new construction industry to work made it impossible for the government to satisfy the *pobladores*' demands, and the only thing it could do was to accept the land invasion and provide some elementary services by putting the squatters in touch with public agencies. In spite of the government's inability, the *pobladores,* including Christian Democrats, collaborated actively with the administration. In contrast, at the end of the second year, when 70,000 housing units were under way and when health, education, and other services started to be delivered, some serious signs of unrest appeared among the *pobladores*. Finally, after October 1972, when the political battle became inevitable, each sector of the *pobladores* aligned with its corresponding political faction, and the squatters movement disappeared as an identifiable entity.

To understand the evolution of this attitude towards the government, we must consider the social class content of the political affiliations in the Unidad Popular as well as the social interests represented by the squatter settlements. In the first year the government, taking advantage of the political confusion in business circles, successfully implemented a series of economic and social reforms which substantially improved the level of production and standard of living. This policy obtained important popular support, as did the preservation of political freedom and social peace. The political debate was kept within institutional boundaries, and the opposition of the Christian Democrats was moderate.

During the second year, however, the economy deteriorated rapidly, due to the sabotage of the economy by the Chilean business sector and landlords, the international boycott, and the end of the benefits· of using formerly idle industrial capacity. The political alliance between the center and the right isolated the Unidad Popular. The radicalization of some popular sectors exacerbated the situation and was used as a pretext for political provocation. International pressure against Allende was reinforced. There was one popular working sector that was particularly sensitive to political alienation, especially in these conditions: workers in small companies, the reason being that, although not included in the nationalized sector, they were asked to restrain their own demands to preserve the alliance with small business. This sector of workers was actually non-unionized since Chilean laws before 1970 prevented union membership in companies with less than 25 employees, with the result that they tended to be politically less conscious than the unionized working class. So the Unidad Popular government asked for a more responsible effort by this less organized and less conscious segment of the working class. The reaction was a series of errant initiatives, sometimes very radical, sometimes conservative, and generally out of tune with mainstream policy of the popular left. Now all surveys show that this sector of small companies' working class—the workers of the 'traditional sector' or 'informal economy'—represented in Chile the most important share of the squatter settlements . . . and therefore the *pobladores'* movement organized the main expression of this social group. But as we have just noted, the squatters' movement was split into political factions and so reflected the disorientation of this group. Instead of becoming factions and so a focus for organization and mobilization the movement only occasionally enabled them to manoeuver against the Unidad Popular government.

During the third year, the left of the Unidad Popular, as well as MIR, tried to build up a people's power base strong enough to oppose the ruling classes by developing territorially based organizations which combined both the *comandos comunales* and *cordones industriales* (Cheetham et al. 1974). But because many people actually leaned towards the center-right, and most workers were following the government, this grassroots movement only gathered a vanguard of industrial workers in some sectors of the big cities, particularly in the area of Cerrillos-Maipu in Santiago. In fact the squatters' movement disappeared as an autonomous entity in the decisive moments of 1973, and was less than ever the unified movement around which the left might have organized some popular sectors, people who in the event supported the center-right as much as the left. Not only did the political influence of the *pobladores* movement wane but, as part of the right wing offensive, the far right organized some middle class neighborhoods along sociopolitical lines (the Proteco organizations), linking the provision of local services to preparations for the military coup. The disappearance of the *pobladores'* movement, then, in 1972–3 was the consequence of the logic of party discipline replacing the search by the left to establish political hegemony.

The only moments of mass participation in the squatters' movement were those where political parties of the left gathered around a clear-cut common cause: the first year of Allende's government, and the mass response to the business strike of October 1972. In both situations, the squatters' movement started to produce a new urban system corresponding to the political transformation of the state by Allende's government. On both occasions, we could observe its potential as a mass movement and as a social movement. As a mass movement, it gathered and organized a larger proportion of people than the left wing parties could. As a social movement it started to produce substantial changes to urban services and the local state institutions because of its capacity to mobilize people. Both moments were exceptional but too short-lived: the unity and the cultural influence of the squatters' movement did not survive the polarization of the political opposition existing inside it.

The squatter movement in Chile was potentially a decisive element in the revolutionary transformation of society, because it could have achieved an alliance of the organized working class with the unorganized and unconscious proletarian sectors, as well as with the petty bourgeoisie in crisis. For the first time in Latin America, the left understood the potential of urban movements, and battled with populist ideology on its own ground, planting the possibilities of political hegemony among urban popular sectors. But the form taken by this political initiative, the overpoliticization from the beginning, and the organizational profile of each political party within the movement, undermined its unity and made the autonomous definition of its goals impossible. Instead of being an instrument for reconstructing people's unity, the *pobladores'* movement became an amplifier of ideological divisions. Yet its memory will last as the most hopeful attempt by Latin American urban masses to improve their social condition and achieve political liberation.

NOTES

1 See the collection of data and analyses presented in Abu-Lughod and Hay (1977).

2 See Leeds and Leeds (1976).

3 See E. Henry (1974, 1978); David Collier (1976).

4 They always start in the periphery of the city, but with the expansion of urban space, some of the early *barriadas* are now located in the core of the metropolitan area.

5 See S. Eckstein (1977).

6 See J. Labastida (1972).

7 For an analysis of the overall process, see M. Castells (1975) and B. Stallings (1978).

8 See M. Castells (1975) and E. Pastrana and M. Threlfall (1974).

9 This according to E. Santos and S. Seelenberger (1968).

10 For quantitative data on the evolution of the squatter movement, see E. Pastrana and M. Threlfall (1974).

11 See *Equipo de Estudios Poblacionales del CIDU* (1973).

12 See *Equipo de Estudios Poblacionales del CIDU* (1973) and F. Vanderschueren (1979).

13 See the extraordinary study by C. Meunier (1976).

14 According to one of the officials in charge of housing policy under Allende, Miguel Lawner. See Lawner and Barrenechea (1983).

15 Alejandro Villalobos actively resisted the military dictatorship in Chile. He was murdered in the street by the junta's political police in 1975, but his death was officially announced as a "clash with guerrillas."

16 Nueva La Habana suffered a fierce repression and became an impoverished shantytown renamed Amanecer (Dawn) by the junta.

17 See C. Cordero, E. Sader and M. Threlfall (1973).

BIBLIOGRAPHY

Abu-Lughod, J. and Hay, R. (eds.) (1977) *Third World Urbanization,* Chicago: Maaroufa Press, and London: Methuen.

Alvarado, L., Cheetham, R. and Rojas, G. (1973) "Movilizacion Social en Torno al Problema de la Vivienda," *Revista Latino-americana de Estudios Urbanos Regionales,* 7.

Bengoa, J. (1972) *Pampa Irigoin: Lucha de Clases y Conciencia de Clases,* Santiago: CESO, Universidad de Chile.

Castells, M. (1971) "L' Urbanisation dependante en Amerique Latine," *Espaces et Societes,* 3.

Castells, M. (1973) "Movimiento de Pobladores y Lucha de Clases en Chile," *Revista Latinoamericana de Estudios Urbanos y Regionales,* 9.

Castells, M. (1975) *La Lucha de Clases en Chile,* Buenos Aires: Siglo XXI.

Cheetham, R. (1971) "El Sector Privado de la construccion: Patron de dominacion," *Revista Latinoamericana de Estudios Urbanos Regionales,* 3.

Cheetham, R., Rodriguez, A. and Rojas, J. (1974) "Comandos Urbanos: Alternativa de Poder Socialista," *Revista Interamericana de Planificacion,* VIII(3).

Collier, D. (1976) *Squatters and Oligarchs: Authoritarian Rule and Policy Change in Peru,* Baltimore: Johns Hopkins University Press.

Collier, D. (1978) *Barriadas y Elites: De Odria a Velasco,* Lima: Instituto de Estudios Peruanos.

Collier, D. (1979) *The New Authoritarianism in Latin America,* Berkeley: University of California Press.

Cordero, C., Sader, E. and Threlfall, M. (1973) *Consejo Comunal de Trabajadores y Cordon Cerrillos-Maipu 1972: Balance y Perspectivas de un Embrion de Poder Popular,* Santiago: CIDU, Working Paper 67.

FIVE

Duque, J. and Pastrana, E. (1973) "Le movilizacion reivindicativa urbana de los sectores populares en Chile: 1964–1972," *Revista Latinoamericana de Ciencias Sociales,* 4.

Eckstein, S. (1977) *The Poverty of Revolution: The State and the Urban Poor in Mexico,* Princeton: Princeton University Press.

Equipo de Estudios Poblacionales del CIDU (1973) "Experiencia de Justicia Popular en Poblaciones," *Cuadernos CEREN,* 8.

Henry, E. (1974) *Urbanisation Dependante et Mouvements Sociaux Urbains,* Paris: Universite de Paris; Ph.D. thesis, 4 vols.

Henry, E. (1977) "Los Asentamientos Urbanos Populares: una Esquema Interpretativo," *Debates* 1: 1.

Henry, E. (1978) *La Escena Urbana,* Lima: Pontifica Universidad Catolica del Peru.

Labastida, J. (1972) "Los Grupos Dominantes Frente e las Alternativas de Cambio," *El Perfil de Mexico en 1980,* Mexico, DF: Siglo XXI, Vol. 3.

Lawner, M. and Barrenechea, A. (1983) "Los Mil Dias de Allende: la Politica de Vivienda del Gobierno Popular en Chile," in M. Castells (Ed), *Estructura Social y Politica Urbana en America Latina,* Mexico, DF: EDICOL.

Leeds, A. and Leeds, E. (1976) "Accounting for Behavioural Differences in Squatter Settlements in Three Political Systems: Brazil, Peru and Chile," in L. Massoni and J. Walton (Eds.), *The City in a Comparative Perspective,* Beverly Hills: Sage.

Meunier, C. (1976) *Revendications Urbaines, Strategie Politique et Transformation Ideologique: le Campamento Nueva La Habana, Santiago, 1970–73,* Paris: Universite de Paris; Ph.D. thesis.

Montano, J. (1976) *Los Pobres de la Ciudad en los Asentamientos Espontaneos,* Mexico City: Siglo XXI.

Navarro, B. and Moctezuma, P. (1981) "Clase Obrera, Ejercito Industrial de Reserva y Movimientos Sociales Urbanos de las Clases Dominadas en Mexico," *Teoria y Politica,* 2.

Pastrana, E. and Threlfall, M. (1974) *Pan, Techo y Poder: el Movimiento de Pobladores en Chile, 1970–3,* Buenos Aires: Ediciones STAP.

Pingeot, F. (1976) *Populisme Urbain et Crise du Centre-Ville dans les Societes Dependantes: Santiago-Du-Chili 1969–7,* Paris: Universite de Paris; Ph.D. thesis.

Pueblo, E. (1982) *Surgimiento de la Coordinadora Nacional del Movimiento Urbano Popular: Las Luchas Urbano-Populares en el Momenta Actual,* Mexico, DF: CONAMUP.

Rodriguez, A. and Riofrio, G. (1974) *Segregacion Social y Movilizacion Residencial: el Caso de Lima,* Buenos Aires: SIAP.

Rojas, J. (1978) *La Participation Urbaine dans les Socieres Dependantes: l'Experience du Mouvement des Pobladores au Chili,* Paris: Universite de Paris; Ph.D. thesis.

Safa, H. (ed.) (1982) *The Political Economy of Urbanization in Third World Countries,* New Delhi: Oxford University Press.

Santos, E. and Seelenberger, S. (1968) *Aspectos de un Diagnostico de fa Problematica Estructural del Sector Vivienda,* Santiago: Escuela de Arquitectura: Universidad Catolica de Chile.

Stallings, B. (1978) Class *Conflict and Economic Development in Chile, 1958–1973,* Stanford: Stanford University Press.

Unikel, L. (1971) *El Desarrollo Urbano de Mexico,* Mexico, DF: El Colegio de Mexico.

Urrutia, C. (1972) *Historia de fa Poblaciones Callampas,* Santiago: Quimantu.

Vanderschueren, F. (1971) "Significado Politico de las Juntas de Vecinos," *Revista Latinoamericana de Estudios Urbanos y Regionales,* 2.

Vanderschueren, F. (1979) *Mouvements Sociaux et Changements Institutionnels: L'Experience des Tribunaux Populaires de Quartier au Chili,* Paris: Université de Paris; Ph. D. thesis.

Villarreal, D. (1979) *Marginalité Urbaine et Politique de l'Etat au Mexico: Enquéte sur les Zones Residentielles Illegales de la Ville de Monterrey,* Paris: Université de Paris, Ecole des Hautes Etudes en Sciences Sociales.

Weisslitz, J. (1978) *Développement, Dépendance, et Structure Urbaine: Analyse Comparative de Villes Péruviennes,* Paris: Universite de Paris; Ph.D. thesis.

"Global Mobility, Shifting Borders and Urban Citizenship"

Tijdschrift voor Economische en Sociale Geografie (2009)

Michael Peter Smith and Luis Eduardo Guarnizo

INTRODUCTION

Global migrants live and work in cities and towns in countries other than those in which they were born. Added to the hypermobility of both skilled and unskilled labor migrants to localities in the global North, increasing numbers of retirees from affluent Northern countries now live and invest in Southern localities where they can afford a more comfortable and enjoyable life than in their homelands. International tourism also has experienced rapid and steady growth, reaching a record 903 million tourists in 2007 (UN World Tourism Organisation 2008). The sheer size, heterogeneity, and multi-scalar scope of all these mobilities and the relationships and social structures they engender across national borders, have encouraged the formation of complex transnational connections between multiple localities across the world, helping transform the local, national, and international contexts in which citizenship is defined, bestowed, and exercised. The once taken for granted correspondence between citizenship, nation, territory, and state has been called into question as new forms of supra-national and sub-national membership and belonging have taken on an increasingly trans-territorial character. The increasingly transnational character of global migration flows, cultural networks, and grassroots political engagement has dramatically changed the discourses and practices of citizenship in the past two decades. Some nation-states are extending the inclusiveness of citizenship well beyond their national territorial jurisdiction; others are trying to erect further barriers of exclusion. Some analysts welcome these changes as progressive transformations, while others categorize them as dangerous and destabilizing developments that should be curtailed. Changes in the institution of citizenship have not been consistent, straightforward and definitive. Rather, they have been inconsistent, complex, tentative and contested. Thus, categorical conclusions about the theoretical and practical implications of these changes should be avoided and replaced by historical, non-ideological and nuanced analyses. This is what we intend to do in this paper.

We begin by considering ... key political economic developments operating at the global scale that have unsettled the established view of the close correspondence between nationhood and citizenship. A key effect of global migration has been that many local residents have moved to places throughout the world in which they lack national citizenship and are thus alienated from the means of democratic participation. This disjuncture between the spaces of citizenship and daily life, in turn, has led to a political devolution of citizenship claims-making from national to urban space. The heart of our analysis of this disjuncture is our assessment of the uses and limits of the increasingly voluble discourse on "the right to the city" as a way to create alternative political spaces where variously excluded groups of globally mobile urban inhabitants may empower themselves in the face of the unsettled terrain of national citizenship.

We offer an analytical framework suitable for overcoming the limits of this perspective that accounts for the causes and effects of the rising demand for urban citizenship as the twenty-first century moves forward.

[. . .]

THE RIGHT TO THE CITY: EMPOWERING AND DISEMPOWERING LOCAL INHABITANTS

Global mobility is the engine of global urbanization. Today, 3.3 billion urban residents constitute more than half of the world's population. By 2030, world urbanization is projected to rise to nearly five billion and three of every five people will live in cities (UN Population Fund 2007). In the next 50 years, two-thirds of the world's population will be urbanized (International Alliance of Inhabitants 2008).

Rural–urban migration, identified by the Chicago School of sociology a century ago as a constitutive element of modernity, has taken a new turn. Rural inhabitants are moving not only within national borders in the global South, but also from the rural South to the urban North. This fact has generated multiple points of view about the consequences of the arrival of new urban inhabitants in the North. These range from optimistic hopes of developing progressive, super-diverse, and inclusive kinds of transnational cities and migrant citizenship, to pessimistic, exclusionary and xenophobic fears of the destruction of preexisting national cultures.

The "right to the city" (RTTC) argument has been advanced as an alternative theoretical construction as well as a slogan motivating a new transnational social movement. While we are very impressed with the effectiveness of this slogan in practice, our concern lies primarily with the shortcomings of the theoretical underpinning of this construction. Urban geographer Mark Purcell (2002, 2003) has most fully explicated its theoretical logic. Drawing upon the pioneering work of Lefebvre (1968, 1973), Purcell deconstructs the RTTC into two main dimensions—the right of city dwellers to participate in decision-making processes that affect the quality of city life, and their right to appropriate and use urban space. Accordingly, inhabitance is privileged as a way to enable city dwellers to promote use values over exchange values in the production and use of urban space.

The RTTC argument is a direct response to the political disenfranchisement of urban residents resulting from global mobility, the neo-liberal restructuring of the global economy, and the rescaling of the state (Purcell 2003). Urban residents have been disenfranchised by new forms of "governance" dubbed public–private partnerships, which privatize major decisions over the development of urban space. A primary goal of city governments is to promote economic development connected to global capital circuits. They no longer seek to directly improve the everyday living conditions of the inhabitants, but rather to create environments attractive to corporate investments in global competition with other cities. To do this, cities have developed a wide array of tools, for example, creating special taxation districts controlled by private interests, reducing demands for social services by expelling low income populations from central city areas and promoting the image of safe and secure downtowns. To stem this imbalance of power, RTTC advocates propose to valorize the power of all city inhabitants, independent of their national citizenship, to shape the decisions regulating the use of urban space. This implies not only the inhabitants' capacity to affect local governmental decision-making, but also private corporate decisions affecting urban space, the built environment and socio-cultural life, including labor markets and environmental impacts.

Bypassing conventional regimes of national citizenship, the RTTC argument incorporates all residents of the city. Inhabitance becomes a privileged status, granting citizens and non-citizens alike a right to participate in public policy-making, as well as in decisions of private corporations affecting urban life chances. Accordingly, Purcell argues that, because the right to the city operates simultaneously at different scales and affects different territories, it thus empowers urban dwellers to affect all corporate and governmental decision-making, even if those decisions are actually taken outside the city proper. Residents of Los Angeles, for example, would be given a seat at the corporate table where decisions are made about the location and relocation of industrial plants into or away from Los Angeles, for these decisions are likely to affect labor markets and living conditions for Los Angelinos.

Lefebvre argued that urban inhabitants should have the right to participate centrally in all decisions that produce urban space. Purcell further pushes this argument several scales. He argues that a hypothetical decision by the Mexican government to change land tenure policies in the state of Oaxaca would be likely to strongly affect immigration patterns from that Mexican state to Los Angeles. Therefore, under the RTTC perspective, inhabitants of Los Angeles would have the right to participate centrally in the Mexican national government's decision-making process concerning Oaxacan land tenure policies because such a decision would likely change Los Angeles's population geography and urban space (Purcell 2002: 104). Left out of this equation are the rights of Mexican citizens residing in Oaxaca to shape their own localities.

In practical terms, the RTTC theoretical discourse has found expression in a global social movement, the International Alliance of Inhabitants (IAI). Founded in Madrid in September 2003, the IAI has brought together a large network of grassroots associations of urban inhabitants from many parts of the world. It seeks to coordinate actions to "jointly stand against the perverse effects of exclusion, poverty, environmental degradation, exploitation, violence, and problems related to transportation, housing and urban governance produced by the neo-liberal globalization" (IAI 2011).

[...]

The RTTC, thus, embodies the normative reinvention of national citizenship as urban citizenship. This is the vision imagined not only by right to the city theorists like Lefebvre and Purcell, but also by more mainstream political theorists like Ranier Bauböck, who has advocated the creation of "a formal status of local citizenship that is based on residence and disconnected from nationality" (Bauböck 2003: 139).

In practice, the RTTC movement seems to provide a mechanism for greater inclusiveness in today's burgeoning multicultural cities, perhaps even offering the first step in the development of a transnational form of multicultural municipal citizenship. Yet its theoretical assumptions are fraught with what Purcell recognizes as radical open-endedness. The RTTC approach, by emplacing global relations in urban space, also runs the risk of essentializing urban life and place-making. Inadvertently, the combination of radical open-endedness and making "place" the basic site of citizenship can result in quite unexpected reconfigurations of power. Consider the following two less than salutary possibilities.

[...]

MAKING SENSE OF CONTESTED CITIZENSHIP

Today, more people, with more or less power, are moving across the globe finding more or less resistance and more or fewer opportunities in the localities to which they move. In practice the RTTC approach is a strong and progressive response to contest the marginalization of southern immigrants in hyper-diverse northern cities, or even poor, racialized rural minorities in urban centers in the South. However, this approach, at least as presently theorized, seems to flatten the actual heterogeneity of the new inhabitants from abroad into a homogeneously poor, marginalized and powerless social grouping. Most fundamentally, the RTTC perspective has a large blind spot that prevents it from detecting and incorporating migrants' persistent transnational interests and engagements that are embodied in complex global sociopolitical, communicative and multifaceted microstructures extending well beyond the city limits.

Tensions between established residents and newcomers, particularly when the power relation between is them is asymmetrical, are unavoidable. The question of "inhabitance" is often fought out along the lines of who has the power to decide who is an established resident, legitimate local actor, or who is acceptable as a new resident and, thus, who has the right to local sociopolitical, cultural, and economic space and who does not. This is a common theme in the cases discussed above, whether it is affluent northern retirees in historic villages in Mexico, poor immigrants in Los Angeles or Providence, or claims to historical presence in South Ossetia, Georgia and the Crimea.

In the last instance, "the politics of municipal citizenship" is a question of power. Who has the power to make place out of space, who contests this power, who wins, who loses, and with what effects? These issues are further complicated in a context of high, multidirectional mobility embedded in transnational relations that may reach a global scale. In this context, empirical research on municipal citizenship needs to address these central questions of power. If it is not possible to regard the RTTC logic of inhabitance as a cure-all to the local effects of globalization and the limits of national citizenship, how can we best understand and map a practical way forward for reconfiguring citizenship in the decades ahead?

Like national "citizenship" itself, "inhabitance" is an abstract status associated with urban belonging. Both formal statuses ignore the actual subject positions that people making claims to citizenship or inhabitance actually occupy, speak from, fight for, and fight over. Urban and national belonging and political participation are shaped by actors situated in particular locations based on social relations of difference such as class, gender, generation, ethno-racial identities, religious affiliation, formal legal status, and so forth. This is precisely what politics is all about.

The approach that we propose is comparative, multi-scalar, and sensitive to the dialectical relationships between micro-, meso- and macro-structures. Our approach starts from a close examination of the role of human agency and historically specific individual and institutional actors in producing urban change. Comparatively, our approach has two dimensions, namely, historical and cross-sectional. Historically, we need to understand the long-term changes in the dominant frameworks for organizing economic, political, and social relations as, for example, in the shift from the post-Second World War world "order" centered on the Bretton Woods global institutions of governance and economic management built around state-centered conceptions of national citizenship, sovereignty, and development to the market-centered neo-liberal project in which corporate domination is enshrined, while state power is diffused, privatized and refocused on new techniques of governmentality, to the current global fiscal crisis where state power over economic management is being reasserted. Cross-sectionally, we need to understand how these ruling ideologies differentially affect different groups of people located in different places around the globe. This implies comparing the experiences and consequences of the different mobilities of Southern, as well as Northern peoples, as we have tried to do in the examples above.

Our framework is thus multi-focal. It recognizes that while keeping one eye on the developing global, national, transnational, and local structural contexts within which human mobility occurs, it is also necessary to focus on the agency of the mobile subjects operating at all these scales. Our view defines transnational politics as the study of processes of incorporation, accommodation, and resistance to changing structural opportunities and constraints. These processes do not operate in a vacuum or a world of pure voluntary action. Rather, mobility is embedded in a web of historically changing opportunities and constraints operating in the context of economic globalization mediated by the institutional practices of nation-states and highly variable local contexts of reception. In this process, new local and transnational political spaces are emerging as myriad state and non-state actors seek to reposition or "replace" themselves in the wider political economic setting of global mobility.

Our approach is consistent with that of Ong who has proposed a "transversal" mode of perception. Her "analytics of assemblage" views the logics underlying the application of neoliberal policies as not taking place uniformly in a single social space. Rather than viewing neoliberalism as an economic tsunami sweeping up everything in its wake, she conceives it as a "migratory set of practices" that articulate diverse situations and engender diverse configurations of possibility both within and across countries for the key actors we have already identified. Accordingly, states, market driven forces, and different civil society actors recalibrate their respective capacities in relation to "the dynamism of global markets" (Ong 2007: 4). While recognizing the multiple logics of neoliberal globalization, the analytics of assemblage is a metaphor for the situated practices that come together in specific places, at particular times in history. It is within this historically contingent mode of analysis that we have framed our specific narratives illustrating the complex, non-linear and often unpredictable realities operating at the intersection between human mobility, neoliberal globalization, and the reconfiguration of national and urban citizenship.

REFERENCES

Bauböck, R. (2003). "Reinventing urban citizenship," *Citizenship Studies* 7, 139–160.

International Alliance of Inhabitants (2008). HTTP available at: http://eng.habitants.org/the_urban_way/word_urban_forum_2008.

International Alliance of Inhabitants (2011). "Who we are." HTTP available at: http://eng.habitants.org/who_we_are/.

Lefebvre, H. (1968). *Le Droit à la Ville*, Paris: Anthropos.

Lefebvre, H. (1973). *Espace et Politique*, Paris: Anthropos.

Ong, A. (2007). "Neoliberalism as a mobile technology," *Transactions of the Institute of British Geographers* 32: 3–8.

Purcell, M. (2002). "Excavating Lefebvre: The right to the city and its urban politics of the inhabitant," *GeoJournal* 58: 99–108.

Purcell, M. (2003). "Citizenship and the right to the global city," *International Journal of Urban and Regional Research* 27: 564–590.

UN Population Fund (2007). *State of the World Population 2007: Unleashing the Potential of Urban Growth*, New York: UNFPA.

UN World Tourism Organisation (2008). *Tourism Highlights*, UNWTO Publications Department. HTTP available at: http://www.unwto.org/facts/menu.html.

"Cyberactivism and Citizen Mobilization in the Streets of Cairo"

Essay written for *Cities of the Global South Reader* (2015)

Sahar Khamis and Katherine Vaughn

Unlike the popular belief that Internet activism, through Facebook posting, blogging and tweeting, gave birth to the Egyptian revolution, the stage for national change had been already set in Egypt by existing protest movements and a network of activist groups that had learned from their previous attempts at affecting change. In fact, January 25, 2011 was not the first anti-government popular protest called in Egypt. The Egyptian people were frustrated with the high levels of corruption, dictatorship, economic distress, and humiliation they had been suffering for a long time. This had given birth to several active protest movements in the Egyptian political arena, including the *Kefaya* movement (which means "enough" in Arabic); the Muslim Brotherhood, which was active for a long time, despite its officially banned status; and Ayman Nour's political party (*Hizb el Ghad*). These movements, however, were not able to achieve public mobilization on a massive scale. Able to attract only a few hundred participants made it easy for the police to crack down on their marches and protests.

The popular protests in Egypt in 2011 prevailed because they were able to bring a much larger number of people into the streets. Using the new media technologies, activists were able to spread the word about planned protests and, thus, ensure a wider base of support for popular mobilizations. While the use of Internet technology allowed the word to spread more, it could not replace the important role of street activism, which preceded the actual revolution and paved the way for it. This is especially important, as Egyptian political activist and member of the "National Coalition for Change" Mohamed Mustafa (2011) explains, because not everyone in Egypt has Internet access. To reach out to those whose sentiments are in support of the revolution but do not have access to the Internet, street activism is irreplaceable.

Moreover, the wider spread of information about protests will not necessarily result in actual protest participation. The limits of social media for democratic movements becomes more evident considering a 2008 incident involving the April 6 Movement. This movement was named for its first effort, before the revolution, to organize a labor strike in the Nile Delta city of *El-Mahalla el-Kubra* on April 6, 2008 (Rosenberg, 2011). The movement used cell phones, blogs, Twitter, Facebook, and YouTube to document police excesses, organize meetings and protests, alert each other to police movements, and get legal help for those who had been arrested (Ishani, 2011; Nelson, 2008). But while Facebook attracted many sympathizers online, it was unable to organize them offline (Rosenberg, 2011). As a result, the demonstrations were disorganized, did not spread widely, and fell apart during a police crackdown.

This story is a sober reminder of the real-world risks and limitations of organizing protests online only, since both online and offline efforts are needed for a protest to succeed. However, the success of protest movements aided by cyberactivism is not just a matter of technology allowing citizens to come together in common cause against the state by knowing how their fellow citizens feel about the regime. Rather, it is a matter of technology allowing people, as Marc Lynch (2011) rightly remarks, to calculate the risks of protesting and weigh the chances of succeeding, based on a more accurate and better informed realization of the sheer volume of popular support and shared sentiments, in deciding how much they were ready to sacrifice for freedom. In other words, a strong political dissatisfaction and resentment of the ruling regime was shared among people at large for a long time, but there was a missing link between public anger and the possibility of actually mobilizing the public to bring about real change. Political activism in the real world, aided by cyberactivism in the virtual world, created this missing link.

THE USES AND LIMITS OF CYBERACTIVISM

To fight against state political oppression, long before the 2011 protests many Egyptian tech-savvy activists had reached out to the international community to educate themselves on new technologies for bypassing state controls. Some activists, like members of the April 6 Movement, received technical advice from the Italian anarchist party on how to use "ghost servers," which "bounce Internet searches to nonexistent servers to confuse any online monitoring, allowing users to share information and continue coordinating their activities in heavily monitored digital and telecom environments" (Ishani, 2011). Others worked with the Kenyan NGO Ushahidi to learn how to securely capture raw video with their cell phones during events and use it to build credible online content. Moreover, some Egyptian activists received training from US-based NGOs to use mapping tools, such as Google Maps and UMapper, to document protests and choose demonstration sites (ibid.).

Different social media tools lend themselves to different types of networks. Facebook, for example, is built on linkages between trusted circles of "friends," whereas Google Moderator and Twitter allow anyone to comment on a subject. Google Moderator allows for commentary and voting on subjects by all users. Twitter allows users to create a subject for discussion and post a comment, or "tweet," about that subject (which could include a link to other content), which can then be picked up by other users and "retweeted" multiple times, until it becomes widespread. Thus, tools like Twitter lead to an environment where the best ideas and content, regardless of who posted it, can spread and gain great influence in a type of "meritocracy" of ideas and information (Maher, 2011). By combining these multiple functions of different types of online media together in one effective communication network during the January 2011 revolution, it is easy to understand how Egyptian political activists won their battle against the regime, both online and, most importantly, offline.

For example, because of security concerns, most of the activists' sensitive planning occurred offline to avoid detection, especially during the early stages of mobilization. Moreover, "when technology was used, it was private and one-to-one (SMS, phone calls, GChat), unlike social media, which is public and many-to-many" (Joyce, 2011). In general, as Brisson et al. (2011, p. 28) report, "technology was only marginally, if at all, used by several factions critical to the revolution. Even mobile phones, while near ubiquitous, were little used in campaigns by the labor movement and the judiciary." This was mainly due to many of the mobilized groups not being Internet users or being fearful of Internet surveillance. This again reminds us of the crucial role played by on-the-street public mobilization, both before and during the revolution.

Once people were in the streets, Facebook and similar platforms were less immediately relevant, but other online tools were still important for coordination, such as maps made with Google tools and SMSs to alert protesters to sniper locations. Twitter was used simultaneously for citizen journalism and on-the-ground moment-by-moment mobilization during street protests. For example, through tweeting locations of gatherings, progress and attacks, protesters succeeded to safely gather people as they marched through back alleys. Moreover, images were posted on Twitter showing satellite maps marked with arrows indicating where protesters could go to avoid pro-government thugs (Meier, 2011).

While the Egyptian activists combined their strong public will and determination for change with the effective utilization of new media to achieve political reform and democratization, the Egyptian government combined its incompetent political strategy with an equally ineffective communication strategy that not only failed to halt political activism, but even fueled it further. The regime realized too late that many, if not most, of the people in the streets were not Internet users (Ishani, 2011). In fact, as Egyptian academic Adel Iskander (2011) rightly remarks, some of the protesters in Tahrir Square have never heard of Facebook before, but they were energized and inspired by the huge numbers of people flooding to the streets each day.

Beside the evident role of civic engagement in the Egyptian uprising, citizen journalism also played a major role. It was inspired by examples from other mobilizations, such as those in Iran and Tunisia, where state-controlled television prevailed before their uprisings. People "no longer had to read stifled accounts in state-run newspapers when they could go on the Internet and hear from ... protesters directly through social networks" (Idle and Nunns, 2011, p. 26).

Moreover, social media's horizontal and non-hierarchical structure was empowering for women, who not only engaged in online activism and citizen journalism through social media, but also effectively and courageously participated in demonstrations and protests. At the peak of the protests in Egypt, for example, roughly one quarter of the million protesters who poured into Tahrir Square each day were women—"Veiled and unveiled women shouted, fought and slept in the streets alongside men, upending traditional expectations of their behavior" (Otterman, 2011). Even after the revolution, Egyptian women are mobilizing to ensure a gender inclusive democracy that provides them with full social and political rights, including the right to run for presidential elections. To achieve their goals, Egyptian women activists are deploying both street activism and cyber-activism in support of their causes, such as insisting on constitutional reforms that safeguard their rights, amid concerns by women activists that the post-revolution committee revising the constitution is all male (Krajeski, 2011).

CONCLUSION

One of the most striking aspects of the Egyptian uprising was its loose structure and lack of identifiable leaders. It was largely a grassroots, across the board, horizontal movement that had a bottom-up, rather than a top-down, structure. In other words, as Beckett (2011) rightly notes, it was a network of activists, not organizations, since it was not organized by conventional opposition parties, nor led by charismatic leaders. This diffuse, horizontal, and diverse nature of the Egyptian uprising gave it a true organic strength that made it very difficult for the regime to break.

It can also be said that protest movements, like the ones witnessed before and during the Egyptian revolution, were more about "processes," rather than "persons." In other words, they were characterized by collective and effective processes of group mobilization, both online and offline, rather than individual acts of leadership by one or more charismatic persons. The way such movements worked was through effectively connecting networks of citizens, both online and offline, through the deployment of social media technologies to catalyse, coordinate and orchestrate protests in the streets. This enabled mass mobilization on a large scale, as people suddenly realized there were many others willing to risk demonstrating against the state, just like them, which enhanced their chances of success and minimized potential risks. The amplifying and snowballing effects of social media prove to be particularly effective in mass mobilization when combined with on-the-ground coordinating and protesting. This is because the virtual networks online activate the personal networks on the ground and mutually reinforce and strengthen each other.

While social media were certainly a powerful instrument in Cairo's street protests and the Egyptian revolution, it would be a mistake to characterize the uprising as a Facebook or Twitter revolution. Social media played a key role in the Egyptian revolution, but it must be stressed that these new media were nothing more than powerful tools and effective catalysts: social media were only effective because of the willingness of large numbers of people to physically engage in, and support, peaceful social protest, sometimes at great personal risk, including grave injuries or even loss of life. In short, social media were not causes of revolution, but new vehicles for citizens' empowerment, which helped amplify, magnify and expedite the process of revolt. Social media enabled the revolution to snowball by turning small protests into a huge challenge to the regime leading to its ultimate demise (Adel Iskander, 2011).

In analysing the Internet's role in the ability of Egyptian citizens to assert and practice their citizenship rights during the Egyptian revolution, one should not exaggerate the power of new media technologies in a way that undermines the social, cultural and economic forces that underlie the movement. It is crucial to acknowledge the role of the actors on the ground, in this case the Egyptian citizens themselves. The ways in which citizens organize and mobilize, however, can be facilitated, or frustrated, by the use of distinct technologies. Citizens could organize at the neighborhood level through pamphleteering or through soup kitchens and residents' associations, or they could organize at the city or national levels through cyberactivism, creating cyber public spheres and cyber public spaces. The use of distinct technologies influences the ways in which citizens wage their struggles for an inclusive citizenship or the right to the city in the contemporary era. The use of new technologies influences, but does not determine, the process or the outcome of citizens' struggles, such as those witnessed in contemporary Cairo. At the end of the day, if it were not for the power and determination of Egyptian citizens to act, organize, and mobilize on the streets, this revolution would have never happened.

REFERENCES

Beckett, C. (2011). "After Tunisia and Egypt: towards a new typology of media and networked political change," *POLIS*, Journalism and Society Think Tank. Available HTTP: http://www.charliebeckett.org/?p=4033 Accessed March 24, 2011.

Brisson, Z. et al. (2011). "Egypt: from revolutions to institutions," New York: Reboot. Available HTTP: http://thereboot.org/wp-content/Egypt/Reboot-Egypt-From-Revolutions-To-Institutions.pdf Accessed April 18, 2011.

Idle, N. and Nunns, A. (Eds.) (2011). *Tweets from Tahrir: Egypt's Revolution as it Unfolded, in the Words of the People Who Made It*, New York: OR Books.

Ishani, M. (2011). "The hopeful network," *Foreign Policy* Available HTTP: http://www.foreignpolicy.com/articles/2011/02/07/the_hopeful_network Accessed March 24, 2011.

Iskander, A. (2011). Adjunct faculty at Georgetown University's Center for Contemporary Arabic Studies. Phone interview on April 24.

Joyce, M. (2011). "The 7 activist uses of digital technology: popular resistance in Egypt," International Center on Nonviolent Conflict. Available HTTP: http://www.nonviolent-conflict.org/images/stories/webinars/joyce_webinar.pdf Accessed April 18, 2011.

Krajeski, J. (2011). "Egyptian women look beyond Tahrir Square," *The New Yorker*, March 14.

Lynch, M. (2011). "Revolution 2.0? The role of the Internet in the uprisings from Tahrir Square and beyond," Paper presented at the conference: *Theorizing the Web* at the University of Maryland, College Park, April 2011.

Maher, K. (2011). Interview with Frog Design. Video available online at: http://www.youtube.com/watch?v=zulsoFZ-Kqs&feature=player_embedded.

Meier, P. (2011). "Maps, activism and technology: check-in's with a purpose," iRevolution. Available HTTP: http://irevolution.net/category/digital-activism/ Accessed March 10, 2011.

Mustafa, M. (2011). Egyptian political activist and coordinator of the National Coalition for Change campaign. Skype interview on April 23.

Nelson, A. (2008). "The web 2.0 revolution—extended version," *Carnegie Reporter*. Available HTTP: http://carnegie.org/publications/carnegie-reporter/single/view/article/item/71/ Accessed March 24, 2011.

Otterman, S. (2011). "Women fight to maintain their role in the building of a new Egypt," *New York Times*. Available HTTP: Accessed http://www.nytimes.com/2011/03/06/world/middleeast/06cairo.html Accessed March 24, 2011.

Rosenberg, T. (2011). "Revolution U: What Egypt learned from the students who overthrew Milosevic," *Foreign Policy*. Available HTTP: http://www.foreignpolicy.com/articles/2011/02/16/revolution_u Accessed March 10, 2011.

SECTION 12
The Transfer of Knowledge and Policy

Editors' Introduction

In this concluding section of the *Reader*, we consider how knowledge (understanding an issue) and policy (thinking about ways to address issues to achieve favorable outcomes) get produced and travel between places and across borders. This is a topic that in one way or another, runs through every section of this *Reader*, particularly the commissioned pieces by Anthony King, Richard Harris, Michael Goldman, and Sudeshna Mitra. Our section introductions to various issues also consistently emphasize global trends, knowledge movements, and resistance in framing debates. We focus on modes of analysis modeled on dominant patterns that originate in very different institutional contexts but are applied with little attention paid to the ways in which they embody socio-political interests or power relations. We use the term institutions broadly to include social, cultural, economic, and political structures that shape and condition everyday life and behavior.

What we know, how we know it, and the practical problem of transferring knowledge from one organization, region or city to another, from one group of individuals or a society to another are issues taken up by a range of disciplines and professions. In the world of urban studies and planning they are approached in a variety of ways. This includes historical studies that seek to understand the transfer of ideas on how to create the infrastructure for the automotive city (Norton 2008) or borrow from best practices of providing low-income housing solutions (Parnreiter 2011). The processes include more issues than we can cover here: the politics of knowledge creation, organizational structures and routines, social reconstructions, policy mechanisms, governmental regimes, activist organizing and the power of protest, technology, and communication systems. Various aspects of these have already been discussed in other articles in this *Reader*. Here we draw attention to two ideas: first, the continual production of knowledge in and of complex urban settings in the global South; second, understanding policy transfer and its outcomes as a consequence of institutional contexts and power relations.

KEY ISSUES

The production of knowledge: where you are looking from matters

In the field of urban studies–as in most any academic or policy field–the disciplinary lens of the observer shapes perspective and the resulting interpretation of conditions and events. We focused on this in the Editors' Introduction to Historical Underpinnings when we asked, "whose history . . . [was] being uncovered and recorded, and by whom?" (page 25). This issue of positionality comes up repeatedly in other Editors' Introductions throughout this volume. However, where one is looking from goes further than academic discipline or professional affiliation.

As feminist theorists have long argued, location or "standpoint" matters and guides both the production of knowledge and political praxis. Standpoints are "embodied in individual knower's spoken experiences and social identities and produced in communities as well as a site of inquiry" (Naples 2003: 197). This multi-dimensional notion of standpoint extends the idea of social or identity markers that shape us beyond class and socio-economic status to include gender, race, caste, ethnicity, indigeneity, sexual orientation, age, or national origin. It is not our intention to promote any sort of essentialism (which privileges one viewpoint and sees it as fundamental to understanding motivations and behaviors), to valorize the local or make claims as to whether one identity marker trumps another. Rather, we emphasize that it is crucial to understand our positions and how they shape our worldviews in all their diversity and contradiction. Thus, to draw on some examples from this *Reader*, we see how city experience differs in the stories of each resident of Tehran in Madanipour's contribution and how the ability to recover from disasters in Mexico City as described by Davis shifts based on where you live, your networks and connections, and how you are governed. In the city, as our introductions to the sections on housing, infrastructure, and basic services have shown, women's needs for housing or mobility are different from their male family members and from women of a different ethnic or religious group. The poor and young children, but especially girls, experience the lack of electricity, safe public spaces, and the ill-effects of dirty water and clogged drains most acutely. There is nothing radical or particularly "Southern" about the fact that the city is experienced and understood differently by different people or that experience builds knowledge. What is important is that this idea of location and positionality reinforces the importance of taking institutions and context seriously.

The production of knowledge, institutions and context matters

In a slim, free-wheeling yet difficult volume, *How Institutions Think*, the anthropologist Mary Douglas (1986) argued that knowledge is essentially social and shaped by institutional context. Marx's key insight that we are a product of social context and structure has also remained deeply influential in many ongoing debates, including those around distributive justice (Fleischacker 2004). Skipping to yet another field, many development economists influenced by Douglass North's (1990) new institutionalism now argue that "institutions rule" and that the quality of regulatory and governance institutions trumps most everything else in determining development outcomes (Rodrik, Subramanian and Trebbi 2004). Elinor Ostrom (1990) drew on a set of cases to explain how institutions evolved to govern common pool resources such as fisheries, irrigation, and communal tenure in meadows and forests. These perspectives offer different insights on how institutions structure, create or constrain human behavior and knowledge creation. We use these examples to point out that the role of institutions has become central to understanding economies, societies, politics, and governance over time. Cities and urban planning and policy are no different.

The vast literature on cities includes case studies of planning processes that are particularly useful to understand how context and institutions work to structure knowledge production and transfer in an urban setting. Anthropologists like James Holston (1989) and Ahmed Kanna (2011, also this volume) or architects like Madhu Sarin (1982) have critiqued the paradoxes inherent in imagining and planning either a modernist utopia (Brasilia for Holston, Chandigarh for Sarin) or the postcolonial, hypermodern architectural spectacle (Dubai for Kanna). Their books, like much of the general literature on cities, focus on the flaws of rational planning practice and on the ways in which designed utopias come crashing down under the weight of the hard gritty realities of power, politics and economics. A smaller group of urban planning scholars also focuses attention on the role of institutions in producing desired outcomes (Verma 2007; Teitz 2007; Sanyal and Fawaz 2007; Srinivas and Sutz, 2008; Kim 2012).

Yet others, like anthropologist Lisa Peattie (1968, 1987), who taught for many years in MIT's planning school, reflected on the planning process for creating Ciudad Guyana in Venezuela, where she worked on the original planning team. Her work is both part of the tradition of critique described above, and part of a tradition that focuses on planning practice to understand the profession (Forester 1999, 2013) and the places and cities being produced (Flyberg 1998; Sanyal 2005; Kudva 2008). Twenty years after working on the planning team in Ciudad Guyana, Peattie reflected on what planners could learn from her insider's view of practice. She wrote:

> The planner who wants to ally with some group or interest cannot simply adopt a set of techniques and forms of representation developed in the context of a *different* social placement and set of alliances, but must rethink the technical issues afresh. (italics in original, Peattie 1987: 171)

FIVE

Peattie's emphasis on rethinking "technical issues afresh" and foregrounding context and institutions is echoed in many urban planning case studies across the world. Oren Yiftachel (2006), a critical planning scholar at Ben Gurion University, Israel, writes about the dominance of the North-West in the processes of knowledge production and stresses the role that English-language planning journals based in Europe and North America play. Mapping the "gatekeepers" of theoretical knowledge he analyzes articles published and editorial board membership at major international planning journals to demonstrate the dominance of Western academia in the production and dissemination of planning knowledge. Bringing to light the location from which planning knowledge is produced, he argues that most theorizing is based on the experience of the global North, which might not speak to the reality of cities and societies in the rest of the world. This pointed critique has resulted in a series of publications and a call for de-colonizing planning thought through theorizing from the perspective of the global South or "seeing from the south" (Miraftab 2009:44). To move away from North-dominated planning paradigms and assumptions Vanessa Watson (2009), a planning academic at the University of Cape Town, emphasizes once again the importance of developing theoretical positions and practical strategies that draw on the varied contexts of cities across the global South, particularly when considering issues such as informality and service provision described in other article selections in this *Reader*.

Watson also writes of the difficulty involved in transferring these new understandings:

> [p]erhaps one of the most difficult problems that have to be faced when trying to draw on understandings or ideas developed in a different context is that of how transferable they are: what is unique to a particular time and place, and what is more general? It is only the depth of contextualizing detail (thick description) that will allow this judgment to be made. Deep situational understanding is essential input when dealing with new problems and circumstances. (Watson 2002: 184)

This takes us to a discussion on the challenge of knowledge and policy transfer.

Knowledge and policy transfer: the consumption of knowledge across borders

An earlier generation of urban professionals disseminated the idea of the Master Plan (King, Part II, this volume) and particular solutions like decentralization and privatization to solve the problem of indebted governments (Kohl and Farthing, this volume; see also Editiors' Introduction to Governance, this volume). The current generation also celebrates and transfers innovative ideas that have emerged in Southern contexts like participatory budgeting (Baiocchi, this volume), bus rapid transit and gondola transit over slums (see Editiors' Introduction to Infrastructure, this volume). What holds largely true across both periods is the lifting of ideas out of the contexts in which they were generated to transplant them elsewhere. How do policy and planning ideas travel? What are the channels and mechanisms that allow diffusion and transfer? What is emulated and adopted in the receiving areas? Do learning mechanisms rely on rational processes or are they driven by coercion and power? Who are the agents involved in such processes? These are issues of abiding scholarly interest across many disciplines.

Two main lines of research are distinguishable in the large literature on international policy transfer and diffusion (Marsh and Sharman 2009). The first line of research, dominant in the international relations field in political science, focuses on structural patterns and trends that lead to policy diffusion and convergence across countries. It is a literature dominated by quantitative studies and large-N data sets. The second line of research frequently appears in the public policy and urban planning literature. This involves case studies that emphasize agency. This literature includes many forms of knowledge and policy transfer through mechanisms such as rational learning, lesson-drawing, transnational norm-making and coercion (Evans and Davies 1999; Evans 2004; Dolowitz and Marsh 2000). The literature emphasizes circumstances, agents, institutions, channels, policy-content and the "industry" that enable transfer. It has produced studies of think tanks, international

NGOs, global consulting firms such as McKinsey and Company, and multilateral institutions such as UN-HABITAT (Ladi 2005; Parnreiter 2011; Stone 2000, 2004, for a fuller discussion on this literature see Kudva and Dharamavaram 2011).

The frameworks used widely in the urban planning literature overlap with these perspectives (for a recent collection see Healey and Upton 2010; also Healey 2011). The emerging literature on South–South transfers of knowledge and policy ideas tend to draw on these frameworks as well. However, much less attention is paid to actual decision-making processes. Drawing on the work of social psychologists Kahneman and Tversky, some researchers consider the reasons why local leaders tend to use "heuristics" and take "inferential short cuts" (timesaving mental shortcuts to reduce complex judgments to simple rules and allocate attributes to a type that seems to match a given category). One result of taking these shortcuts is that leaders and professionals adopt models that seem attractive as Weyland (2004, 2005) describes in an interesting study of pension reform adoption across Latin America and Kudva and Dharamavaram (2011) discuss in their study of how TDR (transfer of development rights) travelled from America to Mumbai, where it was repurposed to allow for slum redevelopment. Others like Appadurai (2001: 30) note that the "tyranny of emergency . . . that characterizes the everyday lives of the urban poor" forces project-based responses that work against patient, long-term work in difficult circumstances. Yet others like Parnreiter (2011) describe how expert-promoted models, which include the consumption of "fast policies" and "off-the-shelf" products, often lead to the depoliticization of local urban development and the reluctance to develop appropriate, contextual solutions in collaboration with those who are the most affected by planning and policy decisions. What continues to be difficult to grasp is how leaders, practitioners and professionals can acquire and learn to act on the "deep situational understanding [that is] essential input when dealing with new problems and circumstances" across the many cities of the global South (Watson 2002: 184).

ARTICLE SELECTIONS

Our article excerpts for this section highlight the two issues we have discussed above: knowledge production and policy transfer. The first excerpt is from an article by Ananya Roy, a prominent scholar and educator at the University of California at Berkeley. She is well known for work on informality (see Editiors' Introduction to Urban Economy, this volume) and her analysis of urban poverty in the global South including her critique of microfinance in the volume *Poverty Capital* (Roy 2010). The article selection included here is a finely written provocation published with the intention of prodding people into thinking of planning in the context of urban informality. In terms of this introduction, it takes on the issue of situational understandings and the impossibility of planning in one context using norms, rules, and skill-sets that have emerged from a very different institutional context. Her work demonstrates a mode of building knowledge based on her own research in Asia as well as drawing on the work of others studying urban development across the global South. As a whole, this perspective builds an argument for a transnational approach–specifically for a critical transnationalism where the notion of justice remains at the core of knowledge transfer and traveling policies (Roy 2011).

The second article is by Richard Tomlinson, a prominent South African planning scholar and practitioner currently Chair of the University of Melbourne's Planning Program in Australia. Grounded in his experience as a planning practitioner in South Africa, Tomlinson's writings are particularly insightful. More recently, he has examined the impacts of mega-projects like the summer Olympics in London and edited a book on Australian cities (Tomlinson 2012). In his earlier work, including the article included below, he examines processes of policy making as they unfold in post-apartheid South Africa. The article excerpt included here is a finely detailed case study of best practices, which follows the role of local government and "international development agencies and globe-hopping consulting firms and academics" in the implementation of urban policies in South Africa. Like Peattie before him (1987), Tomlinson writes from the unique vantage point of being an insider, a participant in the processes of policy transfer and planning that he analyzes. His insights on the limits of best practices–a much touted term in urban studies and policy circles–are useful and the comment from Rebecca Black, an aid official, is particularly chastening: "'best practice' may be the wrong word . . . rather than being driven by real best practice, current policy is driven by informed history of worst practice, with emphatic advocacy for what is the most unlike the past but as yet untested" (Tomlinson 2002: 386). Given the debate on whether the African National Congress (ANC) was coerced or leapt at the opportunity to adopt the pre-packaged neoliberal policies of the Washington Consensus (Bond 1999), Tomlinson's work helps us see how this process involved not just coercion but also persuasion and a conditioned willingness.

REFERENCES

Appadurai, A. (2001) "Deep democracy: Urban governmentality and the horizon of politics," *Environment and Urbanization,* 13(2): 23–43.

Bond, P. (1999) "Basic infrastructure for socio-economic development, environmental protection and geographical desegregation: South Africa's unmet challenge," *Geoforum,* 30: 43–59.

Dolowitz, D. P. and Marsh, D. (2000) "Learning from abroad: The role of policy transfer in contemporary policy-making," *Governance*, 13(1): 5.

Douglas, M. (1986) *How Institutions Think*, Syracuse, NY: Syracuse University Press.

Evans, M. (Ed.) (2004) *Policy Transfer in Global Perspective*, Aldershot: Ashgate.

Evans, M. and Davies, J. (1999) "Understanding policy transfer: A multi-level, multi-disciplinary perspective," *Public Administration,* 77 (2): 361.

Fleischacker, S. (2004) *A Short History of Distributive Justice*, Cambridge, MA: Harvard University Press.

Flyberg, B. (1998) *Rationality and Power: Democracy in Practice*, Chicago: University of Chicago Press.

Forester, J. (1999) *The Deliberative Practitioner: Encouraging Participatory Planning Processes*, Cambridge, MA: The MIT Press.

Forester, J. (2013) *Planning in the Face of Conflict: The Surprising Possibilities of Facilitative Leadership*, Washington, DC: APA Planners Press.

Healey, P. (2011) "The universal and the contingent: Some reflections on the transnational flow of planning ideas and practices," *Planning Theory,* 11(2): 188–207.

Healey, P. and Upton, R. (Eds.) (2010) *Crossing Borders: International Exchange and Planning Practices*, London: Routledge.

Holston, J. (1989) *The Modernist City: An Anthropological Critique of Brasilia,* Chicago: The University of Chicago Press.

Kanna, A. (2011) *Dubai, the City as Corporation*, Minneapolis: University of Minnesota Press.

Kim, A. (2012) "The evolution of the institutional approach to planning," in Randall Crane and Rachel Weber (Eds.) *The Oxford Handbook of Urban Planning*, New York: Oxford University Press, pp. 69–86.

Kudva, N. (2008) "Teaching planning, constructing theory," *Planning Theory and Practice*, 9(3): 363–376.

Kudva, N. and Dharamavaram, S. (2011) "Slum TDR, policy and knowledge transfer from America to Mumbai." Paper presented at the American Collegiate Schools of Planning Annual Conference.

Ladi, S. (2005) *Globalisation, Policy Transfer and Policy Research Institutes*, Cheltenham: E. Elgar.

Marsh, D. and Sharman, J. C. (2009) "Policy diffusion and policy transfer," *Policy Studies,* 30(3): 269–288.

Miraftab, F. (2009) "Insurgent planning: Situating radical planning in the global South," *Planning Theory,* 8(1): 32–50.

Naples, N. (2003) *Feminism and Method: Ethnography, Discourse Analysis, and Activist Research,* New York: Routledge.

North, D. (1990) *Institutions, Institutional Change and Economic Performance*, Cambridge: Cambridge University Press.

Norton, P. D. (2008) *Fighting Traffic: The Dawn of the Motor Age in the American City,* Cambridge, MA: The MIT Press.

Ostrom, E. (1990) *Governing the Commons: The Evolution of Institutions for Collective Action*, Cambridge: Cambridge University Press.

Parnreiter, C. (2011) "Commentary: Towards the making of a transnational urban policy?" *Journal of Planning Education and Research,* 31(4): 416–422.

Peattie, L. (1968) *The View from the Barrio*, Ann Arbor: University of Michigan Press.

Peattie, L. (1987) *Planning, Rethinking Ciudad Guyana*, Ann Arbor: University of Michigan Press.

Rodrik, D., Subramanian, A. and Trebbi, F. (2004) "Institutions rule: The primacy of institutions over geography and integration in economic development," NBER Working Paper No. 9305. Cambridge, MA: National Bureau of Economic Research. HTTP Available at http://www.nber.org/papers/w9305, Accessed June 23, 2013.

Roy, A. (2009) "Why India cannot plan its cities: Informality, insurgence and the idiom of urbanization," *Planning Theory,* 8(1): 76–87.

Roy, A. (2010) *Poverty Capital: Microfinance and the Making of Development*, London and New York: Routledge.

Roy, A. (2011) "Commentary: Placing planning in the world–transnationalism as practice and critique," *Journal of Planning Education and Research,* 31(4): 406–415.

Roy, A. and Ong, A. (Eds.) (2011) *Worlding Cities: Asian Experiments and the Art of Being Global*, Chichester: Wiley Blackwell.

Sanyal, B. (Ed.) (2005) *Comparative Planning Cultures*, New York: Routledge.

Sanyal, B. and Fawaz, M. (2007) "The transformation of an olive grove: An institutionalist perspective from Beirut, Lebanon," in Niraj Verma (Ed.) *Institutions and Planning*, Boston: Elsevier, pp. 207–228.

Sarin, M. (1982) *Urban Planning in the Third World: Chandigarh Experience*, Los Altos: Mansell Publishing.

Srinivas, S. and Sutz, J. (2008) "Developing countries and innovation: Searching for a new analytical approach," *Technology in Society,* 30: 129–140.

Stone, D. L. (2000) "Learning lessons, policy transfer and the international diffusion of policy ideas," Center for the Study of Globalization and Regionalism. February 9.

Stone, D. L. (2004) "Transfer agents and global networks in the 'transnationalization' of policy," *Journal of European Public Policy,* 11(3): 545–566.

Teitz, M. (2007) "Planning and the new institutionalisms," in Niraj Verma (Ed.) *Institutions and Planning*, Boston: Elsevier, pp. 17–36.

Tomlinson, R. (2002) "International best practice, enabling frameworks and the policy process: A South African case study," *International Journal of Urban and Regional Research,* 26(2): 377–388.

Tomlinson, R. (Ed.) (2012) *Australia's Unintended Cities: The Impact of Housing on Urban Development*, Melbourne: CSIRO Publishing.

Verma, N. (2007) "Introduction: Institutions and planning: An analogical inquiry," in Niraj Verma (Ed.) *Institutions and Planning*, Boston: Elsevier, pp. 1–16.

Watson, V. (2002) "Do we learn from planning practice? The contribution of the practice movement to planning theory," *Journal of Planning Education and Research,* 22(2): 178–187.

Watson, V. (2009) "Seeing from the south: Refocusing urban planning on the globe's central urban issues," *Urban Studies,* 46(11): 2259–2275.

Weyland, K. G. (2004) *Learning from Foreign Models in Latin American Policy Reform*, Washington, DC: Woodrow Wilson Center Press; Baltimore: Johns Hopkins University Press.

Weyland, K. G. (2005) "Theories of policy diffusion: Lessons from Latin American pension reform," *World Politics,* 57(2): 262–295.

World Bank (no date) *Program Description for Intergovernmental Relations and Local Financial Management,* Washington, DC: World Bank.

Yiftachel, O. (2006) "Re-engaging planning theory? Towards 'south-eastern' perspectives," *Planning Theory,* 5(3): 211–222.

"Why India Cannot Plan Its Cities: Informality, Insurgence and the Idiom of Urbanization"

Planning Theory (2009)

Ananya Roy

1 TWO SCENES OF INDIAN URBANIZATION

In late May (2008) an article appeared in the *New York Times* about the city of Bangalore in southern India (Sengupta, 2008). Bangalore, often understood as India's Silicon Valley, is a booming metropolis whose economic prosperity has far outstripped its urban infrastructure. The article noted that while the city had built a new airport, modeled after the Zurich airport, a whole set of planning failures accompanied this infrastructural investment. The airport was located 21 miles outside town and the roads connecting the city center to the airport had not been widened sufficiently to accommodate traffic. The city water supply had not reached the airport area and so the shops, office towers, and other developments that were supposed to surround the airport had not yet been built. The article pinpoints various explanations for this planning failure.

[...]

At the far edges of the Calcutta metropolitan region in eastern India, people are indeed being asked to make way for planned development. This is the second scene of Indian urbanization. The rhetoric this time is not of planning failures but rather of the need for planners to take decisive action under the sign of "public purpose." Here, the Left Front, a socialist coalition led by the Communist Party of India-Marxist (CPI-M), is busy acquiring agricultural land for private development. To do so, it has displaced subsistence and smallholder farmers and sharecroppers, often using not only the instrument of eminent domain but also the sheer violence of its political apparatus. The argument has been made thus, by Nirupam Sen, Minister of Industries, West Bengal:

> If a particular industry wants a big chunk of land in a contiguous area for setting up a large plant there, it is not possible for the industry to purchase land from each and every farmer, particularly in West Bengal where fragmentation of land is very high. If a large chunk of land is needed for a very important industrial project, will the State government not acquire it for the project? And, of course, it is a public purpose. Industrialization means employment generation, it means development of society; the entire people of the State will be benefited. Therefore, it is in the interest of public purpose that the land has been acquired. (Chattopadhyay, 2006)

2 THE IDIOM OF URBANIZATION

[...]

The two scenes of Indian urbanization described above when taken together can be seen to present an incontrovertible argument about the failure of planning in India: that informality and insurgence together undermine the possibilities of rational planning, and that therefore India cannot plan its cities. Against this narrative of failed planning, I present the argument that what is at work in the two scenes is an idiom of urbanization. This idiom is peculiar and particular to the Indian political economy and yet can be detected in many other contexts. While this idiom seems to be antithetical to planning, and indeed seems to be anti-planning, it can and must be understood as a planning regime. I also argue that the key feature of this idiom is informality.

[. . .]

3 FOUR PROPOSITIONS ABOUT INFORMALITY

Over two decades ago Janice Perlman (1986) published a now famous essay on the six misconceptions about squatter settlements. Her proposition undermined the "myth of marginality" and established a new common sense about urban marginality. My task here is less ambitious. Nevertheless, I wish to set out four propositions about informality that call into question the ways in which this concept is often used in the study of cities and planning. In particular, I present these ideas in opposition to a dominant viewpoint that conceptualizes informality as a separate and bounded sector of unregulated work, enterprise and settlement. While often sympathetic to the struggle of the "informals," this framework presents informality as an extra-legal domain and thus argues for policy interventions that would integrate the informal into the legal, formal, and planned sectors of political economy. Such a perspective pervades a wide right-to-left spectrum of analytical work, from the neoliberal populism of Hernando de Soto (2000) to the Gramscian conceptualization of subaltern politics by postcolonial theorists, notably Partha Chatterjee (2004). In contrast, I call into question this division between the law and informality and argue that legal norms and forms of regulation are in and of themselves permeated by the logic of informality. In recent times, planning theorists have sought to take up the idea of informality as a feature of planning. For example, an article in the flagship *Journal of the American Planning Association* by Judith Innes, Sarah Connick and David Booher (2007: 207) presents informality as a "valuable strategy of planning." For these authors, the term "informality" signifies planning strategies that are "neither prescribed nor proscribed by any rules . . . The idea of informality also connotes casual and spontaneous interactions and personal affective ties among participants." But such a framework quite drastically depoliticizes the concept of informality by misrecognizing systems of deregulation and unmapping as casual and spontaneous. Indeed, there is nothing casual nor spontaneous about the calculated informality that undergirds the territorial practices of the state. This idiom of state power is structural and is thus a far cry from the "personal affective ties" that Innes et al. seek to highlight. The following propositions make evident the structural nature of informality as a strategy of planning.

3.1 Informality is not synonymous with poverty

The current common sense on informality is that it is synonymous with poverty. Davis (2006) sees the "slum" as the global prototype of a warehousing of the rural–urban poor, marginalized by structural adjustment and deindustrialization. De Soto (1989, 2000) sees informality as a revolution from below, the entrepreneurial strategy or tactical operations of the poor marginalized by bureaucracy and state capitalism. Neither approach is able to pinpoint the ways in which informality is also associated with forms of wealth and power. The splintering of urbanism does not take place at the fissure between formality and informality but rather, in fractal fashion, *within* the informalized production of space. A closer look at the metropolitan regions of much of the world indicates that informal urbanization is as much the purview of wealthy urbanites and suburbanites as it is that of squatters and slum dwellers. With the consolidation of neoliberalism, there has also been a "privatization of informality." While informality was once primarily located on public land and practiced in public space, it is today a crucial mechanism in wholly privatized and marketized urban formations, as in the informal subdivisions that constitute the peri-urbanization of so many cities. These forms of informality are no more legal than squatter settlements and shantytowns. But they are expressions of class power and can thus command infrastructure, services, and legitimacy in a way that marks them as substantially different from the landscape of slums. The important analytical (and political) question to ask in the Indian context, as well in others, is why some forms of informality are criminalized and thus rendered illegal while others enjoy state sanction or are even practices of the state.

3.2 Informality is a deregulated rather than unregulated system

It is commonplace to talk about informality as the lack of regulation. In the classic text, *The Informal Economy*, Castells and Portes (1989) designate the informal as that which is unregulated in an economy where similar activities are regulated. In short, "it is because there is a formal economy that we can speak of an 'informal' one" (Castells and Portes, 1989: 13). This work was revolutionary because two decades ago it departed "from the notions of economic dualism and social marginality" such that the same concept could be applied to "a street seller in Latin America and a software consultant moonlighting in Silicon Valley" (Castells and Portes, 1989: 12). But there is an important distinction between unregulated systems and those that are deregulated. Deregulation indicates a calculated informality, one that involves purposive action and planning, and one where the seeming withdrawal of regulatory power creates a logic of resource allocation, accumulation, and authority. It is in this sense that informality, while a system of deregulation, can be thought of as a mode of regulation. And this is something quite distinct from the failure of planning or the absence of the state. Thus, many scholars, working in the context of development, have pointed to ambiguities of land tenure systems but they have done so to indicate the "fragility of authority," the "Achilles' heel" of resettlement schemes and state-led development (for example, Li, 2007). I argue that such ambiguities are precisely the basis of state authority and serve as modes of sovereignty and discipline.

Two very different examples may help make this point. In the case of Calcutta, I have sought to plot the relationship between the formal plan and unmapped, deregulated territory by reworking the questions we may ask of such a city. My initial questions, in keeping with a traditional understanding of planning as the management of land use and growth were: How can I find the appropriate map? Who owns this piece of land? What uses are planned for it? In their place, I learned to ask: What does it mean to have fluid and contested land boundaries? How does this ambiguity regarding status and use shape processes of urban development? How does this establish the possibilities and limits of participating in such land games (Roy, 2003)? More recently, Naomi Klein (2007) has presented an analysis of the disaster capitalism complex that makes evident how the deregulation of political economies is tied to the deregulation of space. She shows how, in the last decade, there has been the emergence of a parallel, privatized disaster infrastructure that caters exclusively to the wealthy and the "chosen," those who can opt out of the collective system. Here, as in the case of Calcutta, deregulation becomes a logic of resource allocation, accumulation, and authority.

3.3 The state is an informalized entity, or informality from above

As informality is defined as an unregulated domain of activities, so it is often understood to be unplanned. In particular, the informal sector is seen to exist outside "institutionalized regulation" (Castells and Portes, 1989: 12) and is subsequently imagined as extra-legal (de Soto, 1989) or para-legal (Chatterjee, 2004) or as a "shadow city" (Neuwirth, 2004). Informality is thus viewed as the practices of the subaltern (Bayat, 2000), a democracy "from below" (Appadurai, 2002). I argue that informality has to be understood not as a grassroots phenomenon, but rather as a feature of structures of power. In my earlier work, I conceptualized the informal as a site of extra-legal discipline, continuous with formal systems of regulation (Roy, 2003). While I wish to maintain the idea of informality as a mode of discipline, power, and regulation, I now seek to reject the designation of extra-legality. That terminology implies that informality is a system that runs parallel to the formal and the legal. Yet, the formal and the legal are perhaps better understood as fictions, as moments of fixture in otherwise volatile, ambiguous, and uncertain systems of planning. In other words, informality exists at the very heart of the state and is an integral part of the territorial practices of state power. For example, in the Calcutta context, I have argued that it is not sufficient to examine the deployment of eminent domain or "vesting" as an instrument of the state. Instead, it is also necessary to understand the informalization of "vesting" (Roy, 2004). The concept of informal vesting may seem to be an oxymoron. Vesting indicates the legal expropriation of land by the state in the public interest or confiscation of land in excess of land ceilings set by agrarian reforms and the urban land ceiling act. Informality signifies extralegal, and possibly illegal, mechanisms of regulation. But what makes vesting such a powerful instrument in Calcutta is precisely this convergence of legality and extra-legality in the same process. It is the informal vested status of the land that allowed sharecroppers, supported by the Left Front [Ed. Note: An alliance of political parties], to establish de facto use rights; it is this informal vested status that 10 years later made it possible for the Left Front to reclaim this land for the resettlement of central city squatters; and that yet 10 years later allowed the Left Front to displace both squatters and any remaining sharecroppers to make way for peri-urban townships, Special Economic Zones, and other forms of development. It is this territorialized flexibility that allows the state to "future-proof," to make existing land available for new uses, to devalorize current uses and users and to make way for a gentrified future; in short, to plan. It is naïve to designate such processes as extra-legal, for they do not exist outside the law. Rather as practices of the state they are elements of an ensemble of sovereign power and the management of territory. This is informality from above, rather than informality as a subaltern revolution from below.

3.4 Insurgence does not necessarily create a just city

It is tempting to interpret the tactics and struggles of the urban poor in the cities of the global South as instances of rebellion and mobilization. Are these "shadow cities" not revolutionary, examples of a "globalization from below" (Appadurai, 2002; Neuwirth, 2004)? Is not the community organizing work of squatter settlements an inspiring case of the "politics of patience" in the face of the "tyranny of emergency" (Appadurai, 2002)? And if this is not planning—the politics of patience in the face of the tyranny of emergency—if this is not future-proofing, what is? But the relationship between insurgence, informality, and planning is more complicated. As planning is not an antidote to informality, so insurgence is not an antidote to the exclusionary city, particularly not to the types of exclusion that are deepened and maintained through the informalized practices of the state. Here it is important to reflect on the arguments presented by Castells (1983) in the seminal text, *The City and the Grassroots*. Taking a closer look at the insurgence of squatters in various world-regions, Castells argues that most of these are examples of urban populism rather than of radical social movements. Urban populism, according to him, is the "process of establishing political legitimacy on the basis of a popular mobilization supported by and aimed at the delivery of land, housing, and public services" (Castells, 1983: 175). Such forms of urban populism characterize the enfranchisement of squatters and sharecroppers in the Calcutta metropolitan region, allowing the rural–urban poor fragile and tenuous access to shelter and services in exchange for political and electoral loyalties. The fierce and bloody struggles in Nandigram seem to mark a break with such patterns of political dependence. And yet, they can also be understood as yet another instance of populist patronage, one where insurgent peasants are now bound to the electoral calculus of oppositional politics and the protection of the Trinamul Congress [Ed. Note: A regional political party in India]. Such forms of insurgence then do not and often cannot call into question the urban status quo; they can imagine but cannot implement the just city. And most of all, they depend on, and simultaneously perpetuate, the systems of deregulation and unmapping that constitute the idiom of planning. This is the informal city, and it is also an insurgent city, but it is not necessarily a just city. It is a city where access to resources is acquired through various associational forms but where these associations also require obedience, tribute, and contribution and can thus be a "claustrophobic game" (Simone, 2004: 219).

The complex relationship between insurgence and social justice is carefully delineated in the recent work of James Holston (2007). Holston designates the struggles of São Paulo's urban working classes as "insurgent citizenship" and notes the territorial rights that are established through such social mobilizations. The city's auto-constructed peripheries, and their gradual formalization, are a vivid example of insurgence. Such insurgence also transcends the peripheries as it creates a solid base for Brazil's "right to the city" movement and the institutionalization and articulation of such a right in planning processes. Yet Holston's analysis indicates that the insurgent citizenship manifested in the auto-constructed peripheries is a form of propertied citizenship, one where the right to the city is expressed through home ownership and where politics is expressed through neighborhood or homeowner associations. Such propertied citizens are quick to mark the distinctions between their (newly) legal territory and the supposedly illegal territory of more recent squatters. In short, the policing of the arbitrary and fickle boundary between the legal and illegal, formal and informal, is not just the province of the state but also becomes the work of citizens, in this case insurgent citizens. This is an insurgent city, one where the very legal basis of informality has been challenged by the urban poor, and yet it is also an exclusionary city where the poor recreate the margins of legality and formality, imposing new socio-spatial differentiations in the periphery.

4 A CONCLUDING NOTE

The title of this article suggests that India cannot plan its cities. Indian cities serve, in such a framing, as a proxy for the Third World megacity, that which defies all norms of rational planning and "future-proofing." The persistent failure of planning or the splintering of cities through the privatization of planning all seem to be convincing and adequate explanations for the crisis that is the Indian city, or the Third World megacity. Yet, this article has presented a different argument. It has linked India's urban crisis to the idiom, rather than the failure, of planning. In particular it has identified informality as a key feature of this idiom such that Indian planning proceeds through systems of deregulation, unmapping, and exceptionalism. These systems are neither anomalous nor irrational; rather they embody a distinctive form of rationality that underwrites a frontier of metropolitan expansion. And yet, at least in India, urban developmentalism remains damned by the very deregulatory logic that fuels it. It is thus that the territorialized flexibility of the state gives way to the various impasses that mark the two opening scenes of Indian urbanization. Good or better planning cannot "solve" this crisis for planning is implicated in the very production of this crisis. It is in this more fundamental sense that India cannot plan its cities.

REFERENCES

Appadurai, A. (2002). "Deep democracy: Urban governmentality and the horizon of politics," *Public Culture* 14(1): 21–47.

Bayat, A. (2000). "From 'dangerous classes' to 'quiet rebels': The politics of the urban subaltern in the global South," *International Sociology* 15(3): 533–57.

Castells, M. (1983). *The City and the Grassroots*, Berkeley: University of California Press.

Castells, M. and Portes, A. (1989). "World underneath: The origins, dynamics, and effects of the informal economy," in L. Benton et al. (eds), *The Informal Economy*, Baltimore, MD: Johns Hopkins University Press, 11–40.

Chatterjee, P. (2004). *The Politics of the Governed: Reflections on Popular Politics in Most of the World*, New York: Columbia University Press.

Chattopadhyay, S. (2006). "Land reform not an end in itself: Interview with Nirupam Sen," *Frontline* 23(25): 16–29.

Davis, M. (2006). *Planet of Slums*, New York: Verso.

de Soto, H. (1989). *The Other Path: The Invisible Revolution in the Third World*, London: I. B. Taurus.

de Soto, H. (2000). *The Mystery of Capital: Why Capitalism Triumphs in the West and Fails Everywhere Else*, New York: Basic Books.

Holston, J. (2007). *Insurgent Citizenship: Disjunctions of Democracy and Modernity in Brazil*, Princeton, NJ: Princeton University Press.

Innes, J., Connick, S. and Booher, D. (2007). "Informality as planning strategy: Collaborative water management in the CALFED Bay-Delta Program," *Journal of the American Planning Association* 73(2): 195–210.

Klein, N. (2007). *The Shock Doctrine: The Rise of Disaster Capitalism*, New York: Metropolitan Books.

Li, T. (2007). *The Will to Improve: Governmentality, Development, and the Practice of Politics*, Durham, NC: Duke University Press.

Neuwirth, R. (2004). *Shadow Cities: A Billion Squatters, A New Urban World*, New York: Routledge.

Perlman, J. (1986). "Six misconceptions about squatter settlements," *Development* 4: 40–4.

Roy, A. (2003). *City Requiem, Calcutta: Gender and the Politics of Poverty*, Minneapolis: University of Minnesota Press.

Roy, A. (2004). "The gentleman's city," in A. Roy and N. AlSayyad (eds), *Urban Informality: Transnational Perspectives from the Middle East, South Asia, and Latin America*, Lanham, MD: Lexington Books.

Sengupta, S. (2008). "An Indian airport hurries to make its first flight," *New York Times*, 22 May.

Simone, A. (2004). *For the City Yet to Come: Changing African Life in Four Cities*, Durham, NC: Duke University Press.

"International Best Practice, Enabling Frameworks and the Policy Process: A South African Case Study"

International Journal of Urban and Regional Research (2002)

Richard Tomlinson

In this article I analyse urban policies and policy processes that serve the needs of low-income groups in South Africa. I show how conceptions of international best practice and indicators of best practice have shaped the formulation of urban policy in the country, in particular, examining how the government has set out to address housing and municipal services backlogs. In addition, I demonstrate the key role of international development agencies and globe-hopping consulting firms and academics in helping to interpret how international best practice might be locally applied. These points are demonstrated with a case study of municipal services partnerships (MSPs) and the role played by the World Bank, the United States Agency for International Development (USAID) and their consultants in bringing them into being. Lastly, I debate the potential for locally differentiated policies outside the realm of best practice.

The origins of my views lie in my experience managing policy teams for the South African government, during which time I interacted with the World Bank, USAID and various European development agencies, and their consultants that, typically, consisted of large US firms with offices in many cities around the world and projects in very many more. Prominent US and European academics and some independent consultants also contributed, working directly with the development agencies or indirectly through contracts with the large consulting firms. This work has brought me into contact with persons whose previous postings or consulting contracts were, inter alia, in Warsaw, Hanoi and Sao Paulo, and some of whose next postings or contracts were in Moscow, New Delhi, Jakarta, Washington DC and Berlin. I often had occasion to wonder how it was that they might presume to contribute to policy formulation in their "host" countries, until I realized that their confidence is based on conceptions of international best practice.

ADOPTING OR ADAPTING INTERNATIONAL BEST PRACTICE

Two contradictory issues emerge from the preceding pages [Ed. note: which contain an account of policies aimed at urban areas and at the delivery of municipal services]. The first underlies this article, namely that the preparation of policy, the drafts of legislation and regulations, and institutional support were heavily dependent on financial and technical support from USAID and its consultants. The second issue, briefly mentioned, was that the MIIF (Municipal Infrastructure Investment Framework) was intensely negotiated among local stakeholders. What is the potential for locally differentiated policies arising from local dynamics and specificities?

During the course of my experience managing teams and writing and obtaining comments on this article, I found that the answer was viewed as depending on the character of the people concerned and their motivation for adopting best practice. I believe, however, that there is another, perhaps even more fundamental, issue arising from constraints that are intrinsic to the policy formulation process in a context of best practice.

THE ROLE AND MOTIVATION OF THE INDIVIDUAL

In regard to the people concerned, it was frequently observed that best practice means nothing if the South African senior bureaucrats and their political bosses fail to understand the issues concerned and if they are not personally driven, ambitious individuals. Best practice stumbles if there are not "local champions." This view creates the potential for both the effective diffusion of international best practice and its failure. When it comes to adopting and implementing best practice, Bond (1998: 47) has repeatedly observed that it is remarkable how former activists became good technocrats, locked into an urban privatization strategy with "a chain that began with international capital and that was welded together by international development agencies and national and local states." It would seem that there were two reasons for this happening. First, there were immediate political concerns. Gilbert (2001) reports that in the period leading up to and shortly after 1994, leading figures in the ANC confronted the prospect of capital flight and were concerned to allay business fears. Concessions in the role of the private sector in delivering housing should be seen in this light.

Second, the motivation to adopt best practice represents the intellectual hegemony of the neoliberal agenda. Its rationale is as follows. "Unless the partially marketized city [like Sao Paulo or Johannesburg] accepts its linkage to the developed world and increases its ability to adapt to the changes in the global economy, continued deterioration is assured" (Ruble *et al.*, 1996: 13). The World Bank (n.d.: 1) links this position to decentralization to local government and privatized service delivery: the "world-wide trend toward decentralization ... links to overall economic reform and improved governance [and] progress in stabilization, privatization, and poverty alleviation." Thus, Pycroft (2000) argues that, in South Africa, the motivation for decentralization to local government and the stress on the developmental mandate of local government are viewed as necessary in order to meet the challenges of a global economy.

What leeway does this provide the South African government to selectively use the policy insights and recommendations found in best practice and to chart its own course? Thinking back to the intense negotiations that characterized the early stages of the MIIF, these centered on the key role of intellectual, supremely confident former activists, now senior bureaucrats, struggling for power and influence. (Many subsequently left government, to be replaced by less experienced and confident officials.) For example, there came the point that responsibility for revising the MIIF shifted from the Department of Housing to the Department of Provincial and Local Government. Were they also struggling over the interpretation of best practice? Certainly this constituted the platform for debate, but Bond (2000) probably gets it wrong when he asserts that the platform was delimited by a highly prescriptive neoclassical policy agenda. Instead, as revealed in the following quote from a USAID publication (Black *et al.*, 2000: 2), the platform was delimited by the confidence that in certain issues there was a best practice, but also that in others there was not.

The lessons of development are often not absolute. In some cases, we have been able to say with some certainty that such-and-such a path *must* be followed if a sound housing finance market is to develop. In other cases, we can say no more than that a particular path has been followed in Poland or Hungary, and that it appeared to work (or not to work), but that we do not have enough evidence to generalize for other countries and other circumstances. It would seem that the potential for debating best practice increases with declining confidence in what constitutes best practice.

Occasionally, the opportunity arose for discussion surrounding truly alternative policies and programs, where the teams I have managed have included "leftists," but their proposed contributions often have been too far-reaching to be included in the policies that government, in search of best practice, is prepared to contemplate. There has also been a self-defeating element to their contributions that arise from the scope of their ambitions (or the extent of their frustrations). What is one to make of the proposal that the revision of the country's framework for delivering water and sanitation to households should be used as an opportunity for critiquing the country's "unsustainable Minerals-Energy Complex?"

CONSTRAINTS TO ADAPTING BEST PRACTICE

There are a number of constraints to adapting international best practice. To begin, documents such as the *Habitat Agenda* may actually inhibit locally differentiated policies. The reason for this lies in their influence, such that local and international research and policy analysis too often confront immutable policy preconceptions. For example, when managing the team that prepared Swaziland's housing policy, I was given a copy of the *Habitat Agenda*, which was to serve as a guide. The terms of reference did not include examining the effects of HIV/AIDS, which infects about a quarter of the country's adult population and which is causing the disintegration and impoverishment of many families. This is a problem because the *Habitat Agenda* emphasizes that "the design, development and management of human settlements should enhance the role of the family," and presumes that families, if provided with secure tenure and serviced sites, will seek to invest in housing. On the contrary, many families seek to minimize expenditure on housing and services (Tomlinson, 2001a).

The same is true in South Africa, which is a signatory to the *Agenda*. In a context where most low-income families comprise extended families, but also in a context where many are fragmenting due to HIV/AIDS, a policy targeting a western nuclear family is singularly inappropriate (Tomlinson, 2001b)

Thus, Gilbert (2001: 4) notes that "even if the advice available today is neither so dogmatic nor so uniform as it once was, the danger remains that local decision makers will pick the wrong option for their country. Even if they are invited to reinvent locally, this will not help if the basic policy menu on offer is very limited."

The resolution to this potential for choosing the wrong option requires a willingness among senior bureaucrats, their political bosses, development agencies, and consultants, academics and NGOs who are "in the loop," to regularly and critically evaluate their preconceptions. Instead, the reverse is true.

This is because, firstly, signatories to the *Agenda* are required to "monitor the implementation of the commitments of the agreement" (Department of Housing, 2001: 31). In South Africa, this applies particularly to the Department of Housing, which was the signatory for the South African government. The Department's housing capital subsidy is central to the delivery of shelter and services (40% of the housing subsidy is used for investment in services). But, with its monitoring role, the emphasis is on conformity, not innovation.

Secondly, conformity is a precondition to attracting foreign investment in, say, service delivery. It was partly to this end that USAID and the Department of Finance created the Municipal Infrastructure Investment Unit. The Unit's American consultants brought their worldwide experience to South Africa with a view to building a market in service delivery.

And thirdly, for those "in the loop," the constraint to critical review, surely, is profits. Under the competitive impress of schedules and delivering what the client—the development agency rather than the host country—wants, international best practice allows the development practitioner to circumvent thinking too deeply about local circumstances.

Yet change does occur, mostly, it is hypothesized, as a result of the evolution of experience *within* international development agencies rather than among host countries. This requires differentiating between pressures on, say, the World Bank's neoclassical agenda, and policies and programs within that agenda. Rebecca Black, of USAID, explains this evolution. In commenting on this article, she observed that "'best practice' may be the wrong word . . . rather than being driven by real best practice, current policy is driven by informed history of worst practice, with emphatic advocacy for what is the most unlike the past but as yet untested."

CONCLUSION

In this article I set out to show how conceptions of international best practice have shaped the formulation of urban policy in South Africa, to demonstrate the key role of international development agencies and consultants in helping to interpret and apply best practice, and to look at the potential for the formulation of policies that are at odds with best practice.

It is apparent that best practice has guided the formulation of policy for the delivery of services in South Africa and that this has occurred every step along the way: policy, legislation, regulation, guidelines and institutional support. The key role of development agencies is clear in USAID's helping the South African government to create an enabling framework for MSPs and in funding consultants to assist with preparing each of these steps. The significance of institutions such as USAID and the World Bank is out of all proportion to the money they spend in a country like South Africa.

Turning to the potential local stakeholders have for deviating from international best practice, my conclusion is that there are two constraints to this happening. The first is that for countries seeking to be part of the global economy, the neoliberal agenda sets the parameters for policy alternatives. (An economist at my university refers to this as "the straight-jacket for which we should be grateful.") The second is that, within these parameters, the potential for debating best practice increases with declining confidence in what constitutes best practice. When there is great certainty on the part of the international agency about the appropriate policy, then there is minimal room for negotiation.

The danger with this policy process is that it weakens participatory processes and commitment to the policy and, where development institutions and consultants play a leading role, it separates authority from responsibility and accountability. Indeed, this is intrinsic to the policy process described above.

The comments of a consultant to the World Bank help to conclude this paper: "There is a limit to best practice. There is a context ... These guys are so confident. They bring their experiences from one country and impose it . . . in another—short-circuiting participatory policy processes."

REFERENCES

Black, R., Jaszczolt, K. and Lee, M. (2000). *Solving the housing problem: Lessons from Poland and Hungary in creating a new housing finance system*, USAID/Warsaw, Regional Urban Development Office, Central Europe.

Bond, P. (1998). "Privatisation, participation and protest in the restructuring of municipal services," *Urban Forum* 9.1: 37–76.

Bond, P. (2000). *Cities of Gold: Townships of Coal*, Trenton, NJ: Africa World Press.

Department of Housing (2001). *Multi-year provincial housing development plans: guideline version 1.0*, Prepared for the National Department of Housing by the Council for Scientific and Industrial Research, Pretoria.

Gilbert, A. (2001). "Scan globally, reinvent locally: reflecting on the origins of South Africa's Capital Housing Subsidy Policy," Draft manuscript.

Pycroft, C. (2000). "Democracy and delivery: The rationalization of local government in South Africa," *International Review of Administrative Sciences* 66: 1.

Ruble, B. A., Tulchin, J. S. and Garland, A. M. (1996). "Introduction: Globalism and local realities—five paths to the urban future," In M. A. Cohen, B. A. Ruble, J. S. Tulchin and A. M. Garland (eds.), *Preparing for the Urban Future: Global Pressures and Local Forces*, Washington, DC: Woodrow Wilson Press.

Tomlinson, R. (2001a). "Housing policy in a context of HIV/AIDS and globalization," *International Journal of Urban and Regional Research*, 25(3): 649–57.

Tomlinson, R. (2001b). "Subsidies benefit only a few," *Mail & Guardian*, 28 September to 4 October: 51.

Copyright Information

Every effort has been made to contact copyright holders for their permissions to reprint selections in this book. The publishers would be grateful if any copyright holders not acknowledged could get in touch so any errors or omissions can be rectified in future editions of the *Reader*.

I THE CITY EXPERIENCED

MADANIPOUR Madanipour, ALI, "Urban Lives: Stories from Tehran." Text has been written specifically for *Cities of the Global South Reader*. Printed with permission from the author. Photographs © Ali Madanipour, reproduced with permission.

II MAKING THE "THIRD WORLD" CITY

KING King, Anthony D., "Colonialism and Urban Development." Text has been written specifically for *Cities of the Global South Reader.* Printed with permission from the author.

MASSEY Masey, Doreen, "Cities Interlinked." From Doreen Massey, John Allen, and Steve Pile (eds.) *City Worlds*. Routledge, 1998. Reproduced with permission from Taylor and Francis.

GOLDMAN Goldman, Michael, "Development and the City." Text has been written specifically for *Cities of the Global South Reader*. Printed with permission from the author.

ROBINSON Robinson, Jennifer, "World Cities, or a World of Ordinary Cities?" From *Ordinary Cities: Between Modernity and Development*. Routledge, 2006. Reproduced with permission from Taylor and Francis.

III THE CITY LIVED

MAYEKISO Mayekiso, Mzwanele, (1996) *Township Politics Civic Struggles for a New South Africa*, New York: Monthly Review Press.

SIMONE Simone, AbdouMaliq, "The Urbanity of Movement: Dynamic Frontiers in Contemporary Africa." *Journal of Planning Education and Research* 2011; 31(4): 379–391. Reproduced with permission from Sage Publications.

ZHANG Zhang, LI, "Migration and Privatization of Space and Power in Late Socialist China." *American Ethnologist* 2001; 28: 179–205. Reproduced with permission from the American Anthropological Association and the author.

BROMLEY Bromley, Ray, "Working in the Streets of Cali, Colombia: Survival Strategy, Necessity, or Unavoidable Evil?" Excerpted from pp. 124–138 in Josef Gugler (ed.) *Cities in the Developing World: Issues, Theory, and Policy*. Oxford and New York: Oxford University Press, 1997. Earlier version published in 1988 as pp. 161–182 in Josef Gugler (ed.) *The Urbanization of the Third World*. Oxford and New York: Oxford

University Press. Original article published in 1982 as pp. 59–77 in Alan Gilbert, Jorge Hardoy and Ronaldo Ramirez (eds.) *Urbanization in Contemporary Latin America*. Chichester and New York: John Wiley and Sons. Reproduced with permission from John Wiley & Sons and Oxford University Press.

MITRA Mitra, Sudeshna, "Anchoring Transnational Flows: Hypermodern Spaces in the Global South." Text has been written specifically for *Cities of the Global South Reader*. Printed with permission from the author.

MOSER Moser, Caroline, O. N., "Women and Self-Help Housing Projects: A Conceptual Framework for Analysis and Policy Making," in Kosta Mathey (ed.) *Beyond Self-help Housing*. London and New York: Mansell, 1992. Reproduced with permission from Mansell Publishing Limited.

LEAF Leaf, Michael, "The Suburbanization of Jakarta: A Concurrence of Economics and Ideology." *Third World Planning Review* 1994; 16(4): 341–356. Reproduced with permission from Liverpool University Press.

IV THE CITY ENVIROMENT

HARDOY and SATTERTHWAITE Hardoy, Jorge E. and Satterthwaite, David, "Environmental Problems of Third World Cities: A Global Issue Ignored?" *Public Administration and Development* 1991; 11(4): 341–361. Reproduced with permission from John Wiley & Sons.

BEALL, CRANKSHAW and PARNELL Beall, Jo, Crankshaw, Owen and Parnell, Susan, "Victims, Villains and Fixers: The Urban Environment and Johannesburg's Poor." *Journal of Southern African Studies* 2000; 26(4): 833–855. Reproduced with permission from Taylor and Francis.

ASSAAD Assaad, Ragui, "Formalizing the Informal? The Transformation of Cairo's Refuse Collection System." *Journal of Planning Education and Research* 1996; 16(2): 115–126. Reproduced with permission from Sage Publications.

BADAMI Badami, Madhav G., "Urban Transport Policy as if People *and* the Environment Mattered: Pedestrian Accessibility the First Step." *Economic and Political Weekly* 2009; XLIV(33): 43–51. Reproduced with permission from the author.

DE BOECK and PLISSART De Boeck, Filip and Plissart, Marie-Françoise, "Kinshasa and Its (Im)material Infrastructure." From Marie-Françoise Plissart and Filip De Boeck (eds.), *Kinshasha: Tales of the Invisible City.* Ludion, 2005. Reproduced with permission from Ludion.

KANNA Kanna, Ahmed, "'Going South' with the Starchitects: Urbanist Ideology in the Emirati City." From *Dubai, the City as Corporation*. University of Minnesota Press, 2011. Reproduced with permission from University of Minnesota Press.

DAVIS Davis, Diane E., "Reverberations: Mexico City's 1985 Earthquake and the Transformation of the Capital." In Lawrence J. Vale and Thomas J. Campanella (eds), *The Resilient City, How Modern Cities Recover from Disaster.* Oxford University Press, 2005. Reproduced with permission from Oxford University Press.

GRAHAM Graham, Stephen, "Disruption by Design: Urban Infrastructure and Political Violence." In Stephen Graham (ed.) *Disrupted Cities: When Infrastructure Fails*. Routledge, 2009. Reproduced with permission from Taylor and Francis.

BAVISKAR Baviskar, Amita, "Between Violence and Desire: Space, Power, and Identity in the Making of Metropolitan Delhi." *International Social Science Journal* 2003; 55(175): 89–98. Reproduced with permission from John Wiley & Sons.

BARNETT and ADGER Barnett, Jon and Adger, W. Neil, "Climate Dangers and Atoll Countries." *Climate Change* 2003; 61(3): 321–337. Reproduced with permission from Springer.

V PLANNED INTERVENTIONS AND CONTESTATIONS

KOHL and FARTHING Kohl, Ben and Farthing, Linda, "New Spaces, New Contests: Appropriating Decentralization for Political Change in Bolivia." In Victoria Beard, Faranak Miraftab and Christopher

Silver (eds.) *Planning and Decentralization: Contested Spaces for Public Action in the Global South.* Routledge, 2008 Reproduced with permission from Routledge.

APPADURAI Appadurai, Arjun, "Deep Democracy: Urban Governmentality and the Horizon of Politics." *Environment and Urbanization* 2001; 13(2): 23–43. Reproduced with permission from the author.

MASCARENHAS Mascarenhas, Michael, Excerpted from "Sovereignty: Crisis, Humanitarianism, and the Condition of 21st Century Sovereignty" by Michael Mascarenhas, originally appearing in Hilary E. Kahn, (ed.), *Framing the Global.* Foreword by Saskia Sassen. 2014, pp. 296–316. Reproduced with permission from Indiana University Press.

BAIOCCHI Baiocchi, Gianpaolo, "The Citizens of Porto Alegre." *Boston Review* March/April issue 2006; 31(3). Available from http://www.bostonreview.net/gianpaolo-baiocchi-the-citizens-of-porto-alegre. Reproduced with permission from the *Boston Review.*

CORNWALL Cornwall, Andrea, "Whose Voices? Whose Choices? Reflections on Gender and Participatory Development." *World Development* 2003; 31(8): 1325–1342. Reproduced with permission from Elsevier.

CASTELLS Castells, Manuel, "Squatters and the State: The Dialectics between Social Integration and Social Change (Case Studies in Lima, Mexico, and Santiago de Chile)." From *The City and the Grassroots: A Cross-Cultural Theory of Urban Social Movements* (1983). Reproduced with permission from the author.

SMITH and GUARNIZO Smith, Michael Peter and Guarnizo, Luis Eduardo, "Global Mobility, Shifting Borders and Urban Citizenship." *Tijdschrift voor Economische en Sociale Geografie* 2009; 100(5): 610–622. Reproduced with permission from John Wiley & Sons.

KHAMIS and VAUGHN Khamis, Sahar and Vaughn, Katherine, "Cyberactivism and Citizen Mobilization in the Streets of Cairo." Text has been written specifically for *Cities of the Global South Reader.* Printed with permission from the author.

ROY Roy, Ananya, "Why India Cannot Plan Its Cities: Informality, Insurgence and the Idiom of Urbanization." *Planning Theory* 2009; 8(1): 76–87. Reproduced with permission from Sage Publications.

TOMLINSON Tomlinson, Richard, "International Best Practice, Enabling Frameworks and the Policy Process: A South African Case Study." *International Journal of Urban and Regional Research* 2002; 26(2): 377–388. Reproduced with permission from John Wiley & Sons.

Index

Note: Page numbers in **bold** type refer to **figures**
Page numbers in *italic* type refer to *tables*
Page numbers followed by 'n' refer to notes